Data Analysis in Forensic Science

A Bayesian Decision Perspective

Statistics in Practice

Statistics in Practice is an important international series of texts which provide detailed coverage of statistical concepts, methods and worked case studies in specific fields of investigation and study.

With sound motivation and many worked practical examples, the books show in down-to-earth terms how to select and use an appropriate range of statistical techniques in a particular practical field within each title's special topic area.

The books provide statistical support for professionals and research workers across a range of employment fields and research environments. Subject areas covered include medicine and pharmaceutics; industry, finance and commerce; public services; the earth and environmental sciences, and so on.

The books also provide support to students studying statistical courses applied to the above areas. The demand for graduates to be equipped for the work environment has led to such courses becoming increasingly prevalent at universities and colleges.

It is our aim to present judiciously chosen and well-written workbooks to meet everyday practical needs. Feedback of views from readers will be most valuable to monitor the success of this aim.

A complete list of titles in this series appears at the end of the volume.

Data Analysis in Forensic Science

A Bayesian Decision Perspective

Franco Taroni
School of Criminal Justice
University of Lausanne
Switzerland

Silvia Bozza
Department of Statistics
University Ca' Foscari, Venice
Italy

Alex Biedermann
School of Criminal Justice
University of Lausanne
Switzerland

Paolo Garbolino
Faculty of Arts and Design
IUAV University, Venice
Italy

Colin Aitken
School of Mathematics
University of Edinburgh
UK

A John Wiley and Sons, Ltd., Publication

This edition first published 2010

Registered office
John Wiley & Sons Ltd, The Atrium, Southern Gate, Chichester, West Sussex, PO19 8SQ, United Kingdom

For details of our global editorial offices, for customer services and for information about how to apply for permission to reuse the copyright material in this book please see our website at www.wiley.com.

Library of Congress Cataloging-in-Publication Data

Data analysis in forensic science : a Bayesian decision perspective / Franco Taroni ... [et al.].
p. cm.
Includes bibliographical references and index.
ISBN 978-0-470-99835-9
1. Forensic sciences–Statistical methods. 2. Bayesian statistical decision theory. I. Taroni, Franco.
HV8073.D286 2010
363.2501′519542–dc22

2009054415

A catalogue record for this book is available from the British Library.
ISBN: H/bk 978-0-470-99835-9

Typeset in 10/12 Times Roman by Laserwords Private Ltd, Chennai, India.
Printed and bound in Great Britain by T. J. International, Padstow, Cornwall

To Dennis Lindley, who has opened up new horizons in statistics, law and forensic science

‘We should not look for truth, but should only become conscious of our own opinions. We should not question nature but only examine our consciences. At most I can question nature so that it will give me data as elements of my judgments, but the answer is not in the facts; it lies in my state of mind, which the facts cannot compel but which nevertheless can spontaneously feel itself compelled by them.’ (de Finetti 1989, p. 180)

Contents

Foreword

David H. Kaye
Distinguished Professor of Law and Weiss Family Scholar Pennsylvania State University

Statistics and law make strange bedfellows. Both disciplines are concerned with decision making under conditions of uncertainty, but lawyers work with words and pictures, not numbers and graphs. As a result, among lawyers and judges, statistics and statistical reasoning tend to generate feelings of aversion or trepidation rather than comfort and comprehension. A New York trial judge once derided a probability as "betting odds" and the "so-called [verbal] predicate" as "an impressive title ... which ... roughly corresponded to a form chart for pari-mutuel horse racing." Angela B. v. Glenn D., 482 N.Y.S.2d 971 (Fam. Ct. 1984). Although the appellate court regarded the questioned probability as "clear and convincing," and therefore reversed the trial court's determination (502 N.Y.S.2d 19 (App. Div. 1986)), more recent opinions can be found that echo the judge's antagonism toward probabilities. Nevertheless, scientific and statistical evidence has become increasingly important in modern litigation. The trend is irreversible.

The resulting challenge for forensic scientists or statisticians is to produce, process, and present accurate data that will assist legal decision makers in reconstructing past events. The concomitant challenge for the legal system is to achieve an optimal balance between completeness and comprehensibility of quantitative testimony. This book advances this mutual endeavour by describing the tools of modern Bayesian decision theory, by illustrating their application to paradigmatic types of data in forensic science, and by defending the procedures against various objections. The description is technically and philosophically sophisticated. With remarkable rigour, the authors lay out the elements of the theory – probability, likelihood, and utility – and apply them in the form of Bayes' rule and loss functions to the recurrent statistical problems of estimation, classification, and decision. They advocate the personal or subjective theory of probability and frankly face the daunting task of integrating this perspective into a system in which the "ultimate issue," as the rules of evidence characterize it, is the judge's or jury's to resolve.

The book will not be easy going for readers with no exposure to elementary statistics, but forensic scientists and analysts, as well as evidence scholars and practising lawyers, need to become familiar not only with the ideas behind classical statistical inference (as developed by Fisher, Neyman, and Pearson), but also likelihood theory (as defended by Edwards and Royall), and the reasoning pioneered by Bayes in 1764 and elaborated upon by his successors (ably presented in this volume). Data from tests, experiments, and observations may speak loudly, but even the most extreme data cannot be translated directly into statements of certainty. Certainty is an asymptote, not an end point. When gathered appropriately, data can help us choose between hypotheses about the world (or models of it), but these data do not produce probabilities of zero or one, and they do not speak for themselves. They inform forensic scientists, whose task it is to educate investigators, lawyers, and factfinders.

This task is crucial. Naive calculations and loose talk about the meaning of probabilities associated with forensic findings can cause serious problems for the administration of justice. The extensive appeals and postconviction proceedings in cases such as the prosecutions of Sally Clark in England (R v. Clark, 2003, EWCA Crim 1020, 2003, All ER (D) 223 (Apr),CA and 2000, All ER (D) 1219, CA) and Troy Brown in the United States (McDaniel v. Brown, 129 S.Ct. 1038 (2009)) illustrate the disturbing problems that inadequate statistical analysis and interpretation can produce. Likewise, analysts who match patterns and eschew quantitative testimony in favour of claims of absolute identification increasingly have come under attack. Such statistics-avoidance behaviour cannot prevail much longer.

Not everyone will agree with the strong subjectivist perspective advanced here. Rational people can differ on the relative value of frequentist, likelihood, and Bayesian methods of data analysis and exposition. But a wider understanding of the mathematical techniques and the philosophical ideas in this book can and will enhance the contributions of forensic science in criminal and civil investigations and litigation. For these reasons and others, careful study of the chapters that follow will be an illuminating and valuable undertaking.

Preface

This book was motivated by the facts (a) that the use of statistical methods to analyse quantitative data in forensic science has increased considerably over the last few years, (b) that students, researchers and practitioners in forensic science ask questions concerning the relative merits of differing approaches, notably Frequentist and Bayesian, to statistical inference, and (c) that the ideas of decision theory are now being introduced to forensic science.

Moreover, generations of students in the applied sciences experience lectures in statistics as frustrating. This may be in part due to the most commonly advocated approach in this context, which is the frequentist and which has fundamental philosophical flaws in its inferential procedures. Sometimes both Bayesian and frequentist philosophies are proposed but without a careful comparison and discussion of their relative qualities. In addition – as suggested by D'Agostini (2000) – it is also time to stop the negative reaction developed by students and practitioners towards words like 'belief' and 'subjective'. More effort should therefore be placed on clarifying the conceptual underpinnings of these schools of thought.

With regard to this, the present book intends to set forth procedures for data analysis that rely on the Bayesian approach to inference. An emphasis will be made on its foundational philosophical tenets as well as its implications in practice. In effect, the book proposes a range of statistical methods that are useful in the analysis of forensic scientific data. The topics include, for example, the comparison of allele proportions in populations, the choice of sampling size (such as in a seizure of pills) or the classification of items of evidence of unknown origin into predefined populations.

Data are analyzed to aid with decision making. These decisions have consequences the effect of which are measured by factors known as utilities. For this reason, this book advocates the study of the decision-theoretic extension of Bayesian inference. In fact, the application of decision theory to forensic science is both original and novel and much in this area is still unexplored. The book thus aims to introduce students and practitioners to the application, interpretation and summary of data analyses from a Bayesian decision-theoretic perspective. The text also includes selected code written in the statistical package R which offers to readers an additional opportunity to explore the treated subjects. Further illustrations of both Bayesian inference and decision theory are provided through graphical models (Bayesian networks and Bayesian decision networks).

The book is organized in two parts. The first part explains and defines key concepts and methods from historical, philosophical and theoretical points of view. The second part follows a step-by-step approach for taking the reader through applied examples, inspired and motivated by issues that may arise in routine forensic practice. This mode of presentation includes discussions of the arguments and methods invoked at each stage of the analyses, in particular where the Bayesian decision-theoretic aspect intervenes.

In the opinion of the authors, this book could make an interesting addition to the list of publications already offered in the forensic science and statistics collections. It is obviously designed to fit well with previous books written by several of the authors (Aitken and Taroni 2004; Taroni *et al.* 2006a). These existing works focused essentially on scientific evidence evaluation and interpretation whereas rather little mention was made of data analysis (with an exception of Chapter 5 in Aitken and Taroni 2004). In addition, the topic of decision making had not been given the main attention. This fundamental and critical point was mentioned by Professor Dennis Lindley in a personal letter accompanying his Foreword in the Aitken and Taroni (2004) book:

> A general point that arises at several places in the book is that Bayesian analysis incorporates decision-making as well as inference and your near-silence on the former was disappointing, though understandable. Some mention of utility might be desirable.

Inspired by Professors Lindley's fruitful comment, it is the hope of the authors that they have added in this new book an acceptable extension to the decision-analytic perspective. Aware of the list of specialized books dedicated to Bayesian inference and decision, the authors' primary intention was to provide a more introductory and case-based book on Bayesian decision theory that could also be of benefit to scientists, eventually also from outside the traditional forensic disciplines, who are also interested in this kind of data analysis. As such, the book may prepare the reader for more sophisticated books in this area.

From our point of view, an important message of the present book is (a) that subjective probability is a natural concept to quantify the plausibility of events in conditions of uncertainty, (b) that Bayes theorem is a natural way of reasoning in updating probabilities, and (c) that when we wish to decide whether to adopt a particular course of action, our decision clearly depends on the values to us of the possible alternative consequences. This logic of thought leads to compelling results that conform to the precepts of evidential assessment of forensic evidence. It is thus hoped that it will attract and convince both researchers and practitioners.

As in the authors' earlier projects, this new endeavour involved different backgrounds from statistics, forensic science and philosophy. We think that such a group can shape a well-balanced book and combine interdisciplinary aspects in a distinct way. The different academic profiles allow the introduction of new knowledge and stimulate challenges for all individuals interested in data analysis. The three fields – statistics, forensic science and philosophy – have common features and co-operation amongst them may produce results that none of the disciplines may produce separately.

We are very grateful to a number of people who contributed, in one way or another, to the book. In particular we thank Marc Augsburger, Luc Besson, Catherine Evequoz Zufferey, Raymond Marquis, Williams Mazzella, Fabiano Riva, Neil Robinson and Matthieu Schmittbuhl for their support and the permission to use original data from their personal works.

One of us, Alex Biedermann, has been supported by the Swiss National Science Foundation and the Italian National Research Council through grant PIIT1-121282.

Several software packages were used throughout the book: R, a statistical package freely available at www.r-project.org, Xfig, a drawing freeware running under the X Window System and available at www.xfig.org as well as Hugin which allows the development of the graphical models described throughout the book.

Finally, we would like to express our indebtedness for support throughout the whole project of this book to the Universities of Lausanne and of Edinburgh, the Ca' Foscari and IUAV Universities of Venice, and to our families.

F. Taroni, S. Bozza, A. Biedermann, P. Garbolino, C.G.G. Aitken
Lausanne, Venezia, Edinburgh

Part I

The Foundations of Inference and Decision in Forensic Science

1

Introduction

1.1 THE INEVITABILITY OF UNCERTAINTY

Gaius Julius Caesar was assassinated on March 15, 44 BC. Do you know his last words? What is the capital of the South Asian country Bangladesh? What will be the percentage gain of the Dow Jones Industrial Average during the next six months?

Maybe the first of these questions makes you think of Shakespeare's tragedy Julius Caesar from which you might recall from Act three Scene one the widely known first half of the macaronic line 'Et tu, Brute? – Then fall, Caesar!'[1] (Shakespeare 1994, p. 1032). Caesar's last words before dying severely wounded represent a contested topic among historians. How definite is your answer? With regard to the second question, you might be fairly sure about Dhaka being the capital of Bangladesh, but you may still prefer, for whatever reason, to check your answer by consulting an up-to-date almanac. Finally, there are many relevant events to come up yet during half a year – at present almost certainly unknown to you – so that we dare to anticipate that your beliefs about the future state of the stock market will be vague to at least some degree.

Yours, ours, anybody's understanding of the past, the present and the future is naturally incomplete. It is such imperfect knowledge which implies what we commonly refer to as *uncertainty*, the natural state that inevitably attends all human activities. Throughout this book, uncertainty is regarded in a personal way, that is as a notion that describes the relationship between a person issuing a statement of uncertainty and the real world to which that statement relates – and in which

[1]A literal translation is 'And you, Brutus? Then fall, Caesar.' Alternatively, the first part of the expression may also be translated as 'You too, Brutus?' or 'Even you, Brutus?'.

Data Analysis in Forensic Science: A Bayesian Decision Perspective Franco Taroni, Silvia Bozza, Alex Biedermann, Paolo Garbolino and Colin Aitken

that person acts. A unified language for understanding and handling uncertainty, revising beliefs in the light of new information and usage for practical decision making will be the main focus of attention.

Uncertainty is an omnipresent complication in life, and the case of forensic science is no exception. As a distinct discipline of its own, along with the criminal justice system at large, forensic science is typically concerned with past events which are both unique and unreplicable. Knowledge about past occurrences is bound to be partially inaccessible, however, because of spatial and temporal limitations of our sensory capacities. Uncertainty is a fundamental problem underlying all forensic sciences. Increasingly often, it is perceived as discomforting by both scientists and other actors of the criminal justice system, which illustrates the continuing need to give it careful attention.

Notwithstanding, an objection to this attention to uncertainty is immediately possible. Suffice it to mention that past events, notably criminal activities, may generate distinct remaining traces in the form of tangible physical entities (such as blood stains or textile fibres) that can be discovered and examined. These are thought to have a potential for revealing something useful to retrace past events. The informed reader might even refer to the writings of the pioneer forensic scientist Edmond Locard, some of whose now widely quoted words are as follows:

> Nul ne peut agir avec l'intensité que suppose l'action criminelle sans laisser des marques multiples de son passage [...] tantôt le malfaiteur a laissé sur les lieux des marques de son activité, tantôt par une action inverse, il a emporté sur son corps ou sur ses vêtements les indices de son séjour ou de son geste.[2] (Locard 1920)

This quotation is still valid in these times but with the important difference that forensic scientists are now in the privileged position to analyze crime-related material of many more different kinds of nature as well as in much smaller quantities than was possible at the time of Locard's writing. A primary reason for this is that, owing to vast developments made in science and technology, today's forensic scientists have a broad scope of methods and techniques at their disposal.

Although this instrumental support offers vast capacities for providing scientists with valuable information, the outset is essentially paradoxical. On the one hand, systematic analytical testing and observation may lead to abundant quantities of information, whereas, on the other hand, such information will often be substantively lacking in the qualities that would be needed to entail (or make necessary) particular propositions[3] that are maintained by a reasoner. To state this

[2]'No one can proceed with the intensity that the criminal act requires without leaving multiple marks of his passing [. . .] either the wrongdoer has left on the scene marks of his activity, or, on the other hand – by an inverse process – , he has taken away with him on his person (body) or clothes, indications of where he has been or what he has done.' Free translation by the authors.

[3]A proposition is interpreted here as an assertion or a statement that such-and-such is the case (e.g. an outcome or a state of nature of the kind 'the suspect is the source of the crime-stain') and also as a linguistic description of a decision. An important basis for much of the argument developed throughout this book is that it is assumed permissible to assign personal degrees of belief to propositions.

differently, forensic science is not, primarily, about the *mise en œuvre* of machinery and equipment in artifical and controlled laboratory settings. It is the general framework of circumstances within which testing is performed that makes forensic science a very challenging undertaking. For example, material related to crimes and to many real-world events in general may often be affected by contamination. The material may also be degraded and/or of such low quantity that only a single method or measurement can be applied. Sometimes it may not even be known whether the material submitted for analysis is relevant[4]. So, laboratory performance is undoubtedly important, but it is not a general guard against the rise of uncertainty in attempts to reconstruct the past.

In this book we will repeatedly come upon observations, measurements or counts, sometimes also referred to as 'raw data', and consider how such information should – in the light of uncertainty – be used to inform beliefs and support decision making. There is a practical necessity for this because such considerations represent vital steps in guaranteeing that scientific evidence meaningfully serves the purpose of a particular forensic branch of application. In order to comply with this requirement, forensic science needs to enquire about ways that allow one to learn about past events in the light of uncertain information, preferably in a manner that is in some sense rational and internally consistent.

More detailed explanation of what we mean by 'rational' and 'consistent' is delayed to a discussion in Chapter 2. For the time being, we solely note that these ways are intended to provide assistance in examining whether people's opinions about unobserved matters of fact are justified and whether these people actually have the reasons to believe that their decisions in the light of these opinions are optimal.

The very idea of enquiring about how one ought to reason and act sensibly under uncertainty takes such a central role that Ian Evett relied on it for the purpose of providing a definition of forensic science:

> [...] I will settle for a simple premise: forensic *science* is a state of mind, I mean that whether a particular individual is behaving, at a given juncture, as a scientist can be determined by the mental process underlying his/her actions and words. (Evett 1996, p. 121)

1.2 DESIDERATA IN EVIDENTIAL ASSESSMENT

Prior to providing a more formal introduction to the methods we seek to implement throughout this book, it is useful to set forth some very general, practical precepts to which we wish our analytic thought and behaviour to conform. Such precepts will be helpful, for example, to clarify why we will be giving preference to some methods and views rather than to others.

---

[4]In forensic science, 'relevance' is commonly used as a qualifier for material that has a true connection with the offence or the offender (Stoney 1994).

Consideration is given hereafter essentially to six desiderata upon which the majority of current scientific and legal literature and practice converge in their opinion. These desirable properties are balance, transparency, robustness, added value, flexibility and logic. These notions have been advocated and contextualized, to a great extent, by some quarters in forensic science and from jurists from the so-called 'New Evidence Scholarship' (Lempert 1988).

For an inferential process to be *balanced*, or in the words of some authors, impartial (Jackson 2000), attention cannot be restricted to only one side of an argument. Evett (1996, p. 122) has noted, for instance, that 'a scientist cannot speculate about the truth of a proposition without considering at least one alternative proposition. Indeed, an interpretation is without meaning unless the scientist clearly states the alternatives he has considered.' The requirement of considering alternative propositions is a general one that equally applies in many instances of daily life (Lindley 1985), but in legal contexts, its role is fundamental. Evett, during an interview, has expressed this as follows:

> Balance means that when I am doing anything for a court of law, I do it in full knowledge that there are two sides represented in that court. Even though the evidence that I've found appears to favour one or the other of those sides, my view of that evidence is directed not to proving the case for that side, but to helping the court to set that evidence and the views of both teams, prosecution and defence. (Joyce 2005, p. 37)

Note that there is more in this quotation than a sole requirement of considering alternatives. It also states that forensic scientists should primarily be concerned with the evidence and not with the competing propositions that are forwarded to explain it. This distinction is crucial in that it provides for a sharp demarcation of the boundaries of the expert's and the court's areas of competence. Failures in recognizing that distinction are at the heart of pitfalls of intuition that have caused – and continue to cause – much discussion throughout judicial literature and practice.

Besides balance, a forensic scientist's evaluation should also comply with the requirements of:

- *transparency*, that is, in the words of Jackson (2000, p. 84), '[...] explaining in a clear and explicit way what we have done, why we have done it and how we have arrived at our conclusions. We need to expose the reasoning, the rationale, behind our work.'
- *robustness*, which challenges a scientist's ability to explain the grounds for his opinion together with his degree of understanding of the particular evidence type (Jackson 2000).
- *added value*, a descriptor of a forensic deliverable that contributes in some substantial way to a case. Often, added value is a function of time and monetary resources, deployed in a way such as to help solve or clarify specific issues that actually matter with respect to a given client's objectives.

These desiderata characterize primarily the scientist, that is his attitude in evaluating and offering evidence, as well as the product of that activity. The degree to which the scientist succeeds in meeting these criteria depends crucially on the chosen inferential framework, which may be judged by the following two criteria:

- *flexibility*, a criterion that demands a form of reasoning to be generally applicable, that is, not limited to particular subject matter (Robertson and Vignaux 1998).
- *logic*, that is, broadly speaking, a set of principles that qualify as 'rational'. In turn, that rational system must also conform, as will be explained later in Chapter 2, to certain minimum requirements (Robertson and Vignaux 1993).

These last two issues – properties of an inferential method rather than behavioural aspects of the scientist – represent the principal topics to which the subsequent parts of this book thematically connect.

1.3 THE IMPORTANCE OF THE PROPOSITIONAL FRAMEWORK AND THE NATURE OF EVIDENTIAL ASSESSMENT

A few additional remarks are necessary on the requirement of balance, a criterion described so far as one that requires a scientist to consider at least two competing propositions.

First, attention should be drawn to the exact phrasing of propositions, an idea that underlies a concept known in the context as propositional level or hierarchy of propositions (Cook *et al.* 1998). The reasons for this are twofold. On the one hand, a proposition's content crucially affects the degree to which that proposition is helpful for the courts. For example, the pair of propositions 'the suspect (some other person) is the source of the crime stain' (known in the context as a source-level proposition) addresses a potential link between an item of evidence and an individual (that is, a suspect) on a rather general level. Generally, activity-level (e.g. 'the suspect (some other person) attacked the victim') or crime-level (e.g. 'the suspect (some other person) is the offender') propositions tend to meet a court's need more closely. On the other hand, the propositional level defines the extent of circumstantial information that is needed to address a proposition meaningfully. For example, when reasoning from a source- to a crime-level proposition, consideration needs to be given to the relevance of a crime stain (that is, whether or not it has been left by the offender), an aspect that is not necessarily needed when attention is confined to a source-level proposition.

Secondly, given a proposition of interest, forensic scientists usually assess the relative degrees to which evidence is compatible with the various settings (that is, the propositions) under consideration. The question, however, of what the believability of each setting actually ought to be, is not an issue for forensic scientists.

Addressing a target proposition requires – for reasons given later in Chapter 2 – a belief-state prior to the consideration of new facts as well as profound knowledge of circumstantial information. Forensic scientists cannot comply with any of these requirements. Even if they could, their focus on an issue (e.g. a proposition of the kind 'the suspect is the source of the crime stain') rather than on the evidence would amount to usurping the role of the court (Aitken and Taroni 2004).

Thirdly, it is worth insisting on having a well-defined framework of propositions. This is in sharp contrast to occasionally held opinions according to which data should be allowed to 'speak for themselves', a suggestion that evidential value represents some sort of intrinsic attribute. This is viewed cautiously in forensic science, where the following position has been reached:

> In court as elsewhere, the data cannot 'speak for itself'. It has to be interpreted in the light of the competing hypotheses put forward and against a background of knowledge and experience about the world. (Robertson and Vignaux 1993, p. 470)

As may be seen, the concept of propositions is important because it is closely tied to the notion of evidential value. For the time being, we tentatively consider the latter as a personalistic function of the former, in the sense that value is assigned to evidence by a particular individual depending on the propositions among which that individual seeks to discriminate and auxilary contextual information that is available to that individual. Arguably, evidential value is neither seen as an abstract property of the external world nor as one that can be elicited in a uniquely defined way.

Generally, the propositional framework is organized as part of an evaluative procedure, that is a model that specifies the relevant ingredients of an inferential process, their relationships along with rules that state how these elements ought to be used for inference. The issue that relates this brief mention of the propositional framework to the main topic of this book – data analysis – is the fact that the latter is needed in order to enable the former to provide quantitative expressions that are appropriate for the purpose for which a particular inference procedure has been designed.

1.4 FROM DESIDERATA TO APPLICATIONS

Although the general criteria to which we would like evidential assessment to conform may appear intuitively reasonable, it may be far from obvious how to implement them to bridge practical difficulties associated with forensic science as a discipline of reasoning and acting under uncertainty. As for themselves, the stated principles describe desirable, abstract properties rather than explicit ways in which one ought to proceed. The criteria – if met by the scientist – should contribute to the avoidance of the reduction of forensic expertise to ad hoc guesswork and unwarranted claims of 'many years of experience' (Evett 1996). Beyond this, however, the mere statement of the principles also remains insufficient for the need.

Further concepts and discussion are thus needed for examining whether scientists' analyses, evaluations and reportings are trustworthy. Among these is an approach to the description of uncertainty as well as rules that prescribe the combination of expressions of uncertainty. For this purpose, Chapter 2 will outline in detail a method for calculating with beliefs that is part of a package that also contains a procedure to use personal beliefs to inform decision making. As will be seen, these elements represent the fundamental tenets of the discipline of *statistics* (Lindley 2000b).

In this book we argue that statistics is a beneficial resource with important capacities for both clarifying and analyzing a wide range of practical problems. In particular, we will focus on statistical concepts that allow one to make plain and conceptualize the passage from the assessment of uncertainties associated with evidence to the assessment of uncertainties associated with particular explanatory propositions, including consistent choices amongst them. We justify this focus of enquiry by our conviction that these concepts have a substantial potential for the enhancement of the quality of forensic expertise.

Although we will argue that the methods yet to be introduced in later sections and chapters are the most appropriate ones currently available, we will not address the separate issue of how evidence is best presented before trial. This latter issue is a distinct topic of its own which, in the context, is also referred to as the 'presentation problem' (Redmayne 2001). This topic extends to additional complications that touch on discussions beyond the scope of this book. Evett and Weir (1998, p. 29) expressed this point concisely when they wrote that, 'in particular, we are going to take the evidence into the court room, where the proceedings owe no allegiance to the laws of science or mathematics and many of the participants are stubbornly nonnumerate.'

Notwithstanding the above, this book's central points of attention – forensic inference and decision analysis – draw their legitimacy from cogent practical reasons. An illustrative example for this is provided by courts that, typically, seek to reduce their uncertainty about a defendant's true connection with a criminal act (Lindley 2006). Often, part of this effort is thought to be achieved on the basis of evidence as offered by forensic scientists. According to this view, evidential assessment, that is a process of reasoning under uncertainty, constitutes a preliminary to judicial decision making (e.g. deciding if a suspect should be found guilty for the offence for which he has been charged) and taking such assessment seriously reflects the intention of promoting accurate decision making (Fienberg and Schervish 1986; Kaplan 1968; Kaye 1988; Redmayne 2001; Robertson and Vignaux 1993).

In particular, a decision-based approach can help (i) to clarify the fundamental differences between the value of evidence as reported by an expert and the final decision that is to be reached by a customer, and (ii) to provide a means to show a way ahead as to how these two distinct roles can be conceptualized to interface neatly with each other. Both of these are topics that are currently viewed differently rather than in a unified manner. This illustrates the continuing need for research in this area.

1.5 THE BAYESIAN CORE OF FORENSIC SCIENCE

Prior to proceeding with more technical chapters we anticipate at this point – still with the intention of relying on an essentially informal style of presentation – some of the main arguments and topics that will be advocated throughout this book, while reserving elements of logical and philosophical justifications to later discussion.

One of the credences of which we seek to convince the reader is that uncertainties about propositions should be expressed by the concept of probability. To this viewpoint we immediately add, however, that we will be giving preference to probability theory employed in that of its distinct interpretations which views probabilities as degrees of belief, a standpoint commonly known as the subjectivist (or personalist) interpretation of probability. Such degrees of belief are personalized assessments of credibility formed by an individual about something uncertain, given information which the individual knows or can discover. In short, the probability apparatus will be used as a concept of reference to which personalized weights to the possible states of the uncertain world that surrounds us may be attached.

Even though such uncertainty is inevitable, recall from Section 1.1 that we live in a world in which further information may be gained by enquiry, analysis and experimentation. As a consequence of this, some means is required to adjust existing beliefs in the light of new evidence. A second credence which will thus be emphasized here is that the revision of beliefs should be operated according to Bayesian procedures. The term 'Bayesian' stems from a theorem – Bayes' theorem (Section 2.3.1) – that is a logical consequence of the basic rules of probability. We will repeatedly come across the theorem because it is a very important result that helps one to understand how to treat new evidence. As an aside, we note that, although the theorem has about a 250-year history, the attribute 'Bayesian' as a descriptor of a particular class of inference methods appears to have gained more widespread use only since the middle of the twentieth century.

Given a set of beliefs about the unknowable states of the world, the general objective is to identify an available course of action that is logically consistent with a person's personal preferences for consequences. This is an expression of a view according to which one decides on the basis of essentially two ingredients. These are, on the one hand, one's beliefs about past, present or future happenings and, on the other hand, one's valuation of consequences. As noted above, the former will be expressed by probability. The latter will be captured by invoking an additional concept, known as utility. Both concepts can operate within a general theory of decision that involves the practical rule which says that one should select that decision which has the highest expected utility (or, alternatively, which minimizes expected loss). When the class of such operations is based on beliefs that have received a Bayesian updating (statistical inference), then this process is called Bayesian decision analysis.

Both within and outside forensic science, the Bayesian package for inference and decision is considered – with continually increasing agreement – as the currently most appropriate and comprehensive approach to the various issues pertaining to the assessment of scientific evidence. In a legal context, the concept is particularly relevant because of the support it provides in conforming with the principles and requirements set forth in Section 1.2. Subscription to the Bayesian decision approach, however, does not suggest that the approach is perfect, a point that is noted by Evett and Weir (1998, p. 29):

> It is not our claim that Bayesian inference is a panacea for all problems of the legal process. However, we do maintain that it is the best available model for understanding the interpretation of scientific evidence.

Practical applications of patterns of reasoning corresponding to a Bayesian approach can be found, for example, as early as the beginning of the 20th century (Taroni *et al.* 1998). Bayesian ideas for inference then entered legal literature and debates more systematically only in the second half of the twentieth century. Kingston (1965a), Finkelstein and Fairley (1970) and Lindley (1977a) are some of the main reference publications from that period. Later, within the 1990s, specialized textbooks from Aitken and Stoney (1991), Aitken (1995) and Robertson and Vignaux (1995) appeared. During the past decade, further textbooks – along with a regular stream of research papers – focusing on Bayesian evaluations of particular categories of evidence, such as glass (Curran *et al.* 2000) or DNA (Balding 2005; Buckleton *et al.* 2004) were published. Compared to this, decision analysis is a rather sparsely studied area, in particular within forensic science. Some of the few available references are mostly from legal scholars, mentioned earlier in Section 1.4. However, for forensic science applications, decision making (e.g. about target propositions of interest) is a presently latent topic with room for many thought-provoking and interesting issues that ask for explanations and the formulation of effective answers.

It is worth noting that many discussions of probabilistic reasoning applied to forensic and legal matters in general rely on probability as a concept that is defined on propositions, that is linguistic entities of the kind that were presented at the beginning of Section 1.1. A probability statement about such entities – that is to say, some form of proposition or hypothesis – may typically have that character of a personalized expression of belief, as was mentioned earlier in this section.

Besides this, there is an additional facet of probability relevant to forensic applications, known as the set-theoretical development of probability, introduced in the last century by the Soviet mathematician A. N. Kolmogorov (Kolmogorov 1933). In that development, probability is defined on subsets of some given set. It is customary to denote the latter set as one that comprises all elementary possibilities, often referred to as outcomes of an experiment (sample space). This approach lends itself to a series of extensions that makes it of particular interest for the fields of mathematics and statistics (e.g. due to the applicability of the full differential and integral calculus).

The latter development of probability is used essentially in contexts where the main issue is uncertainty about the true value of a parameter[5], such as a mean or a proportion. The aim then is to use the probability calculus to obtain probability statements about such parameters. To return to the above-mentioned argument, these probability statements have a personalized interpretation in terms of degrees of belief and such degrees are revised by the use of Bayes' theorem, operating after the provision of data. Since one is concerned with a set of objects or individuals (a population), one can usually extract a subset of that set and investigate it. The result of this is a set of numbers – the data – for which a statistic (e.g. a mean) may be calculated and used as a basis for revising beliefs about the population parameter held prior to inspection of those data. When in addition to that, the goal is to choose a particular number as an estimate for the parameter, then this can be conceptualized as a decision problem using the ingredients informally introduced above, that is decisions and utilities.

These are, in brief, the main aspects of the Bayesian approach to statistics and decision analysis, a more formal and detailed account of which is given in the next two chapters.

1.6 STRUCTURE OF THE BOOK

In the previous sections of this chapter, a discussion was initiated on Bayes' theorem, presented essentially as a formalization of logic and common sense which makes it a valuable tool for reasoning about evidence in situations involving uncertainty. In fact, forensic practice routinely involves the collection of sets of observations or measurements, but they may be compatible with several distinct hypotheses of interest given and some of them are (a priori) less plausible than others. The problem, as stated, then consists in drawing conclusions in such situations while the idea underlying the inference analysis consists in offering guidance and influencing one's behaviour (Cornfield 1967).

Bayes' theorem plays a central role in quantifying the uncertainty associated with particular conclusions. The forthcoming Chapters, 2 and 3, provide a careful outline of this role and its relation to the process of consistent choice between hypotheses when available evidence is imperfect. The viewpoint will be that of a unification of the theories of (subjective) probabilities (mainly Bayes' theorem) and utility within decision theory in order to set forth the construction of a co-ordinated and structured whole.

Combining both theoretical elements and practical examples, Part II of the book – Chapters 4 to 8 – will proceed with focusing in more detail on the idea

[5]For the time being, a parameter is taken as a characteristic of the distribution of the measurements on, or categorization of, the entire set of members (e.g. individuals or objects) of a target population.

of 'judgements', that is, stated otherwise, assessments, considered decisions or sensible conclusions (not, however, in the sense of decisions of a court or judge). Actually, day-to-day forensic practice involves judgements, in a variety of facets and in contexts. For the purpose of illustration, consider the following:

1. How is one to judge an estimate of, say, the proportion of red car paint flakes found on a victim of a car accident in a given town, or the alcohol concentration given a series of measurements on a blood sample?
2. How is one to assess whether the value of a parameter of interest, such as the colour dye concentration in ecstasy tablets, lies within a given interval?
3. How is one to decide among competing hypotheses according to which, for example, a rate is greater (or not) than a given value, or two series of continuous measurements differ (or do not differ)?
4. How ought the value of a particular item of evidence be assessed?
5. How many samples should be analyzed in a consignment of, say ecstasy pills seized in a criminal investigation?
6. How ought one to proceed when an unknown item needs to be associated with or arranged into one of different specific (and known) classes? In other words, how is one to judge the appropriateness of two (or more) competing models for a given forensic real-world phenomenon?

Questions of this kind will be described, exemplified, analyzed and commented on in the light of Bayesian statistical methodology. Specifically, Chapter 4 approaches problems in point estimation, as given by question 1 above. Credible intervals, addressed by question 2, are treated in Chapter 5. Chapter 6 focuses on hypothesis testing to approach issues to which questions 3 and 4 relate. Question 5 relates to sampling problems, a recurrent topic in many forensic science disciplines, and considered in Chapter 7. Finally, questions such as question 6 are studied in Chapter 8.

The book focuses on a versatile list of statistical questions that may reasonably be encountered in forensic practice. The aim is to show how the Bayesian framework can be developed, understood and practically applied. Although the Bayesian approach has been in place for some time already and it may be tempting to think of it as an 'old tool', actual practice demonstrates that its help is more than ever indispensable for addressing current problems in evaluating forensic evidence. This, then, is an instance where we seek to answer a viewpoint taken by Good (Good 1962, p. 383), according to which it may be beneficial to 'learn to use a little of the language of the theory of rationality to understand what it means to be reasonably consistent. "Rational" is to be interpreted in relation to the theory of rationality, namely Bayes' theory, in which emphasis is on judgments of probabilities, utilities and expected utilities.'

2

Scientific Reasoning and Decision Making

2.1 COHERENT REASONING UNDER UNCERTAINTY

The mathematical theory of probability can be interpreted as a logical calculus for fitting together judgements of uncertainty: as the laws of deductive logic can be used to define formal notions of *coherence* for beliefs entertained with certainty, and provide constraints to deductive reasoning by means of rules of inference, so the laws of probability can be used as *a standard of coherence* for beliefs you entertain to a certain degree only, and can be used as rules of inference for reasoning under uncertainty[1].

> The central idea is the concept of coherence between uncertainty judgments. If You make some probability statements, then others are implied by the calculus of probability, and effectively You have made those as well. (Lindley 1990, p. 53)

The standard of coherence is a *pragmatic* standard. You are not obliged by whatever a priori reasons to have degrees of belief that satisfy the laws of probability calculus, but if your beliefs do not, then they will be potentially *pragmatically self-defeating*. That means that, if your degrees of belief are not coherent according to the probabilistic standard, and you wish to use them as a guide for action, then

[1]Some introductions to the subjective interpretation of probability theory addresses the readers by using the second person: this is a rhetorical artifice to help them to keep in mind that probabilities is their own degrees of belief based upon information they have. We shall follow this usage throughout this chapter but only this chapter.

Data Analysis in Forensic Science: A Bayesian Decision Perspective Franco Taroni, Silvia Bozza, Alex Biedermann, Paolo Garbolino and Colin Aitken

you will be acting in a way that will bring you consequences that are worse than they might have been if your degrees of belief had been coherent.

If probability is interpreted this way, than it provides not only *inference rules* but *decision rules* as well: if your preferences among actions with uncertain consequences are coherent according to the probabilistic standard, then the values ('utilities') these consequences have for you can be measured by probabilities; that is, these 'utilities' obey, like degrees of beliefs, the laws of the probability calculus.

> The Bayesian paradigm is a complete recipe for appreciation of the world by You, and for Your action within it. The central concept is probability as the sole measure of uncertainty and as a means of expressing Your preferences through (expected) utility. Properly appreciated, both measures are adequate for inference and decision-making. (Lindley 1990, p. 55)

The most important probabilistic rule of inference is the *principle of conditionalization*, and the basic rule of decision is the *principle of maximization of expected utility*.

In this chapter arguments are given for justifying the acceptance of the probability calculus as a standard of coherence for degrees of belief, and the acceptance of the principle of conditionalization as a rule for *coherent updating* of degrees of beliefs. Informal discussion is also given as to how this concept of coherent reasoning under uncertainty can be extended to scientific thinking in general and to statistical and forensic science inferences in particular. A formal treatment of utility theory is provided in Chapter 3.

2.1.1 A rational betting policy

A centuries-old practice that connects degrees of beliefs with action is betting. Betting jargon uses *odds* to quantify the *relative chances* for a proposition H and its negation: odds $m : n$ in favour of H correspond to a probability of H equal to $m/(m+n)$.

Suppose your odds for the proposition that the bus involved in a hit-and-run accident yesterday was blue were 4:1. This means that your personal probability for that proposition is $4/5 = 0.8$, and that it should be *indifferent* for you either to pay €4 for buying a wager that pays back €5 in case the bus was blue (with a net gain of €1) or to sell the same wager for exactly the same amount, engaging yourself to pay back €5 in case the bus was blue (with a net loss of €1). Another way of saying the same thing is as follows: if your odds are 4:1 in favour of the proposition that the bus was blue, then you think a price of 80 cents is *fair* for a ticket worth €1 if the bus was blue and nothing if it was not.

Bets have been offered, and they still are, on a wide range of propositions concerning events, from games to political elections, for which there can be statistics. However, even if there are not statistics, betting is a method, practised for a long time, for quantifying personal degrees of belief in connection with choices. Real wagers, of course, must be decided during the lifetime of the bettors: if you make a real bet on the colour of the bus, you must have a method to verify whether the bus involved in the accident was blue. This is why real bets are not made on legal judgements or scientific theories: one can bet on the actual outcome of the trial, the decision of the jury, but one cannot bet if that decision is the *truth* of the fact. Also the truth of some scientific hypotheses or theories will not be known during our lifetime, and maybe not during the life span of mankind either. But decisions are taken on the basis of uncertain hypotheses, in everyday life, in courts of justice, in laboratories. Any help is welcomed on this subtle and important issue, so consider what betting behaviour can teach us about *coherent* decision making.

What is learnt is that all betting policies that violate the laws of probability are *incoherent* in the sense that they will lead to a sure loss, no matter which proposition turns out to be true. The proofs of this fact are referred to as *Dutch Book arguments*: 'In British racing jargon a *book* is the set of bets a bookmaker has accepted, and a book *against* someone – a *Dutch Book* – is one the bookmaker will suffer a net loss on no matter how the race turns out' (Jeffrey 2004, p. 5). There are two kinds of Dutch Book arguments in the literature: a *synchronic* Dutch Book argument providing a pragmatic justification of the laws of probability as standards of coherence for sets of beliefs held at the same time, and a *diachronic* Dutch Book argument providing a pragmatic justification of the principle of conditionalization as a standard of coherence for belief change.

Denote with pr the price you think 'fair' to pay for winning €1 if proposition H is true, and winning nothing if false (it doesn't matter, in that case, if you lose the price pr you have paid in advance, or if you pay that amount pr after H has been verified to be false). So far, no assumption is made about the mathematical properties of this measure pr, but the assumption is made that your degree of belief that the proposition H is true is measured by pr. In this chapter, the symbol pr will always denote the price of a bet, the symbols P and Q will denote probability measures, and the symbols p and q will denote particular probability values.

Any 'reasonable' price must range between 0 and 1 (inclusive):

$$0 \leq pr \leq 1. \tag{2.1}$$

Indeed, you would be 'unreasonable' if you were willing to pay more than the ticket is worth, and no 'reasonable' person can be willing to pay you for playing against him. Moreover, if H is known to be true, the only 'reasonable' behaviour is not to bet, that is, the unique fair price is equal to the ticket's worth:

$$\text{if } H \text{ is true}\ ,\ pr = 1. \tag{2.2}$$

Suppose now that you can buy a ticket worth €1 if H is true, and nothing if false, at price pr_1 and a ticket worth €1 if H is false, and nothing if true, at price pr_2. If you buy both tickets, you'll win €1 for sure; therefore, the only 'reasonable' prices are all those prices for which:

$$pr_1 + pr_2 = 1. \tag{2.3}$$

Indeed, if you buy the two tickets your gains, say G_1 and G_2, will be:

$$G_1 = 1 - pr_1 - pr_2 \text{ if } H \text{ is true;} \tag{2.4}$$
$$G_2 = 1 - pr_1 - pr_2 \text{ if } H \text{ is false.}$$

Then, $G_1 = G_2 = 0$ if and only if (2.3) holds. If your subjective evaluation of the tickets were such that $(pr_1 + pr_2) > 1$, and you were willing to buy both tickets at those prices, then you would incur in a sure loss. And if your subjective evaluation were such that $(pr_1 + pr_2) < 1$, and you were willing to sell both tickets at those prices, again you would incur a sure loss.

Of course, nobody would participate in such a Dutch Book because it is easily recognizable in this case that there are only two mutually exclusive and exhaustive propositions. The Dutch Book argument holds for $H_1, H_2, \ldots, H_n$ mutually exclusive and exhaustive propositions, and under the very general assumption that the values of n tickets are variables $X_1, X_2, \ldots, X_n$. Under these general assumptions, the price at which you will be willing to buy or sell a ticket will be a fraction $pr_i X_i$ of its winning worth X_i. Therefore, your gains are given by the linear set of equations:

$$\begin{aligned} G_1 &= X_1 - pr_1 X_1 - pr_2 X_2 - \cdots - pr_n X_n; \\ G_2 &= X_2 - pr_1 X_1 - pr_2 X_2 - \cdots - pr_n X_n; \\ &\ldots \\ G_n &= X_n - pr_1 X_1 - pr_2 X_2 - \cdots - pr_n X_n; \end{aligned} \tag{2.5}$$

and, in order to be able to make a Dutch Book against you, a bookmaker has to find some $X_1, X_2, \ldots, X_n$ such that all the G_i's are negative values. It can be proved algebrically that a Dutch Book is not possible if and only if:

$$1 - pr_1 - pr_2 - \cdots - pr_n = 0. \tag{2.6}$$

Equations (2.1), (2.2) and (2.6) are the *axioms of finitely additive probability measures*. Therefore, if your quantitative subjective degrees of belief pr_i satisfy these equations, then they *are* finitely additive probabilities, in the sense that they satisfy exactly the mathematical properties of such measures. Moreover, we say that your subjective degrees of belief are *coherent*, in the sense that no Dutch Book can be arranged against you, if and only if they are finitely additive probabilities. The idea that a quantitative measure of degrees of belief is given by the prices of 'fair' bets can be traced back to the seventeenth century fathers of the mathematical theory of probability, Blaise Pascal and Christiaan Huygens, who called *mathematical expectation* the 'fair' price of a bet, even though the synchronic Dutch Book Theorem was explicitly formulated and proved for the first time only in the twentieth century by the mathematician Bruno De Finetti (de Finetti 1937). It was also known by the philosopher and logician Frank Ramsey, although in his

(1931) article he did not provide a formal proof. De Finetti and Ramsey have been the most important representatives, in the last century, of the so-called *subjective*, or *personalist*, interpretation of probability which is embraced in this book.

When the number of propositions is infinite, a Dutch Book argument for additivity can still be given, provided the infinite set of propositions is a countable set[2]. A simple version of the Dutch Book argument for *countable additive probability measures* can be found in Jeffrey (2004, pp. 6–9). An infinite countable bet can be represented by a ticket that reads: 'For each positive whole number n, pay the bearer €1 if proposition H_n is true'. Of course, such a bet would be only a hypothetical one. Therefore, pragmatic justification can be extended to *countable additive probability measures* and, later in this chapter, it shall be seen what the implication is for subjective Bayesians 'to bet' on generalizations ('All x are such that...'), that is, to bet on scientific theories.

2.1.2 A rational policy for combining degrees of belief

In the standard mathematical presentations of probability theory, the *conditional probability* formula, which holds between any two propositions A and E, is given as a definition:

$$P(A \mid E) = \frac{P(A,E)}{P(E)} \text{ provided that } P(E) > 0. \tag{2.7}$$

In words, the probability that the proposition A is true, given that the proposition E is true, is equal to the probability that propositions A and E are both true, divided by the probability that E is true. The so-called *product rule of probability* is merely a restatement of (2.7):

$$P(A,E) = P(A \mid E)P(E). \tag{2.8}$$

From the subjective Bayesian point of view the product rule *is not a definition* but a *fundamental principle of coherent reasoning under uncertainty* which can be justified by a *Synchronic Dutch Book argument*, on a par with the probability axioms: a Dutch Book cannot be arranged against you if and only if your subjective degrees satisfy the product rule. The argument is based on another centuries-old betting practice, that one of accepting *conditional bets*, and runs as follows (Jeffrey 2004, 1988).

You are uncertain about the truth of propositions A and E: this means that you do not know which one of the possible propositions (A,E), $(A,\bar{E})$, $(\bar{A},E)$ and $(\bar{A},\bar{E})$ is true ($\bar{A}$ and $\bar{E}$ denote the logical negations of A and E). Assume now that your

[2]'A countable set is defined as one whose members (if any) can be arranged in a single list, in which each member appears as the n-th item for some finite n. Of course, any finite set is countable in this sense, and some infinite sets are countable. An obvious example of a countable infinite set is the set $I^+ = \{1, 2, 3, \ldots\}$ of all positive whole numbers.' (Jeffrey 2004, p. 7).

degree of belief in the truth of proposition A, on condition that the proposition E turns out to be true, is the price pr_1 for a ticket that is worth €1 if A and E both happen to be true, nothing if E turns out to be true but A is false, and it is worth exactly pr_1 if E happens to be false, no matter what A turns out to be. That is, if the condition of the bet on A does not occur, then the bet is 'called off' and you are given back the amount paid for the ticket, i.e. pr_1: this is called a *conditional bet*, and it was used precisely by Thomas Bayes (Bayes 1763) to define the concept of conditional probability in his *Essay toward solving a problem in the doctrine of chances*.

Suppose you are offered three tickets T_1, T_2 and T_3, the values of which depend upon all the possible logical combinations of A and E as shown in Table 2.1:

Table 2.1 The pay-off matrix.

	(A, E)	$(A, \bar{E})$	$(\bar{A}, E)$	$(\bar{A}, \bar{E})$
T_1	1	pr_1	0	pr_1
T_2	0	pr_1	0	pr_1
T_3	1	0	0	0

It can be seen that ticket T_1 is a conditional bet on A, ticket T_2 is a simple bet with pay-off € pr_1 if $\bar{E}$ is true, and nothing if E is true, and T_3 is a simple bet on A and E whose pay-off is €1. Now, if you buy both tickets T_2 and T_3, this action is equivalent to buying ticket T_1 only, because they are worth together exactly the same as T_1 (see Table 2.2):

Table 2.2 The combined pay-off matrix.

	(A, E)	$(A, \bar{E})$	$(\bar{A}, E)$	$(\bar{A}, \bar{E})$
T_1	1	pr_1	0	pr_1
T_2, T_3	1	pr_1	0	pr_1

Therefore, it would be surely 'irrational' for you to buy T_2 and T_3 at prices pr_2 and pr_3 which did not satisfy the equation:

$$pr_1 = pr_2 + pr_3. \tag{2.9}$$

But, if you are coherent, these prices are your subjective probabilities for winning, respectively, the fraction $P(\bar{E})$ of €pr_1 if proposition $\bar{E}$ is true, and €1 if proposition (A, E) is true:

$$pr_2 = pr_1 P(\bar{E}); \tag{2.10}$$
$$pr_3 = P(A, E).$$

Therefore, by substituting in (2.9), we obtain:

$$pr_1 = pr_1 P(\bar{E}) + P(A,E) = pr_1\,[1 - P(E)] + P(A,E), \qquad (2.11)$$

and, solving for pr_1:

$$pr_1 = \frac{P(A,E)}{P(E)} = P(A \mid E). \qquad (2.12)$$

Violation of the conditional probability rule means 'to place different values in the same commodity bundle in different guises' (Jeffrey 2004, p. 13). What happens if your prices do violate (2.9)? It can be formally proved that a Dutch Book could be made against you (de Finetti 1937).

The Synchronic Dutch Book Argument also provides a pragmatic justification for the claim that the following elementary theorem of probability calculus, the so-called *theorem on total probabilities*, is an all-important rule of reasoning under uncertainty:

$$P(A) = P(A,E) + P(A,\bar{E}) = P(A \mid E)P(E) + P(A \mid \bar{E})P(\bar{E}). \qquad (2.13)$$

Dennis Lindley has called this rule the *theorem of the extension of the conversation*:

> The extension theorem is perhaps the most widely used of all probability results and yet, at first glance, it looks as if it might be rarely useful, because it evaluates $P(A)$ by introducing another event E. What can be the point of adding E to the consideration of the chance of A occurring? The answer is that in many cases consideration of A alone is difficult and it is much easier to break it down into separate parts, given E and given $\bar{E}$. (Lindley 1985, p. 40)

It shall soon be seen that these 'cases' occur precisely in scientific inference in general, and in statistical inference in particular, where the theorem is extended to any number of exclusive and exhaustive uncertain propositions $E_1, E_2, \ldots, E_n$:

$$P(A) = \sum_{i=1}^{n} P(A \mid E_i)P(E_i). \qquad (2.14)$$

2.1.3 A rational policy for changing degrees of belief

As has been seen so far, the probability axioms, the product rule and the extension of conversation rule are rules for combining together degrees of belief based upon a *given* body of evidence. The bets in Table 2.1 are offered simultaneously on the basis of the same information: nothing has been said so far about how to *change* degrees of belief when the body of evidence changes. However, the product rule immediately provides such advice if the so-called *principle of conditionalization* is assumed.

The principle simply says that, if your degrees of belief at a certain time are represented by a probability P, and then you acquire new evidence which makes you know that a proposition E is true, then your *new* degrees of belief Q ought to be equal to your *old* conditional probabilities P given E, that is, for any proposition A such that $P(A) > 0$, and any proposition E such that $P(E) > 0$ and $Q(E) = 1$:

$$Q(A) = P(A \mid E). \tag{2.15}$$

A simple justification of the principle of conditionalization as a rule of coherence can be given as follows (Jeffrey 1988). Suppose you are offered the choice between buying *today* a ticket T_1, which is again a conditional bet worth €1, at the price pr_1 determined by your probability value p of today, and buying *tomorrow* both tickets T_2 and T_3 at prices pr_2 and pr_3, which will be determined by your probabilities q of tomorrow, whatever they will be (see Table 2.3).

Table 2.3 Pay-off matrix for today's and tomorrow's bets.

	(A, E)	$(A, \bar{E})$	$(\bar{A}, E)$	$(\bar{A}, \bar{E})$	Prices
T_1	1	pr_1	0	pr_1	$pr_1 = P(A \mid E)$
T_2	0	pr_1	0	pr_1	$pr_2 = pr_1 Q(\bar{E}) = P(A \mid E)Q(\bar{E})$
T_3	1	0	0	0	$pr_3 = Q(A, E)$

You do not know today what your probabilities will be tomorrow: they shall depend on what is going to happen tomorrow, but what you want to have *today* is a rational policy for changing probabilities in case you will receive tomorrow new evidence about A and E. But the action of buying tomorrow both tickets T_2 and T_3 will bring you the same pay-off as the action of buying today ticket T_1, no matter what your probabilities will be tomorrow: €1 if A and E are true, €pr_1 if it is not the case that E is true, and €0 if it is not the case that both A and E are true.

Therefore, what you do know *today* is that today's probabilities p and your tomorrow's probabilities q must be related by the equation:

$$P(A \mid E) = P(A \mid E)Q(\bar{E}) + Q(A, E). \tag{2.16}$$

Solving for $P(A \mid E)$ as before, we obtain:

$$\begin{aligned}
P(A \mid E) &= P(A \mid E)[1 - Q(E)] + Q(A, E), \\
P(A \mid E) &= P(A \mid E) - P(A \mid E)Q(E) + Q(A, E), \\
P(A \mid E) - P(A \mid E) + P(A \mid E)Q(E) &= Q(A, E), \\
P(A \mid E) &= \frac{Q(A, E)}{Q(E)},
\end{aligned}$$

and thus

$$P(A \mid E) = Q(A \mid E). \tag{2.17}$$

Equation (2.17) yields the rational policy you were looking for: *your today's and tomorrow's conditional probabilities must be the same*. It is easy to see that this policy amounts to obeying the principle of conditionalization.

Indeed, suppose that tomorrow you know that E is true: $Q(E) = 1$. Therefore, your tomorrow's price for ticket T_2 will be zero and your tomorrow's price for ticket T_3 will be:

$$P(A \mid E) = Q(A, E) = Q(A). \tag{2.18}$$

It can be shown that, if your today's and tomorrow's degrees of belief are not related as prescribed by equation (2.17), then a Dutch Book can be made against you. This *Conditional Dutch Book argument* was put in print for the first time by Teller (1973), who attributed it to the philosopher David Lewis, who in turn gave the credit to another philosopher, Hilary Putnam, who mentioned this result as early as in 1963, in a *Voice of America* broadcast, reprinted in Putnam (1975).

A very useful generalization of the principle of conditionalization is easily obtained by equation (2.17). Suppose that the new body of evidence does not allow you to know E with certainty, but it is such that your probability for E changes: that is, $P(E) \neq Q(E)$ and $Q(E) < 1$. We can rewrite (2.17) as follows:

$$\frac{Q(A, E)}{Q(E)} = \frac{P(A, E)}{P(E)}, \tag{2.19}$$

i.e.

$$Q(A, E) = \frac{Q(E)}{P(E)} P(A, E). \tag{2.20}$$

Given that you can update also $Q(A, \bar{E})$ in the same way, by extension of conversation your new probability is:

$$Q(A) = Q(A, E) + Q(A, \bar{E}). \tag{2.21}$$

You have a rule for changing degrees of belief in a coherent way also in case you do not know E for certain. This rule is known in the philosophical literature as *Jeffrey's Rule* or *Jeffrey's conditionalization*, from the name of the philosopher who proposed it (Jeffrey 1983, 1988, 2004). Dutch Book arguments for Jeffrey's Rule have been given by Armendt (1980) and Skyrms (1987).

2.2 COHERENT DECISION MAKING UNDER UNCERTAINTY

Maybe you do not consider betting a serious business, but insurance is based on the same principles. Maybe you have not made a bet on sport games or played a lottery in your life, but your income also depends on the betting behaviour of the financial markets. The acceptance of a bet is essentially the *choice* of an action with an uncertain outcome as an alternative to the choice of another action, i.e. a refusal of the bet and adherence to the *status quo*. The fact that betting in ordinary life is a voluntary activity should not conceal the fact that

> we are all faced with uncertan events and have to act in the reality of that uncertainty. [...] In this sense all of us 'gamble' every day of our lives [...] The essential concept is action in the face of uncertainty. (Lindley 1985, p. 19)

When you have to make choices among alternative actions with uncertain consequences the standard of coherence valid for wagers can be applied, if you think the case is important enough to deserve a careful appraisal of which is the best course of action. There are two main problems in applying the rational betting policy so far outlined to real life: how to *estimate the values* of the possible consequences of alternative decisions, and how to *estimate the probabilities* of propositions in realistic, and complicated, scenarios. Consider the first problem.

2.2.1 A method for measuring the value of consequences

The consequences considered so far were small monetary rewards, equal to or less than €1, and monetary prices have been used as measures of subjective degrees of belief. But when we have to choose among alternative actions in real life the possible consequences of most actions are neither monetary rewards nor can be sensibly compared to monetary rewards. Even when consequences are quantifiable as monetary rewards, their subjective evaluation, the '*utility of money*', usually is not linearly increasing with the increase of the amount of money, as economists know well. The eighteenth century mathematician Daniel Bernoulli was the first to suggest modification of the simple theory of mathematical expectation by the use of what he called '*moral expectation*': the value or the 'utility' for the owner of a monetary gain should not only be proportional to the gain itself, but also inversely proportional to the owner's total income. The intuitive idea behind this suggestion was that a gain of, say, €100 should be more subjectively valuable, and a loss of €100 more painful, for poor people than for rich people. Bernoulli's hypothesis also implied that monetary losses are more strongly felt than gains of the same amount: a 50%–50% gamble is disadvantageous, for although the expected monetary loss is equal to the expected monetary gain, the loss of 'utility' would be greater than the gain of 'utility'. The hypothesis was a big step forward for economics for it provided an explanation of people's observed behaviour like an aversion to gambling and a propension to subscribe to insurances, but it seemed to be bad news for the recommendation to apply rational betting policies to real life problems: through Sections 2.1.1–2.1.3 monetary prices have been set low enough to guarantee that the subjective 'utility' of money might not have a significant impact on your judgements, but if the tickets' values are significant with respect to your income, and if Bernoulli is right, your prices will not only reflect your subjective degrees of belief but also your subjective 'utility' of money. How are these two components of your evaluation to be untangled?

In general, the '*expected utility*' of an action for you (Bernoulli's moral expectation) will depend not only on the probability of obtaining the desired consequence but also on the value for you of that consequence compared with the values of other consequences that could be obtained my taking alternative courses of action. How is the value of consequences to be evaluated?

For a long time after Bernoulli's invention of a mathematical utility fuction, economists believed that it was only possible to ascertain a *qualitative ordering of preference* among consequences, e.g. to ascertain by means of observed choices that you prefer C to A and that, at the same time, prefer C to B, but a numerical estimate of the strength of your preference of C over A and of C over B was thought to be impossible. It was Frank Ramsey (1931), and a few years later, and independently from him, the mathematician John von Neumann together with the economist Oskar Morgenstern, who got the right insight: *gambles* were not the problem, they were the solution! In their book *Theory of Games and Economic Behavior*, first published in 1944, von Neumann and Morgenstern formulated the basic idea in this way:

> Let us for the moment accept the picture of an individual whose system of preferences is all-embracing and complete, i.e. who, for any two objects or rather for any two imagined events, possesses a clear intuition of preference.
>
> More precisely, we expect him, for any two alternative events which are put before him as possibilities, to be able to tell which one of the two he prefers.
>
> It is a very natural extension of this picture to permit such an individual to compare not only two events, but even combinations of events with stated probabilities (indeed this is necessary if he is engaged in economic activities which are explicitly dependent on probability).
>
> [...] Consider three events, C, A, B, for which the order of the individual's preferences is the one stated. Let α be a real number between 0 and 1, such that A is exactly equally desirable with the combined event consisting of a chance of probability $1 - \alpha$ for B and the remaining chance of probability α for C. Then we suggest the use of α as a numerical estimate for the ratio of the preference of A over B to that of C over B. (von Neumann and Morgenstern 1953, pp. 17–18)

Table 2.4 A hypothetical gamble as to whether E or $\bar{E}$ is true. Possible decisions are d_1 and d_2 with utilities $U(A)$, $U(B)$, $U(C)$ for monetary rewards A, B, C, respectively.

	E	$\bar{E}$
d_1	$U(A)$	$U(A)$
d_2	$U(C)$	$U(B)$

Suppose that, given the choice between three non-monetary rewards A, B, C, you prefer C to A and A to B and assume, for the sake of argument, that the 'utilities' these rewards have for you are measured by three non-negative real numbers $U(C)$, $U(A)$, $U(B)$, less than or equal to 1, such that $U(C) > U(A) > U(B)$. Now imagine a possible choice between a decision d_1, which is the decision to do an action which yields as a consequence reward A with certainty, and a decision d_2, which is the decision to do an action yielding reward C in the case a proposition E is true and B in the case E is false (see Table 2.4).

Consider, for example, a hypothetical gamble where a chance device offers a probability α of obtaining C, and $(1 - \alpha)$ of obtaining B, where α is continuously distributed from zero to one. For example, you can imagine drawing a black ball from a urn containing only black and white balls in a known proportion and such that the total number of balls and its composition can be varied so that the value α can be 'fine tuned' at will. The proposition E will be 'a black ball is drawn' and 'a white ball is drawn' will be the proposition $\bar{E}$.

Now imagine that the composition of the urn is changed until a probability α^* is found such that you are *indifferent* between d_1 and d_2. But, *if you are a coherent bettor*, then your fair 'price' of decision d_2 must be given by the sum $\alpha^* U(C) + (1 - \alpha^*)U(B)$. Therefore, the fact that you are *indifferent* between the choice of d_1 and the choice of d_2 means that, for you:

$$U(A) = \alpha^* U(C) + (1 - \alpha^*)U(B). \tag{2.22}$$

The following expression for α^* is then obtained:

$$\alpha^* U(C) - \alpha^* U(B) = U(A) - U(B),$$

$$\alpha^* = \frac{U(A) - U(B)}{U(C) - U(B)}. \tag{2.23}$$

This is von Neumann's and Morgenstern's 'suggestion', and it is easy to see that such a probability α^* can be taken as a measure of your subjective utility $U(A)$. As is explained later, it is possible, and convenient, to put $U(C) = 1$ for the *best* reward C and $U(B) = 0$ for the *worst* reward B; then, it immediately follows that:

$$U(A) = \alpha^*. \tag{2.24}$$

2.2.2 The consequences of rational preferences

In reframing von Neumann's and Morgenstern's quotation the cart has been put before the horse, assuming (i) that a quantitative ordering in the set of real numbers, $U(C) > U(A) > U(B)$, corresponds to the *qualitative ordering of preference* C, A, B, and (ii) that you behave as a coherent bettor when you are faced with a hypothetical gamble involving the three rewards. Now, the facts (i) and (ii) can be rigorously proved to hold true, provided that your ordering of preferences satisfies three conditions. An informal explanation of these conditions and their consequences is given here, while a formal treatment is offered in Chapter 3.

Any real-world decision problem can be analyzed into three basic elements:

1. A set D of *feasible alternative* decisions $d_1, d_2, \ldots, d_m$: only one of them can be taken at one time.
2. A finite set E of exhaustive and mutually exclusive uncertain events or *states of nature* $e_1, e_2, \ldots, e_n$: only one of them will occur.
3. A set C of rewards c_{ij}, which are the consequences of having taken decision d_i when event e_j occurs.

The probabilities of the events E can also depend upon your decision, so that your probability p_{ij} of e_j given decision d_i can be different from your probability p_{kj} of the same event given another decision d_k. You are supposed to follow a rational policy for assigning degrees of belief to states of nature, so that each feasible decision is a 'gamble' $d_i = (c_{i1}, \ldots, c_{in}; p_{i1}, \ldots, p_{in})$ consisting of the consequences $c_{i1}, \ldots, c_{in}$, with probabilities $p_{i1}, \ldots, p_{in}$, such that $\sum_{j=1}^{n} p_{ij} = 1$. You are now asked to consider, in addition to these gambles all the following hypothetical gambles: (i) all the compound gambles $(c, c'; \alpha, 1-\alpha)$ which can be composed of any pair of different elements c, c' in the set C, with probabilities $0 < \alpha < 1$; (ii) all the compound gambles $(d, d'; \alpha, 1-\alpha)$ which can be composed of any pair of different elements d, d' in the set D, with probabilities $0 < \alpha < 1$; (iii) all the 'degenerate' gambles which yield any one of the rewards in C with probability 1. Decisions d_1 and d_2 in Table 2.4 are 'gambles' of this kind: d_1 is an instance of (iii) and d_2 is an instance of (i).

This enlarged set G of gambles $f, g, h, \ldots$, is infinite, even though the sets D and C are finite, because α can take all real values from zero to one, and you are also asked to be able to express a preference between any two gambles in G which are offered as possibilities. This psychological effort must be made in order to have a rich mathematical structure on your preference system: this rich mathematical structure is defined by the following conditions or *axioms*.

The first condition requires that your system of preferences on G is *complete* and *transitive*. That the ordering is complete means, as already explained in von Neumann's and Morgenstern's quotation, that for any two gambles f and g, you are able to tell either which one of the two you prefer or that you are indifferent between them. That the ordering is transitive means that if you prefer gamble f to gamble g and prefer gamble g to gamble h, then you prefer also f to h; and if you are indifferent between f and g and between g and h, then you are also indifferent between f and h.

The second condition requires that your ordering preference is *invariant* with respect to compound gambles. More specifically, it requires that if you prefer gamble f to gamble g (or you are indifferent between them), then, for any other gamble h, and any probability α, you prefer the compound gamble that offers probability α of playing gamble f and probability $(1-\alpha)$ of playing gamble h, to the compound gamble that offers probability α of playing gamble g and probability $(1-\alpha)$ of playing gamble h:

> f is preferred (is indifferent) to g if and only if the gamble $(f, h; \alpha, 1-\alpha)$ is preferred (is indifferent) to the gamble $(g, h; \alpha, 1-\alpha)$, for any h and any α.

The third condition requires the comparability of gambles, in the sense that there are no gambles in the set G that are *infinitely desirable* or *infinitely undesirable* for you:

> let f, g and h be any three gambles such that f is preferred (is indifferent) to g and g is preferred (is indifferent) to h. Then there exists probabilities $\alpha, \beta \in (0, 1)$, such that (i) g is preferred to the gamble $(f, h; \alpha, 1-\alpha)$; (ii) the gamble $(f, h; \beta, 1-\beta)$ is preferred to g.

If (i) does not hold, then you will prefer a chance of obtaining the best consequence f, *no matter how small* it is, to g; that is, you believe that f is infinitely better than g (and h). If (ii) does not hold, then you will prefer g, *no matter how small* is the chance of obtaining the worse consequence h; that is, you believe that h is infinitely worse than g (and f).

If these three axioms are satisfied by your preference system, then the so-called *expected utility theorem* can be proved, that says that there exists a function U on the set C of rewards such that, for any d_j and d_k belonging to the set D, $d_i = (c_{i1}, \ldots, c_{in}; p_{i1}, \ldots, p_{in})$ is preferred (indifferent) to $d_k = (c_{k1}, \ldots, c_{kn}; p_{k1}, \ldots, p_{kn})$ if and only if

$$\sum_{j=1}^{n} p_{ij} U(c_{ij}) \geq \sum_{j=1}^{n} p_{kj} U(c_{kj}). \tag{2.25}$$

This means that your choice depends only on the value of a unique parameter: the expectation or mean value $\sum_{j=1}^{n} p_{ij} U(c_{ij})$, called the *expected utility* of a decision d_i. As a consequence, the best decision for you is the decision for which the value of this parameter is highest, that is, the decision that *maximizes the expected utility*.

Moreover, this function U is unique up to linear transformations: if $U(c_{ij})$ is a utility function then $aU(c_{ij}) + b$ is another utility function which preserves the same ordering of $U(c_{ij})$, that is, the two functions can be equivalently used to represent the same qualitative preference ordering, shifting only the origin of the curves. This invariance under linear transformation is the mathematical property that allows you to assign utility 1 freely to the best reward in the set C, and utility

0 to the worst reward, as we have done in Table 2.4. Then, if your preference ordering satisfies the axioms, it is true that:

(i) the 'price' for the gamble $(C, B; \alpha^*, 1-\alpha^*)$ will be evaluated by you *as if you were calculating its expected utility* $\alpha^* U(C) + (1-\alpha^*)U(B)$.
(ii) for any A, *there exists* a value α^* such that you will be indifferent between the 'degenerate' gamble $(A, A; \alpha^*, 1-\alpha^*)$ and the gamble $(C, B; \alpha^*, 1-\alpha^*)$.

Note that, in practice, this method to measure utility is subject to measurement errors (like any measure of empirical quantities), especially when the value of α^* is very near to either 0 or 1. Methods to improve the measurement can be found in the technical literature (Keeney and Raiffa 1976).

Formal proofs of the expected utility theorem were given for the first time by Ramscy (1931) and von Neumann and Morgenstern (1953). Marschak (1950) provided the axiomatization in terms of 'gambles', or 'prospects' as he called the probability distributions of rewards $(c_{i1}, \ldots, c_{in}; p_{i1}, \ldots, p_{in})$, an axiomatization which has been followed by the informal presentation given in the text. De Groot (1970) provides a modern presentation of this axiomatic approach in term of 'gambles'.

According to von Neumann and Morgenstern (1953) the probabilities p_{ij} which appear in gambles $d_i = (c_{i1}, \ldots, c_{in}; p_{i1}, \ldots, p_{in})$ were *frequencies* in long runs, but they wrote in a footnote that 'if one objects to the frequency interpretation of probability then the two concepts (probability and preference) can be axiomatized together' (von Neumann and Morgenstern 1953, p. 19). When they wrote this passage the joint axiomatization of subjective probability and utility had been already given by Ramsey (1931), but his approach became widely known only after Savage's *The Foundations of Statistics* (1972) was published.

Savage's axiomatization encompasses probabilities and utilities starting from a preference ordering on the set D of decisions $d_i = (c_{i1}, \ldots, c_{in}; e_1, \ldots, e_n)$, that is, without assuming that decision makers have already assessed their subjective probabilities for the states of nature e_j. Savage's axioms include the condition of completeness and transitivity for the preference ordering on D, and what he called the *sure-thing principle*, which is the equivalent of our requirement of invariance of preferences with respect to compound gambles.

Looking at Table 2.5 the principle says that, if d_1 is preferred (is indifferent) to d_2, then d_3 is preferred (is indifferent) to d_4, for any reward c, r. The reason why it has been called the 'sure-thing principle' should be obvious: the pairs of

Table 2.5 The 'sure-thing' principle.

	e	$\bar{e}$
d_1	a	c
d_2	b	c
d_3	a	r
d_4	b	r

decisions (d_1, d_2) and (d_3, d_4) have the same consequence in the case, $\bar{e}$, that the event e is false.

In Savage's approach it can be proved that, if your preferences among decisions D satisfy these two axioms, plus other axioms which play the same role of comparability of gambles, then your degrees of belief on the set E of states of nature are represented by a unique numerical function P that satisfies the axioms of the mathematical probability, like fair betting prices, and your ordering of preferences on the set C of consequences is represented by a utility function U for which the expected utility theorem holds.

2.2.3 Intermezzo: some more thoughts about rational preferences

It is important to understand the meaning and the scope of the expected utility theorem. The theorem is what in the technical jargon of modern measure theory is called a *representation theorem* (Krantz *et al.* 1971): it says that, if your preferences among gambles have a certain non-numerical structure, namely, the structure satisfying the axioms, then your preferences can be represented numerically by an interval function that *agrees* with them in the sense specified by equation (2.25). If your preferences fail to satisfy any one of the axioms, then they cannot be represented by such a numerical function. But why, or in which cases, should your preferences fail?

One must keep in mind that the preference ordering on the set G of 'gambles' is constructed starting from a given decision problem, with sets D, E, and C which are relative to that problem: this means that your utility function U is always relative to *that* framework. In a sense, the Ramsey–von Neumann–Morgenstern utility function U is context-dependent and it is significantly different from Bernoulli's 'utility' and from the concept of 'utility' held by nineteenth century philosophers and economists. Unlike Bernoulli, it is not supposed that all 'rational' men have the same subjective utility function, i.e. that they are risk averters: a utility function inversely proportional to the decision maker's total income is now only one of the possible shapes of the utility function U, when consequences are monetary rewards. The actual shape of your function U relative to a given decision problem will be revealed by your choices among gambles which are combinations of the rewards in *that* decision problem. Nor is it asked of you to have a unique subjective utility function over all possible decisions in your entire life. It is only required that, for particular decision problems under uncertainty which you consider important enough to think about, you can bring your own preferences among the rewards involved in *that* decision into compliance with the conditions of the theorem. If you are successful in doing that, then your preferences for *that* decision problem can be represented by a utility function and the decision that maximizes expected utility picks up what is the best decision according to the beliefs you entertain and to your revealed preferences. If, after careful thought, you arrive at the conclusion that your preference system cannot be brought to agree with the axioms, then the decision that maximizes expected utility shall not be the best decision for you.

Consider possible violations of the first axiom. Transitivity seems to be a straightforward requirement for rational choices: if d_1, d_2 and d_3 are alternative one-time actions and you prefer d_1 to d_2, and d_2 to d_3 but also d_3 to d_1, then you would be entrapped in a vicious circle at the moment of taking a decision. Experiments show that transitivity can be violated. For example, by adding small enough amounts of sugar to successive cups of coffee, I can arrange a sequence of cups of coffee such that you will be indifferent between adjacent cups, but not between the first and the last. You are 'tricked' into a transitivity failure because our senses have a threshold of perception. If it were really important to distinguish between adjacent cups, the problem could be 'reframed' by using chemical tests.

Completeness is a desirable property of rational preferences whose importance increases with the importance of the decision at hand: failure of completeness can be due to a coarse description of the alternatives and you can try to eliminate it by refining the frame. Sometimes what looks like a failure of completeness can be due to the complexity of the reward set C, when more components enter in the specification of the consequences of a decision. Technically, this is called a *multi-attribute decision problem* (Keeney and Raiffa 1976), and what can be done is to try and analyze a reward c into its k components, called *attributes*, and then compare decision d_1 and decision d_2 by a step-by-step comparison of each attribute. An elementary exposition of this method is given in Smith (1988).

Consider now the third axiom, the comparability of gambles. This is not a rationality condition: it is a mathematical condition that must be satisfied in order that the utility function U takes real numbers as its values. Indeed, the real number line is continuous and enjoys the so-called Archimedean property: any pair of positive numbers x and y are comparable, that is, the ratio x/y is not infinite. The axiom requires that you use 'a reward space whose components are not so different in priority that they cannot be compared using lotteries' (Smith 1988, p. 28). There is nothing 'irrational' in violating such a condition: it is an empirical fact that your reward set satisfies it. Here is an example given by Smith (1988, p. 28), involving a multi-attribute decision problem with two attributes, where the condition can fail.

A doctor's reward set is a pair $(x, -y)$ where $x = 1$ if a patient survives a given treatment and $x = 0$ if the patient does not survive, and y is the cost of treatment. Treatment T_1 has probability 1 of success and costs y_1; also treatment T_2 has probability 1 of success but is more expensive: $y_2 > y_1$; treatment T_3 is equally as expensive as T_2 but has only probability 1/2 of success (see Table 2.6).

Table 2.6 The doctor's reward set.

	Patient survives	Patient does not survive
Treatment T_1	$c_{11} = (1, -y_1); p_{11} = 1$	$c_{21} = (0, -y_1); p_{21} = 0$
Treatment T_2	$c_{12} = (1, -y_2); p_{12} = 1$	$c_{22} = (0, -y_2); p_{22} = 0$
Treatment T_3	$c_{13} = (1, -y_2); p_{13} = 1/2$	$c_{23} = (0, -y_2); p_{23} = 1/2$

Clearly, T_1 must be preferred to T_2 and T_2 to T_3. But the doctor may well prefer T_2 to a 'randomized' treatment which gives treatment T_1 with probability α and treatment T_3 with probability $(1 - \alpha)$, regardless of the value of α, because the survival of the patient is infinitely more important than the cost of treatment. The example is an extreme, and unrealistic, case because consequences having probability one are involved: if probabilites of survival were less then one for T_1 and T_2, things might be different. This example is given to make the point that violations of the axiom are empirically possible and that they are not necessarily evidence of 'irrational' preferences but a reflection of a particular scale of values.

The moral is that you must always carefully analyze your reward set and, as a forensic scientist, the reward set of your client, and you must not always take for granted the applicability of expected utility theory to the problem at hand. Indeed, the most fundamental problem to be dealt with in applying expected utility theory to legal decisions has to do exactly with the comparability of decisions which involve deep issues about ethical values of society: is the conviction of an innocent person an infinitely bad outcome? If the answer is yes, then the theory cannot be applied. This is not an issue that is of direct interest for forensic scientists, because their decisions consider the analysis of data and not the final decision of a trial. Something will be said about this issue at the end of the chapter.

Last but not least, consider the second axiom in both of the two versions that have been presented: preferences are invariant with respect to compound gambles and Savage's 'sure-thing principle'. This axiom is a rationality condition: it requires that in your system of preferences compound gambles like $(f, h; \alpha, 1 - \alpha)$ and $(g, h; \alpha, 1 - \alpha)$ be ranked as simple ones like f and g. In other words, it forces you not to consider the anxieties (or the pleasures) associated with the very act of gambling. This is a necessary condition for the existence of a utility function which satisfies the expected utility property: indeed, equation (2.22) requires a hypothetical comparison between a sure reward and a gamble. Notice that what is being considered is not measurable by a Bernoulli utility function, for to measure your subjective utility of money, in a form that could be Bernoullian, one must use equations like (2.22), where A, B, and C are now monetary rewards, and therefore one takes for granted the comparability of sure and uncertain rewards.

Von Neumann and Morgenstern (1953) asked in their book the questions:

> May there not exist in an individual a (positive or negative) utility of the mere act of 'taking a chance', of gambling, which the use of mathematical expectation obliterates? How did our axioms get around this possibility? (von Neumann and Morgenstern 1953, p. 28).

Their answer was:

> As far as we can see, our postulates do not attempt to avoid it. Even that one which gets closest to excluding a 'utility of gambling'[3] seems to be plausible and legitimate, – unless a much more refined system of psychology

[3]The authors are referring to preferences that are invariant with respect to compound gambles.

> is used than the one now available for the purposes of economics. [...] We have practically defined numerical utility as being that thing for which the calculus of mathematical expectation is legitimate. Since [the axioms] secure that the necessary construction can be carried out, concepts like a 'specific utility of gambling' cannot be formulated free of contradiction on this level. (von Neumann and Morgenstern 1953, p. 28)

It might be argued that a difference has to be made between real gambles and the theoretical idealizations used for calibrating utility scales, and the axiom asks you to be detached from a 'specific utility of the mere act of gambling' in those idealizations. That this is difficult to do, even in hypothetical choices, is shown by the famous 'Allais paradox', named from the French economist who presented it (Allais 1953).

You are offered a hypothetical choice between two lotteries with 100 tickets where all tickets are equally likely to win and the pay-offs are given according to Table 2.7.

Table 2.7 The Allais' Paradox: M stands for a million euros. d_1 is the decision to take a gamble that offers 1 million whichever ticket is drawn; d_2 is the decision to take a gamble that offers nothing if ticket number 1 is drawn, 5 million if tickets numbered from 2 to 11 are drawn, 1 million if tickets numbered from 12 to 100 are drawn; d_3 is the decision to take a gamble that offers 1 million if tickets numbered from 1 to 11 are drawn, and nothing if tickets numbered from 12 to 100 are drawn; and d_4 is the decision to take a gamble that offers 5 million if tickets numbered from 2 to 11 are drawn, and nothing if all the other tickets are drawn.

	Ticket number		
	1	2 − 11	12 − 100
d_1	$1M$	$1M$	$1M$
d_2	0	$5M$	$1M$
d_3	$1M$	$1M$	0
d_4	0	$5M$	0

The 'degenerate' gamble d_1 offers €1 million with certainty; the gamble d_2 offers a 1% chance of winning €0, a 10% chance of winning €5 million and a 89% chance of winning €1 million: which one do you prefer?

Now you are offered the choice between gambles d_3 and d_4 (see Table 2.7) which one do you prefer? If you have preferred d_1 to d_2 and d_4 to d_3, then you have violated the 'sure-thing principle' and a utility function does not exist which agrees with your preferences in this case *and* satisfies the expected utility property (2.25).

Allais's counterexample has been the cause of many heated discussions since its appearance. The standard Bayesian response is to claim that expected utility theory is normative, and not descriptive: its axioms are conditions that characterize the preferences of an ideally rational agent, and the fact that empirically observed behaviour violates them is not a valid argument against the hypothesis that they are coherence requirements. The point is: should an 'ideally rational agent' have a 'utility of the mere act of gambling'? Advocates of expected utility theory say 'no'. According to their intuition the 'sure-thing principle' is a condition that an ideally rational agent ought to follow and one should try for it when and where one can. In Allais' case, the behaviour that contravenes the principle is considered like a mistake in reasoning. Adversaries of the theory argue that the principle is not such a compelling condition of rationality, and many alternative proposals have been put forward in the literature (Allais and Hagen 1979).

Any scientific theory, any scientific research programme contains unsolved problems, 'puzzles' and 'anomalies', to use a term made popular by the historian of science Thomas Kuhn, in his famous book *The Structure of Scientific Revolutions* (Kuhn 1970). Allais' paradox is an 'anomaly' of the Bayesian paradigm, but no paradigm, and no theory, is rejected simply because there are anomalies: there must exist an alternative theory which provides satisfactory solutions to some of the anomalies, is able to 'save' in its framework most of the problems solved by the old theory, and possibly formulates new problems and opens new research paths. As it is attempted to be shown in the rest of the chapter, the Bayesian paradigm links decision making and scientific and statistical reasoning in a rich, fruitful and complete rational approach to everyday life. No alternative theory of rational choice that has been proposed so far is as wide in its scope and as capable of producing new results as the Bayesian paradigm.

> It constitutes a much deeper problem to formulate a system, in which gambling has under all conditions a definite utility or disutility, where numerical utilities fulfilling the calculus of mathematical expectations cannot be defined by any process, direct or indirect. [...] Some change of the system [of axioms], at any rate involving the abandonment or at least a radical modification of [the axiom of compound gambles] may perhaps lead to a mathematically complete and satisfactory calculus of utilities, which allows for the possibility of a specific utility or disutility of gambling. It is hoped that a way will be found to achieve this, but the mathematical difficulties seem to be considerable. (von Neumann and Morgenstern 1953, pp. 629–632)

We think it can be fairly said that no such new system is presently available.

2.2.4 The implementation of coherent decision making under uncertainty: Bayesian networks

The fact that we have a standard for coherent reasoning under uncertainty does not make the assessment of subjective probabilites an easy task. Also the judgements of probabilistically well-trained people might happen not to satisfy the probability axioms: there is extensive empirical evidence that people fall for probabilistic fallacies, that they make systematic errors in their explicit probabilistic judgements and that they make incoherent decisions (Baron 2000; Goldstein and Hogarth 1997; Kahneman *et al.* 1982; Tversky and Kahneman 1986). The best way of framing problems of judgement under uncertainty that could help people to reason in a probabilistically correct way is an issue under current debate amongst psychologists (Gigerenzer and Hoffrage 1995; Gigerenzer *et al.* 1999; Girotto and Gonzalez 2001, 2002; Hoffrage *et al.* 2002; Kahneman and Tversky 2000).

If you are willing to obey the probability rules, the estimation of probability values for singular events turns out to be easy, and quite natural, in those cases where you know *frequencies*. If you know the frequency of the occurrence of a given event E and you *don't know anything else about* E, then the Bayesian policy is to use the known frequency as the measure of your subjective probability for the past occurrence of E or for the next occurrence of E. The proviso is important, because it may happen that you know something else about that particular occurrence that is relevant for its probability. For instance, if you know the fraction of 60-year-old people who live to be 80, you can take it as your subjective probability that Peter, who is 60-years-old, will survive up to 80. But if you know also that Peter is a healthy and non-smoking guy, your subjective probability may well be higher than the overall statistical average. On the contrary, if you know that Peter is a heavy smoker, your subjective probability may well be below that average.

When either you do not know frequencies, or you do know but also know additional information such that your subjective probability leans above or below the average, or you are asked to assign a number to a one-off event, you may feel not very confident that you have got the 'right' number. In these cases *scoring rules* can be used to improve people's measurement of uncertainty (Lindley 1982). A scoring rule is a rule which gives to any probability value p assigned to a given proposition a score depending on whether the proposition is true or false. The basic idea behind the use of scoring rules for measuring the 'goodness' of probabilistic judgements is that it seems 'natural to say that the measurement was good if you had thought an event, subsequently shown to be true, had been likely, that is, given a large p; and an event later seen to be false to have been given a small p' (Lindley 1985, p. 22).

A widely used scoring rule is the *quadratic* (or *Brier*) *rule*. The score is thought of as a *penalty* (or *loss*) equal to $(1 - p)^2$ if the event is true and your judgement has been p, and equal to p^2 if it is false: the smaller the loss the better the judgement. In Table 2.8 the losses are shown for some p values.

It can be proved that, if you know the frequency with which event E happens, then the choice of that frequency as your judgement p is the only policy that minimizes the loss calculated with the quadratic rule: no other policy can fare better. In case of sequences of probability forecasts about a particular repeatable event, the rule can be used to measure one's capability of making good probability judgements. We say that a subject is *well-calibrated* if, over a sequence of forecasts for which they give probability p for event E, the frequency of occurrences of E is p. However, even in the case of one-off events, thinking in terms of the rule

Table 2.8 Quadratic scoring rule.

$P(E)$ p	E $(1-p)^2$	$\bar{E}$ p^2
1.0	0	1
0.9	0.01	0.81
0.8	0.04	0.64
0.7	0.09	0.49
0.6	0.16	0.36
0.5	0.25	0.25
0.4	0.36	0.16
0.3	0.49	0.09
0.2	0.64	0.04
0.1	0.81	0.01
0.0	1	0

when assessing your own personal probabilties can help to inprove the assessment. Indeed, Table 2.8 encourages honest and accurate measurement of probabilities in the sense that it penalizes both overbold and overcautious judgements: extreme values of p, near 0 or 1, yield the highest losses if the truth is the opposite to what you think, whereas you are encouraged to 'fine tune' any information you have, because moving from $p = 0.5$, in both directions, might yield a significant improvement in your performance.

Apart from the calibration problem, the satisfaction of the standards of pragmatic coherence can be a difficult task in real life decision problems when chains of events are involved. In these cases you must build a *probabilistic model* representing all the relationships you believe to hold amongst events. The number of such events to be taken into account can be large, and the task of computing the probabilities of the joint occurrence of a finite set of events rapidly becomes very intractable because its complexity increases exponentially with the number of events. However, it is known how to build computationally tractable probabilistic models: *Bayesian Networks*.

A Bayesian network is a *directed graph* without cycles. A directed graph is a mathematical structure whose elements are a finite set of nodes and a finite set of arrows (directed links) between the nodes. If there is an arrow pointing from node X to node Y, it is said that X is a parent of Y and Y is a child of X. A sequence of consecutive arrows connecting two nodes X and Y is a directed path between X and Y. A directed graph without cycles is a Bayesian network where:

1. Nodes are variables with a finite set of mutually exclusive states (Boolean nodes with only two states, t = true, and f = false, represent *propositions* or *events*).
2. Arrows represent *direct probabilistic relevance relationships* among nodes.
3. To each node X with parents $Y_1, \ldots, Y_n$, is associated a *probability table* containing all the conditional probabilities $P(X \mid Y_1, \ldots, Y_n)$. If X has no parents,

then its probability table reduces to probabilities $P(X)$, unconditional on other nodes in the graph.

The directed graph in Figure 2.1 does not contain cycles, and there are arrows from A and from B to C because the state of C is probabilistically dependent upon the states of both A and B. Analogously, the arrow from C to E stands for the fact that the state of E directly depends upon the state of C. The graph becomes a Bayesian network when the conditional probability tables, Tables 2.9, 2.10 and 2.11, are completed:

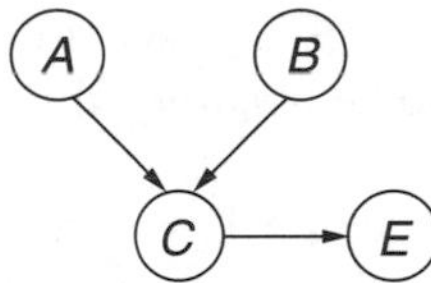

Figure 2.1 An example of a directed acyclic graph with propositions as nodes. A: suspect is the offender; B: blood stain at crime scene comes from offender; C: bloodstain at crime scene comes from suspect; E: suspect's blood sample and crime stain share the same DNA profile. (Adapted from Taroni *et al.* 2006a.)

Table 2.9 Unconditional probabilities assigned to the nodes A and B of Figure 2.1.

A:	t	f
$P(A)$	a	$1-a$

B:	t	f
$P(B)$	r	$1-r$

Table 2.10 Conditional probability table for the node C of Figure 2.1.

	A:	t		f	
	B:	t	f	t	f
C:	$P(C=t \mid A,B)$	1	1	0	p
	$P(C=f \mid A,B)$	0	0	1	$1-p$

Table 2.11 Conditional probability table for the node E of Figure 2.1.

	C:	t	f
E:	$P(E=t \mid C)$	1	γ
	$P(E=f \mid C)$	0	$1-\gamma$

In the tables, assignments of probabilities 1 and 0 are straightforward, probability γ is the so-called random match probability, calculated from statistical data, and a, r and p are subjective probabilities: a should be estimated on the basis of evidence pertaining to the case other than DNA evidence, r should be estimated on the basis of an analysis of the circumstances of the offence, and p, the probability that the stain would have been left by the suspect even though he was innocent, should be estimated on the basis of the alternative hypothesis proposed by the defence.

Notice that there are no arrows from either A or B to E, even though the states of the first two nodes are obviously relevant for the state of the last node. This is so because the influence that the states of A and B have on the state of E is only indirectly transmitted through the state of C, which is directly dependent from A and B, and in turn directly influences E (there are directed paths both from A and B to E). The state of E is said to be *conditionally independent* of the state of both A and B, given the state of C.

For any propositions A, B and C, A is said to be conditionally independent of C given B if and only if the conditional probability of A given B and C is equal to the conditional probability of A given B for all the states of A, B and C (for propositions for which there are only two states, t = true, and f = false):

$$P(A \mid B, C) = P(A \mid B). \tag{2.26}$$

Conditional independence is the fundamental concept in building computationally tractable probabilistic models, because the product rule of probability theory (2.8) can be factorized in Bayesian networks, for *all nodes* X in the network, as the product of their probabilities *conditional only* on the set $\mathbf{PA}(X)$ of their *parents*:

$$P(X_1, \ldots, X_n) = \sum_{i=1}^{n} P(X_i \mid \mathbf{PA}(X_i)). \tag{2.27}$$

For example, the network in Figure 2.1 can be factorized as follows:

$$P(A, B, C, E) = P(A)P(B)P(C \mid A, B)P(E \mid C). \tag{2.28}$$

Networks which can be factorized in this way are said to satisfy the *Markov property*, so called because formula (2.27) can be considered a generalization of a Markov chain.

Judgments of dependence, and conditional independence, among variables reflect your structural analysis of the problem, which is based on your theoretical and practical knowledge, and allow decomposition of complex problems into smaller and more manageable modules. This decomposition also makes the task of measuring uncertainty easier because you are now asked to focus, step-by-step, on conditional probability tables which correspond to structural relationships about which you have information.

Bayesian networks have efficient algorithms to compute formulae like (2.27), and to update probabilities on the states of the nodes, on receipt of new information, by means of a mechanism called *node instantiation*: the state of one or more nodes is exogenously fixed (the nodes are said to be instantiated), and the updating algorithm propagates this information through the network, changing the probabilities of all other nodes. For instance, probability one is assigned to the node E in Figure 2.1, and the algorithm calculates the new probabilities of the nodes A, B, and C, in such a way that the *principle of conditionalization* is satisfied. Bayesian networks obey pragmatic coherence. For a general introduction to Bayesian networks see Cowell *et al.* (1999), Jensen and Nielsen (2007), Kjærulff and Madsen (2008); foundational issues about Bayes networks and the interpretation of probabilistic dependencies as causal relationships are discussed in Glymour (2001), Pearl (2000) and Williamson (2004); Sloman (2005) argues about the psychological plausibility of such probabilistic models; Taroni *et al.* (2006a) provides an introduction to Bayesian networks focused on their use in forensic science.

A graphical method of representing decision problems using *decision trees* was introduced in the 1960s (Raiffa and Schlaifer 1961) and has become widely used since then. In the 1980s a more compact representation than the one provided by decision trees was devised: *influence diagrams* (Howard and Matheson 1984; Shachter 1986). An *influence diagram* is a directed acyclic graph with three types of nodes:

1. *Chance nodes*, represented by circles, which have a finite set E of exhaustive and mutually exclusive events or states of nature;
2. *Decision nodes*, represented by square boxes, which have a finite set D of feasible alternative decisions;
3. *Utility nodes*, represented by diamond boxes. Utility nodes have no states but to each utility node is associated a utility function over its parents: it represents the expected utility given the states of its parents;

and with the following properties:

4. Utility nodes have no children;
5. If there is more than one decision node, then there is a directed path consisting of all decision nodes: this ensures that there is a temporal sequence of decisions.

Bayesian networks can be naturally extended to represent decision problems by adding decision nodes and utility nodes. Decision nodes have no probability tables associated with them, because it is not meaningful to assign probabilities to a variable under the control of the decision-maker, and arrows pointing to decision nodes do not carry quantitative information; they only indicate that the state of the decision node's parents is known prior to the decision. A conditional probability table is associated with each chance node. The arguments of the table now can be not only other chance nodes but also a decision node. An example of an influence diagram for a forensic identification problem is given in Chapter 3 (Example 3.2.4).

The conditional independence relationships embedded in the structure of an influence diagram allow the decomposition of the conditional probability of *all* chance nodes, given the state of *all* decision nodes, as the product of the conditional probability of each chance node given its parents; that is, given the state *only* of the decision node which is its direct predecessor. Let Ω denote the set of all chance nodes and Δ denote the set of all decision nodes; then:

$$P(\Omega \mid \Delta) = \prod_{X \in \Omega} P(X \mid \mathbf{PA}(X)).$$

In the literature there can be found algorithms for solving influence diagrams (Jensen and Nielsen 2007; Kjærulff and Madsen 2008).

2.2.5 The connection between pragmatic and epistemic standards of reasoning

An interesting connection between *pragmatic* principles and *epistemic* principles of reasoning can be pointed out. If (2.17) is written as:

$$\frac{Q(A,E)}{P(A,E)} = \frac{Q(E)}{P(E)}, \tag{2.29}$$

it can be seen that the rational policy for updating the probability of a proposition A, upon new evidence bearing on the probability of a proposition E, can be formulated as the following recommendation: redistribute all your probabilities in such a way that *all the ratios between new and old probabilities are equal to the ratio between the new and the old probability of* E.

This amounts to a rule that redistributes your probabilities among all the propositions, which are possible according to the new evidence, in the *least biased way*: given that your new state of information has changed on learning only the new probability of E, and nothing else ($Q(E) = 1$ in the case of classical conditionalization, and $Q(E) < 1$ in the case of Jeffrey's conditionalization), you have no reason to make a change biased for or against particular propositions, that is, you have no reason to change the *probability ratio* among propositions.

It has been proved (Domotor *et al.* 1980; Williams 1980) that classical conditionalization and Jeffrey's conditionalization amount to a minimization of the 'distance' between the old and new probabilities, p and q, as measured by the *relative entropy*:

$$\sum_{i=1}^{n} \log\left(\frac{q_i}{p_i}\right) \text{ for } n \text{ exclusive and exhaustive propositions.}$$

The concept of relative entropy and its relation to probability updating is described by Diaconis and Zabell (1982). Therefore these rules can be seen as *epistemic principles of minimum change*: changing your degrees of belief according to these rules makes that minimum change in your probabilistic state of knowledge which is needed to take into account the new information, and only that information. It has also been proved (van Fraassen 1989) that they are the unique updating rules which satisfy a very general *symmetry principle* saying, roughly, that one should assign the same probabilities to logically equivalent statements of a problem.

Propagation of evidence in Bayesian networks is implemented using algorithms which allow coherent updating of the probabilities of nodes according to the generalized conditionalization formula (2.20). Therefore, we can say that evidence is propagated in Bayesian networks according to a principle of minimum change: on receipt of new inputs, the propagation algorithm changes all the probabilities of the chance nodes in the net in such a way that the relative entropy between the old and the new probability distributions over the net is minimized (Taroni *et al.* 2006a, pp. 60–66). Notice that during the propagation of evidence in Bayesian networks the probabilities of the nodes do change but the *structure* of the graph does not. Conditional probabilities in the conditional probability tables are kept fixed: they remain the same after one or more nodes have been instantiated, so that the conditionalization principle is satisfied. This fact sheds new light on the principle: it is a rational policy for belief updating whenever the empirical observations are such that they do not influence the form of the theoretical relationships built in the probabilistic model, that is, they do not change the form of the structural equations.

These results show the existence of a subtle link between the *pragmatic standard* of coherence, i.e. avoiding self-defeating betting policies, and *epistemic standards* of rationality, like *minimum principles* and *symmetry principles*, which are highly regarded and have been successfully applied in modern science. The role of symmetry considerations in the foundations of statistical inference is described in Section 2.3.4 below.

2.3 SCIENTIFIC REASONING AS COHERENT DECISION MAKING

Scientific method is a 'technique' of rational reasoning, a basket of intellectual tools which have been forged during centuries of scientific practice in order to correct and reinforce the innate propensity of our perceptual apparatus to organize the inputs from the environment and to form expectations about it. Confronted with data coming from an uncertain and complex environment one develops *hypotheses* and *theories* to explain the data and to serve as guidance to act in that environment. There are hypotheses and theories that range from very simple ones ('the car does not start because the spark plugs are dirty') to highly sophisticated ones such as Schroedinger's wave function. A good 'theory' implies some expectations about future events and related actions: devising and implementing 'experiments' ('check the spark plugs and change them') whose outcomes are highly probable if the 'theory' is true.

> The scientific method is not radically different from the rational attitude in everyday life or in other domains of human knowledge. Historians, detectives, and plumbers – indeed, all human beings – use the same basic method of induction, deduction, and assessment of evidence as do physicists or biochemists. Modern science tries to carry out these operations in a more careful and systematic way, by using controls and statistical tests, insisting on replication, and so forth. Moreover, scientific measurements are often much more precise than everyday observations; they allow us to discover hitherto unknown phenomena; and they often conflict with 'common sense'. (Sokal and Bricmont 1998, p. 56)

Some general epistemological principles can be listed which enjoy a wide acceptance because they have been successful in the past (for example, 'good' theories must predict novel facts because it is always possible to arrange *ad hoc* explanations for known data, but it is impossible to do that for yet unknown data). However 'the scientific method' cannot be defined once and for all, because it is continuously evolving together with scientific results. The fact that scientific method is a work in progress does not impede the recognition and acknowledgement of good scientific practice.

> To illustrate this, let us consider an example that is in a certain sense intermediate between scientific and ordinary knowledge, namely that of criminal investigations. There are some cases in which even the hardest skeptic will find it difficult to doubt, in practice, that the culprit has been really found: one may, after all, possess the weapon, fingerprints, DNA evidence, documents, a motive, and so forth. Nevertheless, the path leading to the discovery of the culprit can be very complicated: the investigator has to make decisions (on the leads to follow, on the evidence to seek) and draw provisional inferences, in situations of incomplete information. Nearly every investigation involves inferring the unobserved (who committed the crime) from the observed. Here, as in science, some inferences are more rational than others. The investigation could have been botched, or the 'evidence' might simply be fabricated by the police. But there is no way to decide a priori, independently of the circumstances, what distinguishes a good investigation from a bad one. Nor can anyone give an absolute guarantee that a particular investigation has yielded the correct result. Moreover, no one can write a definitive treatise on *The Logic of Criminal Investigation*. Nevertheless – and this is the main point – no one doubts that, for some investigations at least (the best ones), the result does indeed correspond to reality. Furthermore, history has permitted us to develop certain rules for conducting an investigation: no one believes anymore in trial by fire, and we doubt the reliability of confessions obtained under torture. It is crucial to compare testimonies, to cross-examine witnesses, to search for physical evidence, etc. Even though there does not exist a methodology based on unquestionable *a priori* reasoning, these rules

(and many others) are not arbitrary. They are rational and are based on a detailed analysis of prior experience. (Sokal and Bricmont 1998, pp. 58–59)

Bayesian epistemology claims that the tools in the basket of scientific method, whatever they might be, must surely satisfy the standard of pragmatic coherence for combining and changing degrees of belief: therefore, the fundamental tool of scientific method is Bayes' theorem (Bovens and Hartmann 2003; Earman 1992; Horwich 1982; Howson and Urbach 1996; Jaynes 2003; Jeffrey 2004; Jeffreys 1961; Maher 1993; Salmon 1990, 1967; Swinburne 2003).

2.3.1 Bayes' theorem

Bayes' theorem is a straightforward consequence of the product rule and the theorem of the extension of the conversation and is so called after Thomas Bayes who first derived it from the definition of a conditional bet (see Section 2.1.2 above):

$$P(A \mid E) = \frac{P(A, E)}{P(E)}$$
$$= \frac{P(E \mid A)P(A)}{P(E \mid A)P(A) + P(E \mid \bar{A})P(\bar{A})}, \text{ provided that } P(E) > 0. \quad (2.30)$$

Used together with the conditionalization principle it yields a simple but powerful formula that captures all the logical relationships between theory and evidence, when probabilities are interpreted as degrees of belief. Consider a set x of observations which will be thought of as evidence, H a hypothesis that has been put forward to explain this set (evidence), and a set y of observations arising from a new experiment devised to test H. Then, the rational policy for changing degrees of beliefs (see Section 2.1.3 above) requires that the new degree of belief in the hypothesis H, after having observed y, must be given by the old conditional probability of H given y and x, and calculated accordingly:

$$P(H) = P(H \mid x, y) = \frac{P(y \mid H, x)P(H \mid x)}{P(y \mid x)}. \quad (2.31)$$

Scientific hypotheses are introduced to simplify a complex web of real-world connections, and this simplification usually means that evidences x and y can be considered conditionally independent given H. Therefore we can rewrite (2.31) as follows:

$$P(H \mid x, y) = \frac{P(y \mid H)P(H \mid x)}{P(y \mid x)}. \quad (2.32)$$

Probability $P(H \mid x, y)$ is conventionally called the *posterior*, or *final*, probability of H and $P(H \mid x)$ is conventionally called the *prior*, or *initial*, probability: of course, probabilities are *a posteriori* and *a priori* only relative to new evidence y. What is prior with respect to y might very well have been obtained by a previous application of the theorem on evidence x: your priors of today are your posteriors of yesterday and your posteriors of today will be your priors of tomorrow. The probability $P(y \mid H)$ is called the *likelihood of H given y*, the *likelihood of H* for short. When H is a scientific hypothesis the numerical value of the likelihood can usually be estimated by means of theoretical equations, and the use of the conditionalization principle is justified because the form of these theoretical relationships does not change from the time when x is observed to the time when y is observed.

If Bayes' theorem is now written as:

$$P(H \mid x,y)P(y \mid x) = P(y \mid H)P(H \mid x), \tag{2.33}$$

it is easy to see that the posterior probability of H, given x and y, is greater than the probability of H, given x, if and only if the likelihood of H is greater than the probability of y, given x:

$$P(H \mid x,y) > P(H \mid x) \text{ if and only if } P(y \mid H) > P(y \mid x). \tag{2.34}$$

In such a case the new evidence y is said to *support*, or *confirm*, hypothesis H; otherwise it is said that it does not support, or disconfirms, H. Evidence y confirms hypothesis H if y is *unexpected*, or *surprising*, *unless* H is true, e.g. $P(y \mid H) > P(y \mid x)$, and evidence y disconfirms H if y is *unexpected*, or *surprising*, if H is true, e.g. $P(y \mid H) < P(y \mid x)$.

The 'qualitative' behaviour of Bayes' theorem shown by equation (2.34) sounds good for both common sense and scientific reasoning. Imagine that y is the result of a scientific experiment devised to test H: in such a case, y is usually a quantitative result predicted by the theory with an uncertainty that depends only on the limits of the theoretical computations, which involve several approximations, and on the limits of the measuring instruments; for the sake of argument suppose that, within these limits, H implies y: $P(y \mid H) = 1$. Then (2.33) becomes:

$$P(H \mid x,y)P(y \mid x) = P(H \mid x) \tag{2.35}$$

and the following important facts may be checked.

(i) The less probable evidence y is, unless H is true, i.e. the lower $P(y \mid x)$, the more informative y is about H; this agrees with Karl Popper's recommendation that good tests of a scientific theory look for improbable predictions, unless the theory is true, and the more improbable the prediction is, the better the test (Popper 1959).

(ii) Scientific hypotheses can never be *verified*, that is, no empirical evidence can raise their probability to 1. Indeed, by the theorem of the extension of the conversation, the value of $P(y \mid x)$ is factorized as:

$$P(y \mid x) = P(y \mid H)P(H \mid x) + P(y \mid \bar{H})P(\bar{H} \mid x). \tag{2.36}$$

It has been assumed that $P(y \mid H) = 1$, therefore the minimum value for $P(y \mid x)$ would be such that $P(y \mid x) = P(H \mid x)$, when $P(y \mid \bar{H}) = 0$. But this minimum, which implies $P(H \mid x, y) = 1$, cannot be attained when H is a high-level scientific hypothesis and y is empirical evidence: there is always a possible alternative explanation of an empirical fact, even though it may not be able to be worked out. Thus $P(y \mid \bar{H}) > 0$. This is what philosophers of science sometimes call the 'catch-all hypothesis'.

The 'catch-all hypothesis' does not disturb current scientific practice: scientists only consider one single theory and they do not take into account alternatives until it is strictly necessary, and they are justified in doing that, according to Bayesian epistemology.

Take x to be a sequence of observations all implied by H, i.e. $x = (x_1, x_2, \ldots, x_n)$, and y to be a further observation implied by H, $y = x_{n+1}$; then, (2.35) can be rewritten as follows:

$$P(H \mid x_1, x_2, \ldots, x_{n+1}) = \frac{P(H \mid x_1, x_2, \ldots, x_n)}{P(x_{n+1} \mid x_1, x_2, \ldots, x_n)}. \qquad (2.37)$$

Now, given that $P(H \mid x_1, x_2, \ldots, x_{n+1})$ cannot exceed 1, and that $P(H \mid x_1, x_2, \ldots, x_n) \neq 0$, then $P(x_{n+1} \mid x_1, x_2, \ldots, x_n)$ must tend to 1 as n increases. Therefore, 'repeated verifications of consequences of a hypothesis will make it practically certain that the next consequence of it will be verified' (Jeffreys 1961, p. 43). Notice that $P(x_{n+1} \mid x_1, x_2, \ldots, x_n)$ involves neither H nor an alternative to H but only the observable x's, and it holds *whether H is true or not*. The 'catch-all hypothesis',

> if it had been thought of, would either (1) have led to consequences $x_1, x_2, \ldots$ or (2) to different consequences at some stage. In the latter case the data would have been enough to dispose of it, and the fact that it was not thought of has done no harm. In the former case the considered and the unconsidered alternatives would have the same consequences, and will presumably continue to have the same consequences. The unconsidered alternative becomes important only when it is explicitly stated and a type of observation can be found where it would lead to different predictions from the old one. The rise into importance of the theory of general relativity is a case in point. Even though we know now that the systems of Euclid and Newton need modification, it was still legitimate to base inferences in them until we knew what particular modification was needed. The theory of probability makes it possible to respect the great men on whose shoulders we stand. (Jeffreys 1961, p. 44)

This is a very important point because it explains why scientific laws are relied upon and are willingly applied with practical certainty in particular instances, even though they are not believed to be true, i.e. $P(H \mid x_1, x_2, \ldots, x_{n+1}) < 1$. Notice also that, after a long sequence of positive observations, the difference $[P(H \mid x_1, x_2, \ldots, x_{n+1}) - P(H \mid x_1, x_2, \ldots, x_n)]$ will be very small, that is, the degree of support provided by the new instance will be negligible. This also provides a first answer to the question raised at the end of Section 2.1.1, that is, what does it mean for subjective Bayesians 'to bet' on generalizations ('All x are such that...') or universal scientific laws: it means to bet in a certain way on next instances of the law. As we have seen, the odds in favour of the next instance can be high, and they will always be equal or higher than the odds in favour of the universal law, because necessarily (see (2.37)):

$$P(x_{n+1} \mid x_1, x_2, \ldots, x_n) \geq P(H \mid x_1, x_2, \ldots, x_n).$$

Scientists are obliged to consider alternatives when a failure in the observation of y *falsifies* H: indeed, if $P(\bar{y} \mid H) = 0$, then $P(H \mid x, y) = 0$. Therefore, falsification of a scientific hypothesis is a particular case of Bayes' theorem that applies when H implies y.

Bayesian epistemology is a *qualified* 'verificationism' that

> does not take the task of scientific methodology to be that of establishing the truth of scientific hypotheses, but to be that of confirming or disconfirming them to degrees that reflect the overall effect of the available evidence – positive, negative, or neutral, as the case may be. (Jeffrey 1975, p. 104)

We know that decision making under uncertainty does not require certainty of knowledge to take rational decisions. Going down from high theories to everyday life the truth of hypotheses like 'spark plugs are dirty' can be verified but it should be remembered that even in most cases of common sense reasoning what can be attained is not certainty but a probability so near to certainty that it can be taken as a practical certainty, 'beyond any reasonable doubt'.

If it is possible to verify some low-level hypotheses, then it is symmetrically possible to falsify other low-level hypotheses, like 'the car is out of fuel', but it is not possible, *pace* Popper, to falsify scientific hypotheses: a high-level, interesting, scientific hypothesis H does not imply observational consequences by its own, but always in conjunction with some *auxiliary hypotheses* A which are usually independent from H and, among other things, 'describe special features of the system under consideration' (Jeffrey 1975, p. 105)[4]. Therefore Bayes' theorem should be written as:

$$P(H, A \mid x, y) = \frac{P(y \mid H, A) P(H \mid x) P(A \mid x)}{P(y \mid x)}. \tag{2.38}$$

This means that, if $P(H, A \mid x, y) = 0$, scientists can choose to put the blame either on H or on A. The history of science shows that scientists' first reaction

[4]Example : 'Newton's law of universal gravitation implies nothing observational about the motions of the planets by itself. It is rather the conjunction of Newton's law of gravitation with the law $F = ma$ and with the hypothesis that all forces except for the gravitational ones are negligible in determining the motions of the planets which has observational consequences about those motions.' (Jeffrey 1975, p. 105)

is to change the auxiliary hypotheses, instead of the beloved H: they shall toil with A, trying modified, or totally new, A'. The planet Uranus was discovered in 1782 and astronomical observations showed that its orbit violated Newton's law of universal gravitation, but the law was not rejected and a new auxiliary hypothesis predicting the existence of a new planet led to dramatic vindication of Newton's theory: the discovery in 1846 of Neptune. Then, another 'falsification' of Newton's universal gravitation popped out: the anomalous Mercury's perihelion was observed for the first time in 1859. Many new auxiliary hypotheses had been floating around since then to save Newton's law (among them the existence of a new planet, *Vulcanus*) until 1915, when the astronomer Karl Schwarzschild calculated exactly the observed value of the perihelion deviation using Einstein's General Relativity (Roseveare 1982). The prior probability of General Relativity could be very low indeed by then, but the probabilities of the *ad hoc* auxiliary hypotheses put forward to save Newton's Laws were also low.

2.3.2 The theories' race

What has been outlined above in Bayesian terms is similar to the story told by Thomas Khun in *The Structure of Scientific Revolutions* (see Section 2.2.3, Kuhn 1970): no particular hypothesis or theory is rejected simply because there are 'anomalies', pieces of evidence disconfirming or even falsifying the theory, but it is rejected only if an alternative is available that is capable of solving the 'anomalies' and of making novel predictions. In the history of science there is never only one runner at a time: there are always two or more serious competitors at the start of the race, that is, hypotheses and theories whose probabilities on the available evidence are greater than zero. This race has been given the name "theories' race". Copernicus put forward his hypothesis against that of Ptolomeus, Galileo defended his hypotheses about motion against Aristotelian theories of motion, Newton's corpuscular theory of light faced Huygens' ondulatory theory, Darwin's theory of natural selection faced Lamarck's evolutionism and eighteenth-century versions of Intelligent Design. The race of scientific theories is not a matter of absolute probabilities but a matter of relative probabilities, a matter of comparative judgements.

The Bayesian way of deciding the winner among alternative hypotheses which are recognized as 'starters' in the race, that is, as worth testing, makes use of the *odds form* of Bayes' theorem. Let H_1 and H_2 be two *mutually exclusive*, but not *exhaustive* hypotheses, that is, it is possible that both are false. This is the typical situation for scientific laws and high-level scientific theories. By the odds on H_1 *against* H_2 is meant the ratio of your probability of H_1 to your probability of H_2. The *prior odds* on H_1 *against* H_2, relative to evidence x, that is the odds based only upon evidence x, are:

$$\frac{P(H_1 \mid x)}{P(H_2 \mid x)}.$$

Note that these odds are different from the odds on H_1 (or H_2) against its *logical negation*, defined in Section 2.1.1 as:

$$\frac{P(H_1 \mid x)}{P(\bar{H}_1 \mid x)},$$

unless, of course, H_1 and H_2 are also exhaustive[5].

The *posterior odds* on H_1 against H_2, relative to evidence y (again, different from the posterior odds on H_1 against its logical negation) are:

$$\frac{P(H_1 \mid x, y)}{P(H_2 \mid x, y)}.$$

Developing this last formula, we obtain:

$$\frac{P(H_1 \mid x, y)}{P(H_2 \mid x, y)} = \frac{\frac{P(y|H_1,x)P(H_1|x)}{P(y|x)}}{\frac{P(y|H_2,x)P(H_2|x)}{P(y|x)}} = \frac{P(y \mid H_1, x)}{P(y \mid H_2, x)} \times \frac{P(H_1 \mid x)}{P(H_2 \mid x)}. \quad (2.39)$$

The ratio $P(y \mid H_1, x)/P(y \mid H_2, x)$ is called the *likelihood ratio*: therefore, the posterior odds equal the likelihood ratio times the prior odds. The likelihood ratio is also called the *Bayes factor* (see Section 6.2.1) and is a measure of the *relative strength of support* which evidence y gives to H_1 as against H_2, given x. If the relative strength of support is greater than 1, it is said that evidence y supports H_1 more than H_2; if it is less than 1, it is said that evidence y supports H_2 more than H_1, and if it is equal to 1, it is said that evidence y is not relevant for the hypotheses at stake.

Suppose that, prior to the observation of evidence y, hypothesis H_1 is believed from the relevant scientific community, on the basis of evidence x, more credible than H_2; Bayesian epistemology dictates that they should reverse their preference if and only if, after observation of y, the overall belief in H_2 is greater than that in H_1. It is easy to see from (2.39) that:

$$P(H_2 \mid x, y) > P(H_1 \mid x, y) \text{ if and only if } \frac{P(y \mid H_2, x)}{P(y \mid H_1, x)} > \frac{P(H_1 \mid x)}{P(H_2 \mid x)}. \quad (2.40)$$

Preferences should be changed from H_1 to H_2 if and only if the likelihood ratio of H_2 to H_1 is greater than the prior odds of H_1 in favour of H_2.

The odds form of Bayes' theorem answers the problem related to the 'catch-all hypothesis', namely, that no likelihood can be calculated for the 'catch-all hypothesis': the denial of a scientific hypothesis is not a scientific hypothesis, and it is void of any explanatory or predictive power. Now, the term $P(y \mid x)$ does not appear in (2.40), so that no reference is made to the 'catch-all hypothesis'. There is a price to be paid for that, and it consists in the fact that no precise quantitative

[5]The formula for translating the posterior odds on H against its negation $\bar{H}$ into probabilities is the following, with O denoting the odds value: $P(H \mid x, y) = O/(O + 1)$.

posterior probabilities can be calculated for H_1 and H_2 when they are exclusive but *not* exhaustive hypotheses. One cannot pretend to be able to execute such a calculation for genuine scientific theories: an estimate of the *posterior odds* on H_1 *against* H_2 is the best that can be aimed for, if quantitative likelihoods can be calculated, and an estimate of the prior odds can be made. This comparison is often sufficient to decide the winner of the race.

> In fact, most applications of Bayesian standpoint in everyday life, in scientific guessing, and often also in statistics, do not require any mathematical tool nor numerical evaluations of probabilities; a qualitative adjustment of beliefs to changes in the relevant information is all that may be meaningfully performed. (de Finetti 1974, p. 117)

In cases where competitors can be reduced to a list of exclusive *and* exhaustive hypotheses, the odds form of the theorem is well suited to a decision-theoretic approach to the choice of hypotheses. Consider again the simplest case where H_1 and H_2 are exhaustive hypotheses: one of them is the 'true' hypothesis, and suppose interest is only in the truth about the world, so that 'utility' 1 is assigned to the consequences of choosing the 'true' hypothesis, and 'utility' 0 is assigned to anything else. Then, the associated decision matrix is given in Table 2.12 and the 'expected utility' EU of choosing hypothesis H_i on data x,y is equal to the probability of H_i given those data:

$$\begin{aligned} EU(d_i \mid x,y) &= U(d_i,H_i \mid x,y)P(H_i \mid x,y) + U(d_i,H_j \mid x,y)P(H_j \mid x,y) \\ &= P(H_i \mid x,y). \end{aligned}$$

Therefore, maximizing 'expected utility' yields the same decision as (2.40).

A more interesting utility structure can be obtained if a *loss function* is used. The concept will be formally defined in Chapter 3 (Sections 3.2.3 and 3.4.2). For the purposes of this chapter it will be sufficient to consider a loss function as a *negative utility*, so that a certain 'loss' L is incurred when the 'false' hypothesis is chosen, and there is no 'loss' when the 'true' hypothesis is chosen (Table 2.13).

Table 2.12 A decision matrix using utilities.

	H_1	H_2
d_1: choosing H_1	1	0
d_2: choosing H_2	0	1

Table 2.13 A decision matrix using losses.

	H_1	H_2
d_1: choosing H_1	0	L_{12}
d_2: choosing H_2	L_{21}	0

There can be a *symmetric* loss $L_{12} = L_{21}$, which reduces to the previous case, and an *asymmetric* loss if $L_{12} \neq L_{21}$. In this second case, the 'expected loss' EL of choosing hypothesis H_i on data x,y will be:

$$EL(d_i \mid x, y) = L_{ij} P(H_j \mid x, y). \tag{2.41}$$

The rational decision is to take the action with the lowest 'expected loss'. Given that

$$EL(d_2 \mid x, y) < EL(d_1 \mid x, y) \text{ if and only if } L_{21} P(H_1 \mid x, y) < L_{12} P(H_2 \mid x, y),$$

the choice involves a comparison of posterior odds with the 'losses' possibly incurred in choosing the wrong hypothesis. Then H_2 will be preferred to H_1 if and only if:

$$\frac{P(H_1 \mid x, y)}{P(H_2 \mid x, y)} < \frac{L_{12}}{L_{21}}$$

or, equivalently:

$$\frac{P(H_2 \mid x, y)}{P(H_1 \mid x, y)} > \frac{L_{21}}{L_{12}}. \tag{2.42}$$

Expression (2.42) can be compared with (2.40). Given that

$$\frac{P(H_2 \mid x, y)}{P(H_1 \mid x, y)} = \frac{P(y \mid H_2, x)}{P(y \mid H_1, x)} \times \frac{P(H_2 \mid x)}{P(H_1 \mid x)},$$

it can be seen that hypothesis H_2 will be preferred to hypothesis H_1 if and only if:

$$\frac{P(y \mid H_2, x)}{P(y \mid H_1, x)} \times \frac{P(H_2 \mid x)}{P(H_1 \mid x)} > \frac{L_{21}}{L_{12}}.$$

Rewriting this, H_2 will be preferred to H_1 if and only if:

$$\frac{P(y \mid H_2, x)}{P(y \mid H_1, x)} > \frac{L_{21} P(H_1 \mid x)}{L_{12} P(H_2 \mid x)}. \tag{2.43}$$

Therefore, (2.40) comes out to be the limiting case of (2.43), when $L_{12} = L_{21}$. The loss ratio in (2.42) fixes a threshold for posterior odds, sometimes called the *posterior odds cutoff*: equation (2.42) says that if the posterior odds for H_2 exceeds the loss from incorrectly choosing action d_2 divided by the loss from incorrectly choosing action d_1, then you should choose d_2. Hypothesis testing as a special case of decision making will constitute the topic of Chapter 6, where more technical details will be given.

This account of the theories' race provides also an answer to the point that genuine scientific generalizations can only have a degree of belief proximate to zero: 'certainly, it seems mad to imagine that we shall ever find ourselves attributing more than a very small positive probability to the whole corpus of physical theory, as found in the textbooks of the day' (Jeffrey 1975, p. 112), or to the whole corpus of life science and social science theories as well. Bayes' theory gives us methodological advice to cope with a world in which human minds can only make fallible guesses.

> The generally accepted scientific hypotheses of the day – the 'laws' – are those serious competitors whose probabilities on the available evidence are orders of magnitude greater than those of the recognized alternatives. But the probabilities of such 'laws' must nevertheless be expected to be orders of magnitude less than those of their denials. (Remember that the denial of a scientific hypothesis is surely no scientific law – not because that denial is improbable but rather because it is not 'lawlike', e.g. it will be essentially devoid of explanatory or predictive content.) (Jeffrey 1975, p. 113)

The winner of the race needs neither to be believed with certainty nor with high probability, but it is either the most probable of the starters on the ground of all the available evidence, if the danger of choosing the wrong theory is the same, or that competitor which minimizes the expected 'loss', if the danger is different (think of the scientific hypotheses about the greenhouse effect). For scientific reasoning as well as for everyday life, 'decision-making under uncertainty – this is the home ground of Bayesianism' (Jeffrey 1975, p. 112).

2.3.3 Statistical reasoning: the models' race

The basket of tools of scientific reasoning today contains *statistical* methods for estimating errors of measurements and for making inferences on statistical data. The Bayesian paradigm in statistical inference will be developed in full in Chapter 3. Here the discussion is limited to showing that statistical inference in particular follows the same standard of coherence as scientific reasoning in general.

The fundamental problem of statistical inference is to use a set $x = (x_1, x_2, \ldots, x_m)$ of random quantities, which are the result of observations made before a decision has to be taken, to make predictions about another set $y = (x_{m+1}, x_{m+2}, \ldots, x_n)$ of random quantities, as yet unknown, which shall influence the outcome of our decisions. It is usually very difficult to try to calculate *predictive probabilities*, $P(y \mid x)$, directly because y is not independent of x. Therefore *probabilistic models* of the data are introduced which simplify the computation of the required probabilities by extending the conversation. A *probabilistic model* is basically a hypothesis about the mathematical form of the *probability distribution* that provides the likelihoods of the observed results; this form contains one or more characteristics of the population, called *parameters*, which are not known. The model enables calculation of the likelihood for different values of the parameters. A formal treatment of the concepts of a probability distribution and a probabilistic model will be given in Chapter 3. In the following generic notation $P(x \mid \theta)$ is used to denote the likelihood of data x, given a particular value of parameter θ. A probabilistic model can also state that all the observations are *independent*, conditional on each particular value of the parameter[6]. In generic notation, we can write

[6]Correlated data such as quantities of drugs on adjacent notes in a bundle associated with drug dealing are not considered. This is beyond the scope of the book.

this as follows:

$$P(x, y \mid \theta) = P(x_1, x_2, \ldots, x_n \mid \theta) = \prod_{i=1}^{n} P(x_i \mid \theta). \tag{2.44}$$

In particular, future observations are taken to be independent from past observations, given the parameter:

$$P(y \mid \theta, x) = P(y \mid \theta). \tag{2.45}$$

By saying that the values of the parameter are unknown, it is meant that they are theoretical quantities, postulated to explain the data, and which can be estimated from the data. Imagine a very large urn containing an unknown proportion of black and white balls (a *population*): the urn is so large that, for all practical purposes, is impossible to draw all the balls in the urn. Estimation of the proportion of black balls in the urn (the 'parameter') is made only on the basis of a finite, and not very long, sequence of drawings (a *sample*): the probability of any sequence of drawings is given by a probability distribution (the hypergeometric distribution if the drawings are without replacement, or the binomial distribution if they are with replacement)[7].

Suppose, for the sake of simplicity, that the *parameter space* Θ of the problem of interest is a discrete set, taking its values over a finite set of values, as in the example of the urn. Then, by extending the conversation over the elements $\theta \in \Theta$, your *predictive probability* will be given by:

$$P(y \mid x) = \sum_{\Theta} P(y, \theta \mid x) = \sum_{\Theta} P(y \mid \theta, x) P(\theta \mid x).$$

This formula can be simplified, by (2.45), as follows:

$$P(y \mid x) = \sum_{\Theta} P(y \mid \theta) P(\theta \mid x). \tag{2.46}$$

The numerical value of the first term of the sum on the right-hand side in (2.46) can be easily calculated *via* (2.44). The second term is a conditional probability that can be calculated *via* Bayes' theorem, with your beliefs about the possible values of θ, before observing x, being represented by an assessment of prior probabilities $P(\theta)$:

$$P(\theta \mid x) = \frac{P(x \mid \theta) P(\theta)}{\sum_{\Theta} P(x \mid \theta) P(\theta)}. \tag{2.47}$$

This is also the basic formula for *parameter estimation*, i.e. in the Bayesian framework, it is the formula for updating your beliefs about the 'true' value of the parameter on data x.

---

[7]See Appendix A for mathematical details about binomial and hypergeometric distributions.

Bayesian parameter estimation and statistical hypothesis testing are a matter of comparison of different values of θ, exactly as happens with theories of large scope and scientific 'laws'. At the start of the 'race' in parameter estimation there may be many competitors, that is, the priors $P(\theta)$ are distributed over the set Θ, and, after observation of x, the posteriors $P(\theta \mid x)$ may be 'peaked' or around only a few values of θ. In the case of statistical hypothesis testing, the same theoretical framework, Bayes' theorem, encompasses all the possible cases of model comparison. One can have a comparison of *simple* versus *simple* hypotheses, if the models assign probability one to specific values θ_0 and θ_1:

$$\frac{P(\theta_0 \mid x, y)}{P(\theta_1 \mid x, y)} = \frac{P(y \mid \theta_0)P(\theta_0 \mid x)}{P(y \mid \theta_1)P(\theta_1 \mid x)}.$$

> The data affect the change of belief through the likelihood ratio, comparing the probabilities of the data on θ_0 and on θ_1. This is in contrast with a sampling-theory (or tail-area) significance test where only the null hypothesis (say θ_0) is considered by the user of the test. Of course, attempts to justify these tests had to consider the alternatives but the user is freed from this necessity. (Lindley 1990, pp. 48–49)

Significance tests can lead to rejection of the 'null' hypothesis even though the evidence actually *supports* the 'null' hypothesis, because the evidence would be even more unlikely given a plausible alternative hypothesis. The failure to consider alternatives *is not* good scientific practice for, as we have seen in the previous section, a scientific hypothesis can be evaluated only in comparison with at least one genuine alternative hypothesis: for Bayesian epistemology, what is good for scientific laws must be good also for statistical hypotheses[8].

One can have a comparison of a *simple* hypothesis ($\theta = \theta_0$) versus a *composite* hypothesis, $\theta \in \Theta_1$, where $\Theta_1 = (\theta_1, \ldots, \theta_k)$ (a finite discrete space) with probability $P_j(\theta)$ assigned to $\theta_j (j = 1, \ldots, k)$. In this case, the likelihood ratio is:

$$\frac{P(y \mid \theta_0)}{\sum_{j=1}^{k} P(y \mid \theta_j)P(\theta_j \mid x)}.$$

For comparison of a *composite* hypothesis ($\theta \in \Theta_0$) versus a *composite* hypothesis ($\theta \in \Theta_1$) (with Θ_0, Θ_1 discrete and exclusive) the likelihood ratio will be:

$$\frac{\sum_{\Theta_0} P(y \mid \theta)P_0(\theta \mid x)}{\sum_{\Theta_1} P(y \mid \theta)P_1(\theta \mid x)}.$$

The previous comments about the minimization of the expected loss in choosing between alternative theories, hold as well for probabilistic models. The choice of methods to use in order to make decisions using probabilistic models will be the main topic of this book.

[8]This point is particularly important because significance tests have been used, and are still used, in forensic science. The problem of rejection of a hypothesis actually supported by evidence has been raised and discussed in the literature (Evett 1991; Lindley 1977b). This situation will be fully discussed in Chapter 6.

> An essential activity of all life is to make judgments about as yet unobserved data y on the basis of observed data x. This is the problem of inference or inductive logic. The Bayesian paradigm requires that this be done solely and entirely within the calculus of probability; in particular, the above judgment is $P(y \mid x)$. The calculation of such probabilities is substantially assisted by the consideration of theories including hypotheses H, and models incorporating a parameter θ. Independence, conditional on H or θ, appears to be basic to the calculations. Statistical practice ought therefore to start from the data, x and y, and regard the analysis, involving theories and models, as means of evaluating the probabilities. (Lindley 1990, p. 50)

2.3.4 Probabilistic model building: betting on random quantities

According to the subjective or personalist Bayesian paradigm, probabilities are always someone's betting ratios. This is not to deny that there are numbers in the world that you may adopt as your personal probabilities, if you knew them, like *observed relative frequencies* and *observed means*. But parameters of probabilistic models are not observed numbers, they are 'theoretical quantities'. A parameter of a probabilistic model is, for example, 'the *limit* of the relative frequency of occurrence of a certain event as the number of observations grows to infinity': but the 'limit of an infinite sequence of observations' cannot be observed, it is a theoretical ideal. Parameters are not the 'numbers' that are out there, in the world, but they are conceptual constructs which constitute the 'bridge' allowing you to pass from your judgements about observed numbers to other judgements about observable but as yet unobserved numbers. It is important to know when it is justifiable to build this 'bridge'. There are some very important results in theoretical statistics which give a precise answer to this question, an answer that can be formulated in terms *only of actual betting ratios about observable events*, that is, in terms of how you would bet on the truth of statements concerning sequences of observations $(x_1, \ldots, x_n)$. These results show how subjective Bayesians can 'recycle' observed relative frequencies as betting odds for unobserved instances, without any need to *identify* probabilities with mathematical limits of relative frequencies.

It has been said before, in Section 2.3.1, that Bayesians can legitimately have opinions about hypotheses that, typically, contain theoretical entities which are not directly observable but do have observable consequences: to have opinions about scientific laws means to bet in a certain way on particular instances of the laws. The same is true for probabilistic models: to have opinions about unobservable parameters means to bet in a certain way on particular sequences of observations. The 'certain way' has a name, *exchangeability*. The concept of exchangeability is explained as follows.

Imagine you are thinking beforehand about all the possible results of an experiment consisting of repeated observations of a certain real-valued random quantity X; i.e. you are thinking about a probability assignment p over all the possible sequences $(X_1 = x_1, \ldots, X_n = x_n)$. Suppose your state of information is such that your p is *symmetric*, or *invariant* for each n-tuple of real numbers $x_1, \ldots, x_n$, and each possible permutation of the individual observations X_i, that is:

$$P(X_1 = x_1, \ldots, X_n = x_n) = P(X_1 = x_{\pi(1)}, \ldots, X_n = x_{\pi(n)}) \tag{2.48}$$

for all permutations π on the set of individual subscripts $\{1, \ldots, n\}$.

This means that only observed numbers matter, and not their particular order, the place of the sequence in which they have been observed. If condition (2.48) does hold for you, then the actual observed sequence $(x_1, \ldots, x_n)$ is called *finitely exchangeable*, and p is called an *exchangeable probability assignment*: intuitively, the name comes from the fact that probabilities do not change by 'exchanging' the place of individual observations in the sequence.

A sequence of tosses from a standard coin, where the random quantity X can take only two values $\{0, 1\}$, or a sequence of drawings from an urn, known to be performed under constant conditions, are typically considered exchangeable sequences. A sequence of measurements of some physical or chemical property made in the same laboratory on the same sample can be judged exchangeable. But suppose you know that some faults occurred in the standard measurement procedure for a subset of the sequence: then, you might not judge the complete sequence exchangeable, but only the subset in which the procedure has been properly accomplished. Alternatively imagine the sequence of measurements of the same substance as a combination coming from different laboratories: then you might not consider the complete sequence exchangeable, even if you believe it to be exchangeable for the sequences coming from the different laboratories; then, if you know from which laboratory each individual measurement comes, you can subdivide the complete sequence into exchangeable sequences, one for each laboratory.

It should be evident from these examples that the notion of exchangeability replaces, for subjectivist Bayesians, the classical notion of 'conditional independence given a parameter', and the related notion of a *random sample*. This also means that judgements of exchangeability overrule *randomization*. One can use randomization devices as practical and economical means to create samples from a population, or to form treatment and control groups in controlled experiments, but their use does not free one from the burden of checking whether exchangeability is satisfied in the samples, given all the available information, because improbable events can occur.

Now, a finite exchangeable sequence $(x_1, \ldots, x_n)$ is said to be *extendible* if it is part of a longer sequence $(x_1, \ldots, x_N)$ $(N > n)$ which is still exchangeable. Actual exchangeable sequences are of this kind: the outcome of an experiment can be considered as part of a larger, but finite, sequence of observations which will be obtained by replications of the experiment. An *infinitely exchangeable* sequence is a sequence *indefinitely extendible*, i.e. extendible for all $N > n$, and such that every finite subsequence is judged exchangeable.

The fundamental results about exchangeability, which are called *representation theorems*, can be informally summarized as follows: if p is your exchangeable probability assignment about a sequence of random quantities that can be indefinitely extended, then your degree of belief can be represented *as if* it were obtained by extending the conversation to a countable set of probabilistic models with a (subjective) probability assignment μ over this set[9].

The most general theorem which can be proved says, more formally, that:

> (*General representation theorem*) if p is an exchangeable probability assignment, for you, on an indefinitely extendible sequence of real-valued random quantities $(x_1, x_2, \ldots)$, then there exists a prior probability measure μ in the space $\Im$ of all the probability distributions F on the set of real numbers such that, for any finite sequence of observations $(x_1, \ldots, x_n)$
>
> $$P(x_1, x_2, \ldots, x_n) = \int_{\Im} \prod_{i=1}^{n} F(x_i)\mu(F)dF$$
>
> where $\mu(F) = lim_{n\to\infty}P(F_n)$, F_n is the empirical distribution function given by the observed sequence $(x_1, \ldots, x_n)$, and μ represents your beliefs about what would be the form of this distribution for n tending to infinity.

In other words, the theorem says that, for you, the probability distribution F is mathematically well defined, and the observed sequence $(x_1, \ldots, x_n)$ is *as if* it were a sequence of independent observations conditional on this (unknown) distribution F:

> thus, 'at a stroke', we establish a justification for the conventional model building procedure of combining likelihood and prior. The likelihood is defined in terms of an assumption of conditional independence of observations given a parameter; the latter, and its associated prior distribution, acquire an operational interpretation in terms of a limiting average of observables (in this case a limiting frequency). (Bernardo and Smith 2000, p. 174).

The theorem provides no clues to the form of the distribution F, and the space of all probability distributions on the real line is so big that other conditions, in addition to exchangeability, are needed in order to be able to specify a particular form. We say that the 'parameter' F is of *infinite dimension* and, conventionally, in statistical literature, models involving infinite dimensional parameters are called *nonparametric models*, whereas those involving *finite dimensional* parameters are called *parametric models*. Fortunately, it is possible to impose some more restrictive conditions which allows the proof of representation theorems for more tractable

---

[9]For the definition of 'countable set' see footnote 2 above.

and familiar classes of distributions such as *binomial*, *Normal*, *Poisson* and *exponential* distributions, among others (for a survey of these results, see Bernardo and Smith 2000, pp. 172–229)[10].

The requirement that an exchangeable sequence of length n be infinitely extendible is necessary to prove the representation theorems, but it might seem contrary to the spirit of the Bayesian paradigm with its emphasis on observables. Appropriate theorems on finite approximation of infinite exchangeability have been proved which show that, if an exchangeable sequence of length n is extendible to sequences of length $N > n$, for N very large but finite, then 'no important distortion will occurr in quantifying uncertainties' (Bernardo and Smith 2000, p. 227).

In the following, details are given only of the representation theorem for the binomial distribution, because it is a mathematically simple but very important particular case. The first reason why it is so important is historical: it was the subject of the first representation theorem for exchangeable random quantities, proved by Bruno de Finetti in 1930 (de Finetti 1930, 1937). The second reason is philosophical: it is the theorem that creates the 'bridge' between observed frequencies and betting ratios, and that clarifies why and how probabilities, which are *subjective degrees of beliefs*, can be estimated given knowledge of *objective relative frequencies*.

The theorem can be stated informally as follows: if the class of observable random quantities is restricted to *binary* quantities, which take only values $\{1, 0\}$ (i.e. 'true' and 'false' or 'success' and 'failure'), then exchangeability is sufficient to specify the form of the distribution F, and it is exactly the binomial or Bernoulli distribution.

(*De Finetti's representation theorem*) if p is an exchangeable probability assignment, for you, on an indefinitely extendible sequence of binary random quantities $(x_1, x_2, \ldots)$, then there exists a prior probability measure μ such that, for any finite sequence of observations $(x_1, \ldots, x_n)$

$$P(x_1, x_2, \ldots, x_n) = \int_0^1 \prod_{i=1}^{n} \theta^{x_i}(1-\theta)^{1-x_i}\mu(\theta)d\theta$$

where the parameter θ is the limit of the relative frequency of 1's (of 'successes'), as n tends to infinite (this limit exists by the strong law of large numbers). Therefore, μ represents your beliefs about the value of this limit.

In other words, if you assign the same probability to all the sequences of length n with the same number of ones, that is, only the number of ones in n trials

[10]In this book only parametric models are presented (see Section 3.2). See Appendix A and Appendix B for the description of distributions.

matters, and not the order in which they are observed, then any finite sequence of observations can be considered as a unique 'weighted' mixture of draws with replacement from a possibly infinite set of urns, containing balls marked 1 and 0 in different proportions θ (these fictional 'proportions' are mathematical limits but, as such, the theorem proves that they exist). The 'weights' are the probabilities μ's expressing your beliefs about the 'true' urn from which the drawing is made.

Applying the theorem to both the numerator and denominator in (2.47), you can update your beliefs about the 'true' urn on the basis of past trials, and then calculate (by 2.46) your probability of 'success' in the next trial, or your probability of observing a fixed relative frequency of 'successes' in a future sequence of trials, *together with* every other piece of information you have (your probabilities μ). A practical example of the calculations involved will be given in Section 3.3.

The fact that only the number of 'successes' (the sample frequency) is informative means that this number is what is called a *sufficient statistic*, that is, a function of the data that summarizes all the relevant information (the *sample mean* in a binomial distribution is another example of a sufficient statistic)[11]. The topic cannot be dealt with here, but this simple example can give a hint about the relationship between exchangeability and the classical concept of sufficient statistics: in many cases, the existence of a sufficient statistic that summarizes the data is a consequence of a judgement of exchangeability.

It has been shown that the use of probabilistic models can be formally justified as a logical consequence of the fact that our beliefs satisfy a very general *symmetry* property, namely, exchangeability. Having said that, this section ends with a *caveat* that will be followed in this book.

> In practice, of course, there are often less formal, more *pragmatic*, reasons for choosing to work with a particular parametric model [...] In particular, specific parametric models are often suggested by an *exploratory data analysis* (typically involving graphical techniques to identify plausible distributional shapes [...]), or by *experience* (i.e. historical reference to 'similar' situations, where a given model seemed to 'work') or by *scientific theory* (which determines that a specific mathematical relationship must 'hold', in accordance with an assumed 'law'). In each case, of course, the choice involves subjective judgments; for example, regarding such things as the 'straightness' of a graphical normal plot, the 'similarity' between a current and a previous trial, and the 'applicability' of a theory to the situation under study. From the standpoint of the general representation theorem, such judgments correspond to acting as if one has a μ[12] which concentrates on a subset of $\mathfrak{I}$ defined in terms of a finite-dimensional labelling parameter. (Bernardo and Smith 2000, p. 229)

[11]A formal definition of a sufficient statistic will be given in Section 3.3.2

[12]The letter Q in the original text has been replaced with the letter μ for the sake of consistency with the rest of the notation in this book.

2.4 FORENSIC REASONING AS COHERENT DECISION MAKING

According to the subjectivist Bayesian paradigm, probability always refers to a *single*, well-*specified*, event. We have seen above how subjective coherent degrees of belief concerning singular events, like a particular sequence of observations, are combined with knowledge concerning what happens 'in the long run', like past observed frequencies, and how this knowledge is used to estimate probabilities for unique events, like the outcome of the next observation. The Bayesian paradigm thus offers a powerful framework for the combination of personal judgements of probability in a *coherent* way, and it fits well with the goals of forensic scientists, who are typically asked to estimate the probabilities of the occurrence of unique events, to evaluate evidence on the light of alternative hypotheses, and in doing that they necessarily must combine judgements based upon both statistical and non-statistical knowledge.

2.4.1 Likelihood ratios and the 'weight of evidence'

The graph in Figure 2.1 above, with its probability tables (Tables 2.9, 2.10 and 2.11), shows a classical 'blend' of the different kinds of probabilistic judgements a forensic scientist is required to make.

Node B represents the so-called *relevance* term (Stoney 1991), and r is the probability that the blood stain is relevant for the case, a subjective probability that the scientist should estimate on the basis of the information available about the circumstances of the fact, information that can be also of a statistical nature (this means that node B could be, in turn, further analyzed: see Aitken *et al.* 2003). Probability γ, the *random match* probability of the DNA profile in the relevant population (Balding and Nichols 1994), is calculated from statistical data; p is another subjective probability, the probability that the stain has been left by the suspect even though he is innocent, and it should be estimated on the basis of the alternative hypothesis proposed by the defence. The probability a of node A is the probability that the suspect is the offender given the available evidence different from the DNA evidence. Probability can be considered the probability held, before considering the DNA evidence, by anyone whose task is to evaluate all the evidence (scientific and non-scientific) pertaining to the case, namely, the judge or the jury. As such, it is not the task of the forensic scientist to give an estimate of this probability, for the scientist is not asked to express an opinion on the question which the court had to decide, the so-called 'ultimate issue'. The scientist's task is to evaluate the *likelihood ratio*:

$$LR = \frac{P(E \mid A)}{P(E \mid \bar{A})}.$$

The *LR* measures the *relative strength of support* which evidence E gives to A as against $\bar{A}$. The judgements of conditional independence contained in the Bayesian network of Figure 2.1, together with the probability assignments in Tables 2.10 and 2.11, allow the decomposition of this complex likelihood ratio according to factorization (2.28), and its reduction to the following manageable formula (Taroni *et al.* 2006a, pp. 97–101):

$$\frac{P(E \mid A)}{P(E \mid \bar{A})} = \frac{P(B) + P(\bar{B})P(E \mid \bar{C})}{P(F \mid \bar{A}, \bar{B})P(\bar{B}) + \left[P(B) + P(\bar{F} \mid \bar{A}, \bar{B})P(\bar{B})\right] P(E \mid \bar{C})}.$$

Substitution of the numerical estimates provided by the forensic scientist, gives:

$$\frac{P(E \mid A)}{P(E \mid \bar{A})} = \frac{r + (1-r)\gamma}{p(1-r) + r\gamma + (1-p)(1-r)\gamma} = \frac{r + (1-r)\gamma}{r\gamma + (1-r)\left[p + (1-p)\gamma\right]}. \tag{2.49}$$

Therefore, adoption of the Bayesian approach provides forensic scientists with the appropriate formula in order to calculate the relative support that E gives to the two parties' propositions and to communicate the results as *likelihood ratios* (Aitken and Taroni 2004; Balding 2005; Evett 1993; Jackson *et al.* 2006; Redmayne 2001; Robertson and Vignaux 1993, 1995).

> Forensic scientists are most helpful in assisting the probabilistic evaluation of evidence if they express their findings in terms of a likelihood ratio which results from a comparison between the probability of the evidence given the competing propositions in a case at hand. This is what in this context is sometimes referred to as the 'likelihood ratio approach'. One of the main arguments for this standpoint is that a likelihood ratio is – due to its convenient multiplicative properties – directly and unambigously amenable for belief updating by a recipient of expert evidence. (Biedermann *et al.* 2007a, p. 86)

The forensic scientists can use in their assessments prior probabilities for the propositions proposed by the parties, provided they make clear either that they are using the parties' prior probabilities, or that they are assuming certain priors in order to carry on simulation of the effect of evidence on the parties' propositions. In the DNA example of Figure 2.1, it is the scientist's task to estimate the probability r of the 'relevance' of the stain, a 'prior' probability with respect to DNA evidence, because 'relevance' is not the 'ultimate issue' in this context, whereas it is the court's task, and that of the parties, to estimate the 'prior' probability a that the accused person is the offender, because the identity of the offender is, in this context, the 'ultimate issue'[13].

A useful function of the likelihood ratio is its logarithm, called by Good the *weight of evidence* (Good 1950, 1985, 1988).

[13] A verbal scale for interpreting likelihood ratio can be found in Section 6.2.1

Let H denote a hypothesis, such as that an accused person is guilty, and let E denote some evidence, such as that presented by a specific witness. We ask how should we define $W(H : E \mid I)$[14], the weight of evidence in favour of H provided by E when background knowledge I is regarded as given or previously taken into account. It is natural to assume that the new evidence converts the prior probability into its posterior probability, that is, that $P(H \mid E, I)$ is a mathematical function of $P(H \mid I)$ and of the weight of evidence. Moreover, $W(H : E \mid I)$ should depend only on (i) the probability of E given that the accused is guilty, and (ii) the probability of E given that he is innocent [...] These desiderata lead to the conclusion that $W(H : E \mid I)$ must be a monotonic function of the Bayes factor and we may as well take the logarithm of the *Bayes factor* as our explicatum because this leads to desirable additive properties. (Good 1988, p. 389)[15]

In our notation, the posterior odds:

$$\frac{P(H \mid E, I)}{P(\bar{H} \mid E, I)} = \frac{P(E \mid H, I)}{P(E \mid \bar{H}, I)} \times \frac{P(H \mid I)}{P(\bar{H} \mid I)}$$

can be rewritten as follows:

$$\log\left[\frac{P(H \mid E, I)}{P(\bar{H} \mid E, I)}\right] = \log\left[\frac{P(E \mid H, I)}{P(E \mid \bar{H}, I)}\right] + \log\left[\frac{P(H \mid I)}{P(\bar{H} \mid I)}\right]. \tag{2.50}$$

The weight of evidence $\log\left[P(E \mid H, I)/P(E \mid \bar{H}, I)\right]$ is the additive change to the log odds of H, due to evidence E only, without taking into account the initial probability of H. Odds vary from 0 to ∞, while their logarithms vary from $-\infty$ to $+\infty$ (see Table 2.14).

Going from odds to log odds means to treat high and low probabilities symmetrically, and one unit of 'weight' corresponds to an increment (or to a decrement) of one decimal place in the odds. This can help handle very low probabilities which often appear when evaluating scientific evidence like DNA evidence, and formula (2.50) makes it easy to calculate how much evidence one needs to turn the scale of the balance, starting from given priors: for example, passing from prior log odds -6 to posterior log odds of 4 requires a weight of evidence equal to 10 ($-6 + 10 = 4$) (Aitken and Taroni 2004).

[14]The letter G in the original text has been replaced with the letter I for the sake of consistency with earlier works of the authors.

[15]The idea of using the log odds as a measure of the probative force of evidence goes back to the American philosopher and mathematician Charles Sanders Peirce: his definition, given in a paper published in 1878, *The probability of induction*, applies only to the special case where the prior odds are 1, i.e. $P(H \mid I) = 0.5$. In such a case, the prior log odds are 0 and the weight of evidence is equal to the posterior log odds. The idea was 'rediscovered' by the great mathematician Alan Turing, while he was working for the British intelligence to break the German code 'Enigma' during the Second World War (see Good (1979)). Moreover, for a formal discussion about *Bayes factor* and likelihood ratio, see Section 6.2.1.

Table 2.14 Logarithmic scale for odds.

Odds	Logarithm to base 10	Odds	Logarithm to base 10
1:100.000.000	−8	1	0
1:10.000.000	−7	10	1
1:1.000.000	−6	100	2
1:100.000	−5	1.000	3
1:10.000	−4	10.000	4
1:1.000	−3	100.000	5
1:100	−2	1.000.000	6
1:10	−1	10.000.000	7

The weights of two pieces of evidence E and F add in the same way (background knowledge I is omitted for ease of notation):

$$\log\left[\frac{P(H \mid E,F)}{P(\bar{H} \mid E,F)}\right] = \log\left[\frac{P(F \mid E,H)}{P(F \mid E,\bar{H})}\right] + \log\left[\frac{P(E \mid H)}{P(E \mid \bar{H})}\right] + \log\left[\frac{P(H)}{P(\bar{H})}\right].$$

When E and F are conditionally independent given H, the formula reduces to:

$$\log\left[\frac{P(H \mid E,F)}{P(\bar{H} \mid E,F)}\right] = \log\left[\frac{P(F \mid H)}{P(F \mid \bar{H})}\right] + \log\left[\frac{P(E \mid H)}{P(E \mid \bar{H})}\right] + \log\left[\frac{P(H)}{P(\bar{H})}\right].$$

The same technique which allows for the computation of probabilities for complex bodies of evidence, namely analyzing them by means of Bayesian networks which enjoy the Markov property (Section 2.2.4) makes it feasible to think about the aggregate weight of evidence. If the full body of evidence can be divided into different networks conditionally independent given the basic hypotheses of the parties, or can be analyzed by a network which can be subdivided into conditionally independent clusters, then the aggregate weight is the sum of the individual weights calculated for each network, or cluster, with regard taken for positive and negative signs.

For example, in a criminal case, the DNA evidence might be evaluated independently from other kinds of evidence, by means of the network in Figure 2.1 (but not, of course, independently by the evidence of witnesses that bears on the probabilities p and r; on the contrary, evidence that bears on the probability a can be considered, in evaluating the DNA evidence as background knowledge I). Therefore, for that piece of evidence, the weight will be given by the logarithm of formula (2.49):

$$\log\left[\frac{r + (1-r)\gamma}{r\gamma + (1-r)\left[p + (1-p)\gamma\right]}\right].$$

It must be emphasised that the same piece of evidence has *as many different weights as there are pairs of alternative hypotheses*: formula (2.50) yields the posterior odds for a hypothesis H and its negation and, if the hypotheses are not exhaustive, 'true' posterior odds cannot be calculated (recall the difference between the odds on H_1 against H_2 and the odds on H against $\bar{H}$ discussed in Section 2.3.2). For the evaluation of a likelihood ratio, the starting point is always the defence's proposition, that is, the explanation put forward by the defence for the existence of that particular physical trace. However, from a strictly logical point of view, that explanation would not be the only possible alternative to the prosecution's proposition. It is true that the basic defence's proposition, 'the suspect did not commit the fact', is the logical negation of the basic prosecution's proposition, but in order to be able to evaluate likelihood ratios both at the 'activity level' and at the 'source level', more specific propositions are needed to provide an alternative explanation of data[16].

There will be always alternative hypotheses available to explain the occurrence of *empirical* facts, for example, the production of physical traces, such as blood stains on the ground, or cellular stimulations on the retina of an eye-witness, so that, in principle, the problem of the 'catch-all hypothesis' does exist in legal settings. In this setting, though, usually the probability of the 'catch-all hypothesis' can be considered to be so low that the prosecution and the defence propositions can be safely taken *as if* they were exhaustive, and on that assumption, and *only* on that assumption, it is permissible to pass from likelihood ratios and weights of evidence to posterior odds.

> The aspiration of the legal system is to approach an assessment of odds. The means by which this is done in the vast majority of cases is to consider the two parties' respective hypotheses. It can be shown that a good approximation of the probability of a hypothesis can usually be attained by comparing it with the next most likely hypothesis. On the assumption that the two most likely hypotheses in a legal case are those advanced by the respective parties, this is what the legal system does. The adversarial process is thus an efficient fact-finding process to the extent (and only to the extent) that this assumption is true. (Robertson and Vignaux 1993, pp. 471–472)

2.4.2 The expected value of information

Information is not usually cost free: even when there is not a monetary price to be paid, it takes time to obtain new data, and loss of time is a cost that sometimes can be quantified. If information was costless, and there were no deadlines for the taking of decisions, then it would always be rational to decide to search for new information. However, this is not the case, and a typical problem scientists face is to decide whether it is worth having new evidence, if it is worth performing a new 'experiment', like another laboratory analysis, or to ask the police to search for new witnesses. Given that this decision has to be taken before the data are

[16]For the concepts of 'source level', 'activity level', and 'crime level', see Cook *et al.* (1998).

available, the problem is to calculate the expected gain of these new data, so that the gain can be compared with the cost of the search. Provided that the utilities, or the losses, of the outcomes of our decisions may be quantified in such a way that they can be compared to the cost of the experiment, Bayesian decision theory explains how the *expected value of information* may be calculated.

Imagine you are advising a client who has to decide which one of two alternative hypotheses H_1 and H_2 to use, and suppose also that these two propositions can be considered as if they were exhaustive. A loss will be incurred in using a 'false' hypothesis, so that the situation may be illustrated by Table 2.15.

Table 2.15 A decision matrix of losses.

	H_1	H_2
d_1: choosing H_1	0	L_{12}
d_2: choosing H_2	L_{21}	0

You know that the best decision is that one for which the expected loss is the minimum. In symbols (I denotes any evidence you already have about the two hypotheses), you should take the decision d_i $(i = 1, 2)$ as the one corresponding to the i for which

$$\sum_{j=1}^{2} L_{ij} P(H_j \mid I) \tag{2.51}$$

is minimized.

You must decide whether to perform an 'experiment' whose possible result is E, and so your problem is to know how much should be paid for that 'experiment'. As a first step towards the solution of the problem, it shall be shown how to calculate the *expected value of perfect information*, that is, how much it would be worth to know with certainty which hypothesis is true.

If you knew the true hypothesis, you want to make the decision with the smallest loss in the column corresponding to that hypothesis. Therefore, to calculate the expected loss *with perfect information* you must multiply the minimum loss for each hypothesis by the probability of that hypothesis, and sum all these products:

$$\sum_{j=1}^{2} (\min_i L_{ij}) P(H_j \mid I). \tag{2.52}$$

In this example, the calculus is very easy, for the minimum for each hypothesis is, of course, zero, so that the expected loss with perfect information is zero.

Your choice before knowing the truth was given by formula (2.51); therefore, the difference between (2.51) and (2.52) is the measure of *the reduction in the expected loss* or, equivalently, the increase of the expected gain that could be obtained with perfect information. In other words, it is the measure of the expected value of perfect information (with the sign inverted, given that we are using negative 'utilities', instead of 'utilities'):

$$\min_i \sum_{j=1}^{2} L_{ij} P(H_j \mid I) - \sum_{j=1}^{2} (\min_i L_{ij}) P(H_j \mid I). \tag{2.53}$$

This is also the *maximum price that one should be willing to pay* for having that perfect information. Notice that the expected value of perfect information will always be greater than zero: indeed, whatever decision is taken without perfect information, the value of (2.51) will be greater than the value of (2.52), since every loss L_{ij} in the former is replaced by a loss ($\min_i L_{ij}$) in the latter which cannot be greater. This does not mean, of course, that perfect information always reduces the expected loss of the best decision (or, equivalently, raises the expected utility of the best decision): it can reduce your expectation, it can be 'good' or 'bad' news, but it always has an informative value, because 'bad' news also increases your body of knowledge.

If your 'experiment' were of such a kind to tell you the truth, and all the truth, this would be the end of the story. Unfortunately, it will provide only *partial information*, changing your probabilities for H_1 and H_2 without letting them to go to 1 and 0. The best decision after having observed the result of the 'experiment' (E or $\bar{E}$) will be the decision that minimizes the expected loss calculated using the posterior probabilities. Equation (2.54) is also called the *Bayes risk* (see Section 7.4.1).

$$\min_i \sum_{j=1}^{2} L_{ij} P(H_j \mid E, I). \tag{2.54}$$

You don't know what will be the outcome of the 'experiment' but you do know the likelihoods; therefore, you can calculate the probabilities for the data by extending the conversation:

$$P(E) = \sum_{j=1}^{2} P(E \mid H_j, I) P(H_j \mid I).$$

Now, given that you should choose the action that minimizes the loss for any possible result of the 'experiment', your expected loss *with partial information* is calculated by multiplying the minimum expected loss for each possible result of the 'experiment' (2.54) by the probability of that result, and sum all these products:

$$\sum_E \min_i \sum_{j=1}^{2} L_{ij} P(H_j \mid E, I) P(E \mid I). \tag{2.55}$$

Application of the formula of Bayes' theorem in (2.55), gives:

$$\sum_E \min_i \sum_{j=1}^{2} L_{ij} \left[\frac{P(E \mid H_j, I) P(H_j \mid I)}{P(E \mid I)} \right] P(E \mid I)$$

$$\sum_E \min_i \sum_{j=1}^{2} L_{ij} P(E \mid H_j, I) P(H_j \mid I). \tag{2.56}$$

Again, your choice before the 'experiment' was given by formula (2.51); therefore the difference between (2.51) and (2.56) is the measure of the expected value of partial information (with the sign inverted):

$$\min_i \sum_{j=1}^{2} L_{ij} P(H_j \mid I) - \sum_E \min_i \sum_{j=1}^{2} L_{ij} P(E \mid H_j, I) P(H_j \mid I). \tag{2.57}$$

This is also the maximum price that one should be willing to pay for having that partial information. Thus, the original problem has been solved: if you can meaningfully compare the value (2.57) with a given cost of the 'experiment', you will be able to decide whether it is worthwhile to conduct the 'experiment'.

There now follow some numerical examples using a particular case. Consider again the choice between two alternative and exhaustive hypotheses (Table 2.15) and assume that the loss is symmetric: this is not a highly unrealistic assumption, because in some cases it seems sensible to believe that the damage one shall incur in choosing the false hypothesis will be the same regardless of which one that is.

This case has the interesting feature that, without loss of generality, setting $L_{ij} = 1$, the expected loss is a probability. Suppose that $P(H_1 \mid I) < P(H_2 \mid I)$: then the best decision will be d_2. The expected loss with perfect information is zero and the expected value of perfect information will be equal to $P(H_1 \mid I)$. Also the expected value of partial information will be a probability, and this allows a qualitative judgement to be made of the order of magnitude of the reduction in the expected loss provided by the evidence, even though the reduction in the expected loss cannot be directly compared to a quantitative cost of the 'experiment'. The calculations are shown in Table 2.16[17]. The decision matrix, the prior probabilities and the likelihoods are entered in the first two numerical columns.

For any decision d_i and 'experimental' result E part of formula (2.56), namely

$$\sum_{j=1}^{2} L_{ij} P(E \mid H_j, I) P(H_j \mid I),$$

may be calculated as follows.

[17]This tabulation method is taken from Lindley (1985, p. 130).

Table 2.16 Calculation of the expected losses of an experiment with two possible mutually exclusive and exhaustive outcomes E and $\bar{E}$ and a choice (d_1 or d_2) between two mutually exclusive and exhaustive hypotheses H_1 and H_2 with a symmetric (0–1) loss function and different prior probabilities for H_1 and H_2 and likelihoods $P(E \mid H_j, I)$ and $P(\bar{E} \mid H_j, I)$ for $j = 1, 2$.

	H_1	H_2	E	$\bar{E}$
	Decision matrix		Expected loss for decisions d_1 and d_2 and outcomes E and $\bar{E}$	
d_1: choosing H_1	0	1	0.57	0.03
d_2: choosing H_2	1	0	0.004	0.396
	Prior probabilities			
$P(H_j \mid I)$	0.4	0.6		
	Likelihoods			
$P(E \mid H_j, I)$	0.01	0.95		
$P(\bar{E} \mid H_j, I)$	0.99	0.05		

Consider each of the first two columns (i.e. each hypothesis H_j), multiply together the values for L_{ij}, the prior and the likelihood for E, and sum the products over the two hypotheses. For example, for d_1 and E, we have:

$$(0 \times 0.01 \times 0.4) + (1 \times 0.95 \times 0.6) = 0.57,$$

and, for d_1 and $\bar{E}$, we have:

$$0 \times 0.99 \times 0.4) + (1 \times 0.05 \times 0.6) = 0.03.$$

These values are entered on the right-hand-side of the table, in the row corresponding to d_1, and in the columns labelled E and $\bar{E}$, respectively. The values for d_2 are similarly entered. Then, the smallest values of columns E and $\bar{E}$ are selected and added together, to give the expected loss with partial information (2.56). The value for Table 2.16 is $(0.03 + 0.004) = 0.034$. Given that the minimum expected loss without this information is 0.4, there will be a reduction of the expected loss from the 'experiment' from 0.4 to 0.034.

If the uncertainty before the experiment were higher, say the two hypotheses were equiprobable, the reduction in the expected loss from the same 'experiment' would also be higher: it is left as an exercise to the readers to check that this is true for $P(H_1 \mid I) = P(H_2 \mid I) = 0.5$, and the same likelihoods as in Table 2.16.

This section concludes with another case that allows some interesting considerations. Assume again that the loss is symmetric, and suppose now that the weight of new evidence E is such that it yields equal posterior probabilities, as in Table 2.17.

Table 2.17 Calculation of the expected losses of an experiment with two possible mutually exclusive and exhaustive outcomes E and $\bar{E}$ and a choice (d_1 or d_2) between two mutually exclusive and exhaustive hypotheses H_1 and H_2 with a symmetric (0–1) loss function and equal posterior probabilities for $P(E \mid H_1, I)$ and $P(\bar{E} \mid H_2, I)$.

	H_1	H_2	E	$\bar{E}$
	Decision matrix		Expected loss for decisions d_1 and d_2 and outcomes E and $\bar{E}$	
d_1: choosing H_1	0	1	0.0099	0.9801
d_2: choosing H_2	1	0	0.0099	0.0001
	Prior probabilities			
$P(H_j \mid I)$	0.01	0.99		
	Likelihoods			
$P(E \mid H_j, I)$	0.99	0.01		
$P(\bar{E} \mid H_j, I)$	0.01	0.99		

The expected loss with this information is $(0.0099 + 0.0001) = 0.01$; the minimum expected loss without is 0.01: therefore, the reduction of the expected loss is *zero*. This result might seem, at first sight, counterintuitive: it says that the 'experiment' is worthwhile to be carried out *only if* it is for free, while, on the other hand, the 'experiment' looks like a good test for the 'underdog' H_1, because E is a highly improbable prediction, unless H_1 is true (see Section 2.3.1). The explanation for the Bayesian result is as follows.

It is true that, if the observed result were E, it would be a surprising and, in a sense, an 'informative' result, because:

$$\frac{P(H_1 \mid E, I)}{P(H_2 \mid E, I)} = \frac{0.99}{0.01} \times \frac{0.01}{0.99} = 1.$$

But consider that, on one hand, the result might be $\bar{E}$, and, in this case, it would be much less 'informative', since H_2 is already highly probable, and, on the other hand, the result E would leave you in a situation of complete uncertainty about the *choice* between H_1 and H_2, since the loss is perfectly *symmetric*. 'Symmetry' in this case means that you are interested only in the truth, and nothing but the truth: which hypothesis is true does not make any difference with respect to what is valuable for you. If this is really the case, then what the Bayesian result says is that it is more convenient for you to look for another 'experiment' which is better able to discriminate between the two hypotheses. If the 'symmetry' is broken, then this experiment has an expected value greater than zero.

Suppose that you 'prefer' that hypothesis H_1 were true. This 'preference' can be perfectly honest: all scientists have their preferred theories and they may well think of a hypothesis that 'it would be a pity if it were not true, because it is mathematically beautiful (it fits well with other theories, it has interesting consequences, and so on)'. These preferences can be represented in terms of 'utilities' and 'losses': if the 'beautiful' hypothesis were not true, they were losing something, their favourite ideas about how the fabric of the Cosmos should look like, if nothing else. So make, for the sake of argument, the simplistic assumption that you would prefer H_1 were true ten times more than H_2. Then the new calculations are as shown in Table 2.18.

Table 2.18 Calculation of the expected losses of an experiment with two possible mutually exclusive and exhaustive outcomes E and $\bar{E}$ and a choice (d_1 or d_2) between two mutually exclusive and exhaustive hypotheses H_1 and H_2 with an asymmetric loss function and equal posterior probabilities for $P(E \mid H_1, I)$ and $P(\bar{E} \mid H_2, I)$.

	H_1	H_2	E	$\bar{E}$
	Decision matrix		Expected loss for decisions d_1 and d_2 and outcomes E and $\bar{E}$	
d_1: choosing H_1	0	1	0.0099	0.9801
d_2: choosing H_2	10	0	0.099	0.001
	Prior probabilities			
$P(H_j \mid I)$	0.01	0.99		
	Likelihoods			
$P(E \mid H_j, I)$	0.99	0.01		
$P(\bar{E} \mid H_j, I)$	0.01	0.99		

Now, the expected loss with information is $(0.0099 + 0.001) = 0.01$, while the minimum expected loss without is 0.1, and, this time, there is a reduction of the expected loss by a factor 10. Thus, the answer is that, if you are a 'supporter' of hypothesis H_1, then it can be convenient (it will depend on the cost) to perform that 'experiment'; if you are a 'supporter' of hypothesis H_2, or if you are 'neutral' between them, it is not.

This example shows how the Bayesian paradigm allows a deep understanding of the real dynamics of scientific work. It is a matter of fact that scientists, in any particular scientific branch, do not work in the same projects, do not share the same opinions about which experiments are worth doing. Their actions are oriented by their 'preferences' about theories, and not by a 'disinterested' love of truth that would be reflected by a perfect symmetrical loss function. But there is nothing wrong in this 'interest': it legitimately influences which experiments will be carried out but it does not influence which one, among several possible experimental results, will occur (of course, manipulation of the results is not a legitimate consequence of having certain theoretical preferences; it is cheating).

The Bayesian analysis emphasizes the importance of taking into due account the consequences of acting according to a certain hypothesis, not only in deciding for *which kind* of evidence it is worthwhile to search, but also in deciding *how much* evidence one needs in order to be able to make a rational decision. If the former is the question that forensic scientists very often address, the latter is fundamental in the law, as shall be seen in the last section of this chapter.

2.4.3 The hypotheses' race in the law

As we have already stressed, forensic scientists are not concerned with 'ultimate issues'. Their decisions have, of course, an influence on decision making about 'ultimate issues', so that the question whether the use of expected utility theory is appropriate is meaningful, as part of the more general question as to whether or not it is legitimate to apply expected utility theory to legal decisions. The debate about this question originated with a paper by Kaplan (1968).

> Within the past decade and a half, statisticians, mathematicians, and economists have developed a new field of learning, statistical decision theory, that has already found its way into use in the highly pragmatic business world. [...] This article will attempt to apply some of the basic tools of decision theory to the factfinding process, particularly that of the criminal trial. (Kaplan 1968, p. 1065)

The proposal aroused a wave of criticism, maybe the most influential of which has been Tribe (1971), and the debate has been going on for 40 years. It will probably be endless, because it concerns the very foundations of our society. Among the advocates of the Bayesian side are Edwards (1991); Egglestone (1983); Fienberg and Finkelstein (1996); Fienberg and Kadane (1983); Fienberg and Schervish (1986); Finkelstein and Fairley (1970); Kaye (1999); Koehler (1992); Lempert (1977); Lindley (1977a); Robertson and Vignaux (1993, 1995). On the side of the critics, there was a philosopher whose ideas were quite popular at the end of the last century (Cohen 1977), and a recent book from a philosopher of science (Laudan 2006). It is not our goal to participate in this debate, therefore we shall limit ourselves to presentation of the standard abstract decision matrix proposed and discussed in the literature mentioned above, and to the provision of an argument in defence of the thesis that expected utility theory is, *in principle*, applicable in that decision matrix.

The 'ultimate issue' loss matrix (Table 2.19) has the following general aspect, where H_d stands for the defendant's proposition, and H_p for the prosecution's,

Table 2.19 The 'ultimate issue'. H_d stands for the defendant's proposition, and H_p for the prosecution's, or the plaintiff's, proposition, and X, Y, Z, and W denote the outcomes.

	H_d	H_p
d_1: decision for H_d	X	Y
d_2: decision for H_p	Z	W

or the plaintiff's, proposition, and X, Y, Z, and W denote the outcomes; no assumptions, at this point, have been made about the 'value' of these outcomes.

Now consider which assumptions can be reasonably made about this set of *uncertain* outcomes: there is an ordering of preferences among them, and there is at least one of them which is the best, and one which is the worst: X and W are better than Y and Z, and, depending on the issue, X can be better than W, or conversely, and Y can be better than Z, or conversely. Assume the 'ultimate issue' concerns a criminal case: as a consequence, nobody will deny that Y (acquitting a guilty accused) is better than Z (convicting an innocent accused). The order of preference between X and W is arguable but, for the sake of simplicity, and without loss of generality, assume that X is indifferent to W: X and W are at the top of the preference ordering with $X = W$, the bottom is Z, and an intermediate outcome is Y.

Given that a Ramsey–von Neumann–Morgenstern utility function is unique up to linear transformations, the origin does not matter (Section 2.2.2), the problem is to put Y somewhere on the utility scale between X and Z (assuming $X \neq W$ would simply introduce two intermediate outcomes, without modifying the argument). This can be done if and only if there exists probabilities $\alpha, \beta \in (0, 1)$, such that:

(i) Y is preferred to the gamble $\alpha X + (1 - \alpha)Z$;
(ii) the gamble $\beta X + (1 - \beta)Z$ is preferred to Y.

If either (i) or (ii), or both, is violated, then X, Y and Z can be ordered but cannot be compared on a utility scale.

Gamble (i) offers a probability α of acquittal of an innocent person and a probability $(1 - \alpha)$ of conviction of an innocent person. Violation of (i) means that, for any probability α, one would prefer to gamble: *no matter how high the probability is of convicting an innocent person*, one is willing to run that risk. However, courts of justice do not behave in that way: (i) is satisfied for some α.

Given that we have assumed that acquittal of an innocent person is equivalent to conviction of a guilty person, i.e. $X = W$, one can substitute W for X in gamble (ii) and obtain an equivalent gamble. Therefore, gamble (ii) offers a probability β of conviction of a guilty person and a probability $(1 - \beta)$ of convicting an innocent person. Violation of (ii) means that, for any probability β, one would prefer to acquit a guilty person: *no matter how high the probability is of convicting a guilty person*, the verdict will be always acquittal. However, courts of justice do not behave that way: (ii) is satisfied for some β.

Therefore, *in principle*, there exists a Ramsey–von Neumann–Morgenstern utility function which represents the preference ordering among the outcomes of the 'ultimate issue'. This is reflected in the law by the '*reasonable doubt*' standard. The problem is that, *in practice*, there are no means to measure the 'utility' of Y in a meaningful way and, by consequence, no means to quantify the '*reasonable doubt*' threshold.

Having established the legitimacy, in principle, of thinking in terms of expected utilities, and expected losses, the problem may be reframed using a loss matrix (the symbols L_d and L_p are self-explanatory):

It is easy to see that, if the losses were measurable, the 'reasonable doubt' threshold would be the *posterior odds cutoff*, fixed by the loss ratio (see Section 2.3.2); E denotes the total evidence pertaining to the case, and I denotes the background knowledge different from E:

$$EL(d_2 \mid E,I) < EL(d_1 \mid E,I) \text{ if and only if } L_pP(H_d \mid E,I) < L_dP(H_p \mid E,I).$$

Therefore, a guilty verdict should be issued if and only if:

$$\frac{P(H_p \mid E,I)}{P(H_d \mid E,I)} > \frac{L_p}{L_d}. \tag{2.58}$$

Development of (2.58) permits a specification as to the weight of evidence necessary to justify a guilty verdict. Given that:

$$\frac{P(E \mid H_p,I)}{P(E \mid H_d,I)} \times \frac{P(H_p \mid I)}{P(H_d \mid I)} > \frac{L_p}{L_d},$$

it holds that

$$\frac{P(E \mid H_p,I)}{P(E \mid H_d,I)} > \frac{L_pP(H_d \mid I)}{L_dP(H_p \mid I)}. \tag{2.59}$$

The chapter is finished with an exercise whose purpose is to show how the Bayesian 'scales of justice' should work.

> In criminal cases, guilt is to be proved 'beyond a reasonable doubt'. [...] Intuition is hesitant. Still, language such as 'it is better to set ten guilty persons free than to convict one innocent one' provides a basis for inferring a numerical standard [...] The principle that decisions should be made by maximizing expected utility permits translation of such a value judgment into an inequality limiting the posterior odds cutoff. [...] we have fairly pinned

down the notion of reasonable doubt. (Edwards 1991, pp. 1063–1064)[18]

If we assume that the acquittal of n guilty people is preferable to the conviction of one innocent person, then we might put $L_d = 1$ and $L_p = n$ in Table 2.20. Therefore, (2.58) says that, if $n = 10$, the 'reasonable doubt' would correspond to a posterior probability of guilty equal to 0.91; if $n = 1000$, it would correspond to a posterior probability of 0.999[19].

Table 2.20 The 'ultimate issue' loss matrix where H_d stands for the defendant's proposition, and H_p for the prosecution's, or the plaintiff's, proposition.

	H_d	H_p
d_1: decision for H_d	0	L_d
d_2: decision for H_p	L_p	0

Suppose then that the prior probability of guilt is $1/m$, (2.59) says that evidence sufficient to a guilty verdict must be such that:

$$LR > n \times (m - 1). \tag{2.60}$$

The threshold (2.58) should hold in civil trials also, and in this case it might present a problem. Whereas, in criminal cases, both n and m should be reasonably high, so to make very high the required value of the weight of evidence, in civil cases we have no clear intuition about the order of magnitude of the loss ratio, and of the prior odds.

> The apparently simpler issue of standard of proof in civil cases is actually more difficult. The phrase 'proof by the preponderance of evidence' is often taken to imply a posterior odds cutoff of 1:1, though the phrase seems to refer to evidence, and so to an aggregate likelihood ratio, rather than to posterior odds. The difficulty is that no standard, vague or otherwise, is proposed for prior odds in civil cases. In the absence of such a standard, a posterior odds cutoff is simply meaningless. (Edwards 1991, p. 1066)

Sometimes it has been proposed to take a prior odds of 1:1 for civil trials: in that case, provided that also the loss ratio is 1:1, the 'preponderance of evidence'

[18] Laudan (2006, p. 63) gives an interesting list of historical opinions about the issue: 'It is better that five guilty persons should escape punishment than one person should die (Matthew Hale 1678). It is better that ten guilty persons should escape [punishment] than that one innocent suffer (William Blackstone 1769). I should, indeed, prefer twenty guilty men to escape death through mercy, than one innocent to be condemned unjustly (John Fortesque 1471). It is better a hundred guilty persons should escape than one innocent person should suffer (Benjamin Franklin 1785). It is better to acquit a thousand guilty persons than to put a single innocent man to death (Moses Maimonides, living from 1135 to 1204).'

[19] Remember that posterior probability equals $O/(O + 1)$.

standard would require an aggregate likelihood ratio simply greater than unity. The problem here is that that weight of evidence seems quite small. It can be argued that likelihood ratios measure the *relative* weight of evidence, making a comparison between two parties' propositions, and they do not measure the 'absolute' amount of evidence, but the most reasonable answer seems to be that, in civil cases, the assumption of equal losses on both sides is too simplistic. The values of the losses reasonably depend on the importance of the issue, and in a litigation between neighbours less is at stake than, say, in an environmental pollution case. This would make necessary an evaluation, by the court, of the overall possible losses in 'utility' the parties would suffer. Fortunately, this is not a problem forensic scientists are asked to solve, and this topic may be left where it stands.

3

Concepts of Statistical Science and Decision Theory

3.1 RANDOM VARIABLES AND DISTRIBUTION FUNCTIONS

3.1.1 Univariate random variables

A statistician draws conclusions about a population. This may be done by conducting an experiment or by studying a population. The possible outcomes of the experiment or of the study are known as the *sample space* Ω. The outcomes are unknown in advance of the experiment or study and hence are said to be variable. It is possible to model the uncertainty of these outcomes because of the randomness associated with them. Thus the outcome is considered as a *random variable*. Each random variable has a set of possible states which might be categorical (the random variable has category labels as possible values) or numerical (the random variable has numbers as possible values). Examples of category labels might be 'positive' or 'negative' in an experiment involving inspection of pills suspected to contain drugs, or 'guilt' or 'innocent' in a verdict. Category labels might also be more than two, such as hair colours.

In many experiments it is easier to deal with a summary than with the original observations. For example, suppose it is of interest to know the proportion of individuals (or objects) in a relevant population that have a certain characteristic. Then, if the size of the population is, for example $n = 100$, the sample space (Ω) for this experiment would be composed of 2^{100} elements, each an ordered string of

Data Analysis in Forensic Science: A Bayesian Decision Perspective Franco Taroni, Silvia Bozza, Alex Biedermann, Paolo Garbolino and Colin Aitken

1s (if the observed item has the characteristic of interest), and 0s (if the observed item does not have the characteristic of interest), of length 100. However, the only quantity which is really of interest is the number (X) of individuals (objects) that present this characteristic. The quantity X is defined as a function (called a *random variable*) from the original sample space Ω to a new sample space $\mathcal{X}$ (the possible states of the random variable), usually the set of real numbers $\mathbb{R}$.

Example 3.1.1 *(Black toners). Consider a population of printed documents. Documents are printed with black dry toner which may be one of two types: single-component (S) or bi-component (B) toners. A sample of three documents is extracted and analyzed. A sample point for this experiment is one which indicates the result of each draw. For example, SSB indicates that two documents printed with black toner of type S have been sampled and then a third one printed with black toner of type B. The sample space for this experiment has eight points denoted ω_j, $j = 1, 2, \ldots, 8$, namely*

$$\Omega = \{SSS, SSB, SBS, BSS, BBS, BSB, SBB, BBB\}.$$

Suppose we are interested in knowing the proportion of documents printed with black toner of type S, and define a random variable X = 'number of observed documents printed with black toner of type S'. In defining the random variable X, we have also defined a new sample space, $\mathcal{X} = \{0, 1, 2, 3\}$, the set of possible values x_i, $i = 1, \ldots, 4$, the variable can take (range of the random variable). This is the usual distinction between a random variable, an upper case letter, and the value that it takes, a lower case letter, here X and x. Note that $X = x_i$ is observed if and only if the outcome of the random experiment is a point $\omega_j \in \Omega$ such that $X(\omega_j) = x_i$. For example, if the outcome is SSB (ω_2), then $X(\omega_2) = 2$ (x_3). The values assumed by the random variable for each point ω_j of the sample space are given in Table 3.1.

Table 3.1 Values (x_i) for the random variable X for each point ω_j of the sample space Ω.

j	1	2	3	4	5	6	7	8
ω_j	*SSS*	*SSB*	*SBS*	*BSS*	*BBS*	*BSB*	*BSS*	*BBB*
$X(\omega_j) = x_i$	3	2	2	2	1	1	1	0
i	4	3	3	3	2	2	2	1

A probability distribution P_X on $\mathcal{X}$ is a function which shows how the probabilities associated with the random variable are spread over the sample space. Thus, considered the random variable X with range $\mathcal{X} = \{0, 1, 2, 3\}$ in Example 3.1.1, the probability distribution is given by

$$P_X(X = x_i) = P_\Omega\left(\left\{\omega_j \in \Omega : X(\omega_j) = x_i\right\}\right), \qquad i = 1, \ldots, 4\,;\ j = 1, \ldots 8,$$

where $\left\{\omega_j \in \Omega : X(\omega_j) = x_i\right\}$ is read as 'the set of ω_j belonging to ($\in$) Ω such that (:) $X(\omega_j) = x_i$'.

To each random variable X is associated a function called the *cumulative distribution function (cdf)*, denoted by $F_X(x)$, and defined by:

$$F_X(x) = P_X(X \leq x), \qquad \text{for all } x.$$

Numerical random variables can be either discrete (the random variable takes values in a finite or countable set, e.g. 1, 2, ...) or continuous (the random variable takes values in a continuum, e.g. the real numbers). A random variable X is *discrete* if $F_X(x)$ is a step function of x. An example of a discrete random variable is given in Example 3.1.1. The associated probability distribution of a discrete random variable is called a *probability mass function (pmf)* and is given by

$$f_X(x) = P_X(X = x), \qquad \text{for all } x.$$

The *pmf* is always non-negative and is such that $\sum_{\mathcal{X}} f_X(x) = 1$.

Example 3.1.2 *(Black toners – continued). Assume that, for the sake of simplicity, the occurrences of the two types of black toners on printed documents are equiprobable ($P(S) = P(B) = \frac{1}{2}$). Then, the probability mass function and the cumulative distribution function are given in Table 3.2. The step function $F_X(x)$ is illustrated in Figure 3.1.*

Table 3.2 Probability mass function and cumulative distribution function of the random variable X = 'number (x) of observed documents printed with black toner of type S' in Example 3.1.1.

x	0	1	2	3
$P_X(X = x)$	1/8	3/8	3/8	1/8
$P_X(X \leq x)$	1/8	4/8	7/8	1

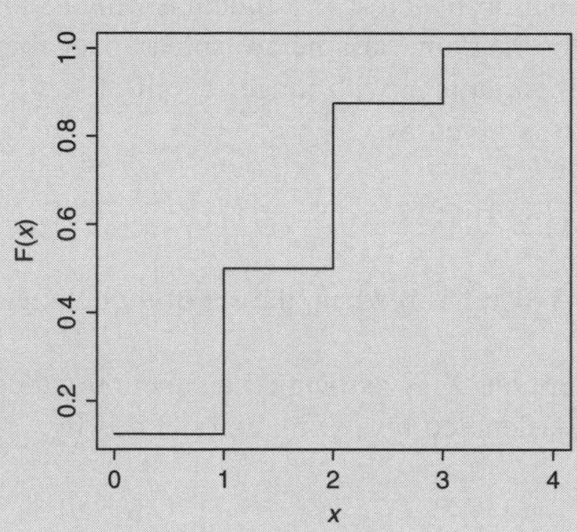

Figure 3.1 Cumulative distribution function of Example 3.1.1.

A random variable X is *continuous* if $F_X(x)$ is a continuous function of x. In this case the sample space Ω is given by a countable infinity of events: the task consists of assigning a probability to the events in the sample space.

Example 3.1.3 *(Weight of seized pills). A typical example for a continuous random variable is the weight of seized pills with illicit content (e.g. ecstasy pills). Observations on items of a given population are collected into classes defined by weight. The relative frequency of each class is reported. The probability density is the relative frequency per unit weight; i.e., relative frequency divided by the range of class (Table 3.3).*

Table 3.3 Weight measurements (x) and distribution in classes.

Weight classes (mg)	Relative frequency	Range of class (mg)	Probability density
[155, 160)	0.0065	5	0.0013
[160, 165)	0.0495	5	0.0099
[165, 170)	0.1555	5	0.0311
[170, 175)	0.2850	5	0.0570
[175, 180)	0.2935	5	0.0587
[180, 185)	0.1560	5	0.0312
[185, 190)	0.0450	5	0.0090
[190, 195)	0.0090	5	0.0018
Total	1		

The limits of each class are specified with two numbers separated by a comma and enclosed in brackets. The inclusion or non-inclusion of the limits in the class follows the standard mathematical conventions, i.e. *the square bracket [means that the lower value is included, the round bracket) means that the upper value is excluded. So, the class* [155, 160) *includes all weight measurements* $155 \leq x < 160$.

The weight classes in Example 3.1.3 can be seen as the elements constituting the sample space, whose probability is given by the relative frequency. Suppose now a single item is extracted from the population, and its weight measured. The probability distribution for weights can be represented by means of a histogram that assigns to each element in a given class a constant value equal to the ratio between the relative frequency and the range of the class (Figure 3.2 (left)). This quantity is known as *probability density* and, assuming the probability is equally distributed in the class, is the probability of an interval of unitary range (in this case an interval of 1 mg). The underlying idea is that the probability is equivalent to an area. Consequently the probability of observing exactly a single value x is null since it is the area of a rectangle with 0 base. It is also evident that the total area of rectangles must be 1 since the total area represents the sum of the probabilities of the elements of the sample space. Imagine a continuous curve that approximates the shape of the histogram (Figure 3.2 (right)). This function is always positive. The underlying area is equal to 1, and is called a *probability density function (pdf)*.

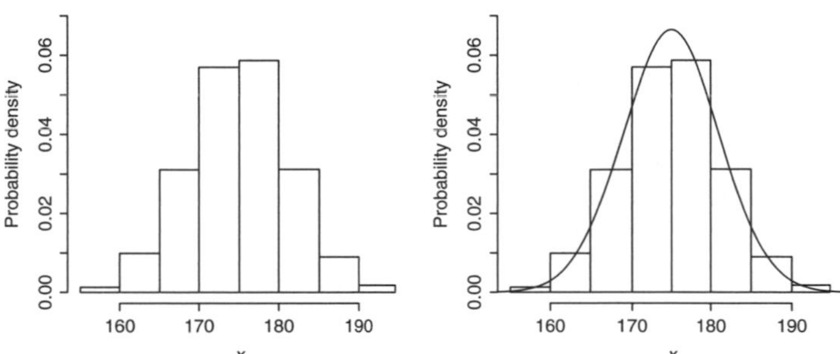

Figure 3.2 Histogram of the distribution summarized in Table 3.3 (left); Histogram of the distribution summarized in Table 3.3 with a probability density function overlaid (right).

A probability mass function (pmf) or a probability density function (pdf) is any function $f_X(x)$ satisfying the following two requirements:

1. $f_X(x) \geq 0$, for all x
2. $\sum_x f_X(x) = 1$ if X discrete; $\int_{-\infty}^{\infty} f_X(x)\mathrm{d}x = 1$ if X continuous.

The probability that the continuous random variable X takes values in any interval $[a, b]$ is the area underlying the probability density function:

$$P(a \leq X \leq b) = \int_a^b f_X(x)\mathrm{d}x.$$

This is illustrated in Figure 3.3 with reference to Example 3.1.3. The shaded area is equal to the probability that the weight of a single pill from the relevant population lies in the interval $[180, 186]$.

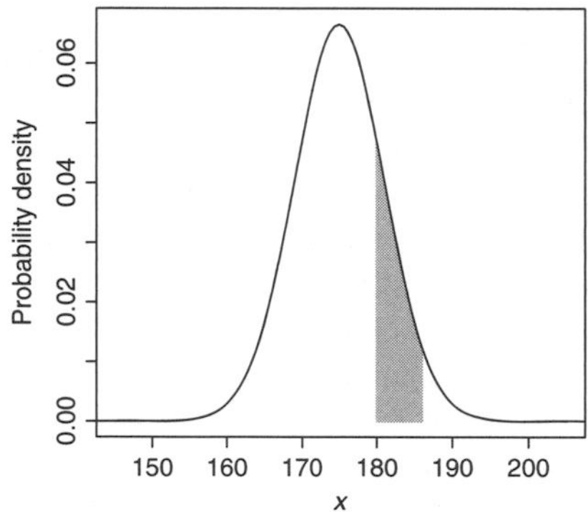

Figure 3.3 Probability of the interval $[180, 186]$ in Example 3.1.3.

Note that, as stated above, $P(X = a) = 0$, and $P(a \leq X \leq b) = P(a < X < b)$.

The cumulative distribution function $F_X(x)$ of a continuous random variable X satisfies:

$$F_X(x) = P(X \leq x) = \int_{-\infty}^{x} f_X(t)\mathrm{d}t, \qquad -\infty < x < \infty.$$

For a single value x of the random variable X, the cumulative distribution $F_X(x)$ gives the probability of the interval $(-\infty, x]$. This is illustrated in Figure 3.4; for

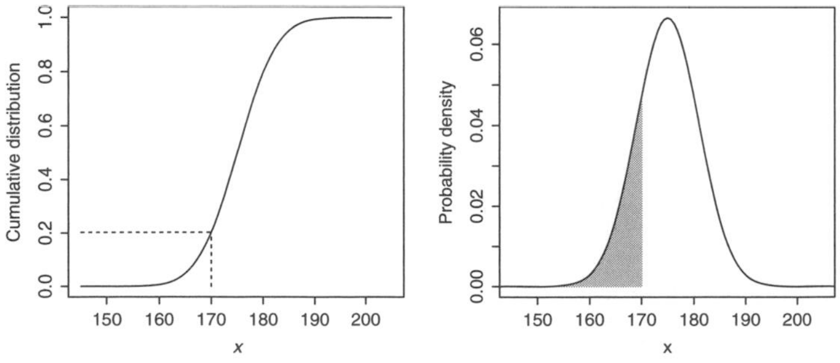

Figure 3.4 Probability of the interval $(-\infty, 170]$ in Example 3.1.3.

$x = 170$. The cumulative distribution is equal to 0.2 (left), which is the shaded area underlying the probability density function (right).

3.1.2 Measures of location and variability

The *mean*, or expected value, of a random variable X is a measure of central tendency that summarizes the centre of the distribution. It is denoted by $E(X)$ and is given by:

$$E(X) = \begin{cases} \sum_x x f_X(x) & \text{if } X \text{ discrete} \\ \int_{-\infty}^{\infty} x f_X(x) \mathrm{d}x & \text{if } X \text{ continuous} \end{cases}$$

The *quantile of order p* (for any p between 0 and 1) of a discrete random variable X is the smallest value of X, x_p, such that

$$F_X(x_p) = P(X \leq x_p) > p.$$

If x_h is a value in the range of X such that $F_X(x_h) = p$ exactly, then the quantile of order p is conventionally taken to be $x_p = (x_h + x_{h+1})/2$. The quantiles of order $p = 0.25$, $p = 0.5$, $p = 0.75$, are commonly identified as the *lower quartile*, the *median*, and the *upper quartile*. For the ease of illustration, consider Example 3.1.1. The lower quartile is $x_{0.25} = 1$, the median is $x_{0.5} = 1.5$, the upper quartile is $x_{0.75} = 2$ see Table 3.2. For a continuous random variable the quantile of order p is that value x_p that satisfies exactly $F_X(x_p) = p$.

The *mode* of a probability distribution is that value x of the random variable X at which its pmf takes its maximum value if X is discrete, or its pdf attains its maximum value if X is continuous (the mode is at the peak). The mode is not necessarily unique; when a density function has multiple local maxima, it is called multimodal. The *mode* and the *median* represent further measures of central tendency of a distribution.

The *variance* of a random variable X is defined by $Var(X) = E\,(X - E(X))^2$, and gives a measure of the degree of spread of a distribution around its mean. The positive square root of the variance, $\sigma_X = \sqrt{Var(X)}$, is called the *standard deviation* of X.

Common families of discrete and continuous random variables, together with their means and variances, are given in Appendices A and B, respectively.

3.1.3 Multiple random variables

In many experimental situations, the value of more than one random variable may be observed. Multiple observations could arise because several characteristics are measured on each item or person. Considering again a population of individual pills. It can be reasonably imagined that more characteristics are observed than just the weight (X). For instance, one may also measure a pill's diameter. A new variable Y may thus be defined. Consequently, one has a two-dimensional random vector (X, Y) that associates to each item a couple of measurements, that is its weight and its diameter. This corresponds to a point in the space $\mathbb{R}^2$.

Definition 3.1.4 *A p-dimensional random vector is a function from a sample space Ω into the p-dimensional Euclidean space $\mathbb{R}^p$.*

Let (X, Y) be a discrete two-dimensional random vector (also called a bivariate random variable). The *joint probability mass function* of (X, Y) is the function from $\mathbb{R}^2$ to $\mathbb{R}$ defined by

$$f_{X,Y}(x, y) = P(X = x, Y = y).$$

This function is always non-negative and such that $\sum_{(x,y)\in\mathbb{R}^2} f(x, y) = 1$, analogously to the univariate case. The marginal probability mass functions of X and Y, $f_X(x) = P(X = x)$ and $f_Y(y) = P(Y = y)$ are given by

$$f_X(x) = \sum_{\mathcal{Y}} f_{X,Y}(x, y) \quad \text{and} \quad f_Y(y) = \sum_{\mathcal{X}} f_{X,Y}(x, y),$$

and are illustrated in Table 3.4, where $\mathcal{X} = \{x_1, \ldots, x_h\}$ and $\mathcal{Y} = \{y_1, \ldots, y_k\}$.

Table 3.4 Joint probability mass function and marginal probability mass functions f of a bivariate discrete random variable (X, Y).

Y		y_1	$\ldots$	y_j	$\ldots$	y_k	$f(x)$
X	x_1	$f(x_1, y_1)$	$\ldots$	$f(x_1, y_j)$	$\ldots$	$f(x_1, y_k)$	$f(x_1)$
	$\vdots$	$\vdots$		$\vdots$		$\vdots$	$\vdots$
	x_i	$f(x_i, y_1)$	$\ldots$	$f(x_i, y_j)$	$\ldots$	$f(x_i, y_k)$	$f(x_i)$
	$\vdots$	$\vdots$		$\vdots$		$\vdots$	$\vdots$
	x_h	$f(x_h, y_1)$	$\ldots$	$f(x_h, y_j)$	$\ldots$	$f(x_h, y_k)$	$f(x_h)$
	$f(y)$	$f(y_1)$	$\ldots$	$f(y_j)$	$\ldots$	$f(y_k)$	1

The marginal distributions of X and Y by themselves do not completely describe the joint distribution of X and Y. There are many different joint distributions that have the same marginal distributions. The joint distribution contains additional information about the distribution of (X, Y) that is not contained in the marginal distributions.

The joint probability distribution of (X, Y) can be completely described with the *joint cumulative distribution function (cdf)*, $F(x, y)$, defined by:

$$F(x,y) = P\left(X \leq x, Y \leq y\right), \qquad \forall (x,y) \in \mathbb{R}^2.$$

Example 3.1.5 *(Black toners – continued). Consider now a new experiment in which only two documents are examined. Suppose that for each document one also notes the resin group contained in each toner sample. For the ease of argument, let this resin group be one of two possible kinds, called Epoxy A and Epoxy C, each equally likely. Then, a new variable Y = 'number of observed documents printed with black toner with resin group Epoxy A' is defined. The random variable X = 'number of observed documents printed with black toner of type S' defined in Example 3.1.1 has now range $\mathcal{X} = \{0, 1, 2\}$, in the same way the random variable Y has range $\mathcal{Y} = \{0, 1, 2\}$, see Table 3.5. The sample space defined by the observation of two characters, the type of toner and the resin group, is given by a total of 16 sample points. For the sample point (S_A, S_C), where $S_{A(C)}$ states for 'black toner of type S and resin group Epoxy A (C)', the random variable X takes value 2, while the random variable Y takes value 1. For each of the 16 sample points the values of X and Y are computed, and the bivariate random variable (X, Y) is defined. The probabilities of events defined in terms of X and Y are just defined in terms of the probabilities of the corresponding events in the sample space. So, the probability of $(X = 0, Y = 0)$ is $1/16$ because there is only one sample point that yields $(X = 0, Y = 0)$, that is (B_C, B_C), both documents are printed with black toner of type B and resin group Epoxy C. The joint pmf of (X, Y) and the marginals are given in Table 3.6. The marginal pmf of X (Y) is computed summing over the possible values of Y (X), for each possible value of X (Y).*

Table 3.5 Random variables X and Y in Example 3.1.5.

Type of toner		X	Resin group		Y
Document 1	Document 2		Document 1	Document 2	
S	S	2	A	A	2
S	B	1	A	C	1
B	S	1	C	A	1
B	B	0	C	C	0

Table 3.6 Joint probability mass function and marginals in Example 3.1.5.

Y		0	1	2	$f(x)$
X	0	1/16	2/16	1/16	4/16
	1	2/16	4/16	2/16	8/16
	2	1/16	2/16	1/16	4/16
	$f(y)$	4/16	8/16	4/16	1

Let (X, Y) be a continuous bivariate random variable. A function $f(x, y)$ from $\mathbb{R}^2$ to $\mathbb{R}$ is called a *joint probability density function* if for every $A \subset \mathbb{R}^2$

$$P\left((X, Y) \in A\right) = \int_A \int f(x, y)\mathrm{d}x\mathrm{d}y.$$

The *marginal probability density function* of X and Y are given by

$$f_X(x) = \int_{-\infty}^{\infty} f(x, y)\mathrm{d}y, \qquad -\infty < x < \infty,$$

$$f_Y(y) = \int_{-\infty}^{\infty} f(x, y)\mathrm{d}x, \qquad -\infty < y < \infty.$$

Note that the important relationship illustrated in the univariate (continuous) case still holds:

$$F(x, y) = P(X \le x, Y \le y) = \int_{-\infty}^{x} \int_{-\infty}^{y} f(s, t)\mathrm{d}s\mathrm{d}t.$$

When two random variables are observed, the values of the two variables are often related. Let (X, Y) be a discrete bivariate random variable with joint pmf $f(x, y)$ and marginal pmfs $f_X(x)$ and $f_Y(y)$. For any x such that $f_X(x) > 0$, the *conditional pmf* of Y given that $X = x$ is the function of y denoted by $f(y \mid x)$ and defined by

$$f(y \mid x) = P\left(Y = y \mid X = x\right) = \frac{f(x, y)}{f_X(x)}.$$

This is the proportion of those values that have the value $X = x$ for which the value of Y is equal to y.

For any y such that $f_Y(y) > 0$, the *conditional pmf* of X given that $Y = y$ is the function of x denoted by $f(x \mid y)$ and defined by

$$f(x \mid y) = P(X = x \mid Y = y) = \frac{f(x, y)}{f_Y(y)}.$$

This is the proportion of those values that have the value $Y = y$ such that the value of X is equal to x.

Let (X, Y) be a continuous bivariate random variable with joint pdf $f(x, y)$ and marginal pdfs $f_X(x)$ and $f_Y(y)$. For any x such that $f_X(x) > 0$, the *conditional pdf* of Y given that $X = x$ is the function of y denoted by $f(y \mid x)$ that is defined by

$$f(y \mid x) = \frac{f(x, y)}{f_X(x)}.$$

(The value $X = x$ is assumed fixed and may be thought of as a parameter.)

For any y such that $f_Y(y) > 0$, the *conditional pdf* of X given that $Y = y$ is the function of x denoted by $f(x \mid y)$ that is defined by

$$f(x \mid y) = \frac{f(x, y)}{f_Y(y)}.$$

(Note that the definitions of the conditional probability density functions involve marginal functions.)

Let (X, Y) be a bivariate random variable with joint pdf or pmf $f(x, y)$ and marginal pdfs or pmfs $f_X(x)$ and $f_Y(y)$. Then X and Y are called *independent random variables* if, for every $x \in \mathbb{R}$ and $y \in \mathbb{R}$,

$$f(x, y) = f_X(x) f_Y(y).$$

If X and Y are independent, the conditional pdf of Y given $X = x$ is (for $f_X(x) \neq 0$)

$$f(y \mid x) = \frac{f(x, y)}{f_X(x)} = \frac{f_X(x) f_Y(y)}{f_X(x)} = f_Y(y),$$

regardless of the value of x. The knowledge that $X = x$ gives no additional information about Y. It can be easily verified that variables X and Y in the black toner example are independent. Take for example $X = 0$, it can be observed

$$f(X = 0 \mid Y = 0) = \frac{1/16}{4/16} = \frac{1}{4},$$

$$f(X = 0 \mid Y = 1) = \frac{2/16}{8/16} = \frac{1}{4},$$

$$f(X = 0 \mid Y = 2) = \frac{1/16}{4/16} = \frac{1}{4},$$

that is, given the knowledge of Y, the conditional probability of $X = 0$ is equal to the marginal $f_X(0)$. This has to be verified for all values of X and Y. Alternatively, variables X and Y might not be independent, as in Example 3.1.6.

***Example 3.1.6** (Black toners – continued). Imagine that the single component toner could be either of resin group Epoxy A or Epoxy C as before, however the bi-component toner (B) could be either of resin group say Epoxy C or Epoxy D, different from A or C. If then a sample of two documents is extracted, one can calculate the values of X (number of observed documents printed with black toner of type S) and Y (number of observed documents printed with black toner with resin group Epoxy A) for the 16 sample points. The joint pmf of the bivariate random variable (X, Y) and the marginals are given in Table 3.7. It can be easily verified that the variables are not independent, in that $P(X = x, Y = y) \neq P(X = x)P(Y = y)$.*

Table 3.7 Joint probability mass function and marginals in Example 3.1.6.

Y		0	1	2	$f(x)$
X	0	4/16	-	-	4/16
	1	4/16	4/16	-	8/16
	2	1/16	2/16	1/16	4/16
	$f(y)$	9/16	6/16	1/16	1

The strength of the (linear) relationship between two random variables is measured through the covariance and the correlation. The *covariance* of X and Y is defined by

$$\text{Cov}(X, Y) = E\left[(X - E(X))(Y - E(Y))\right].$$

The *correlation* of X and Y is defined by

$$\rho_{X,Y} = \frac{\text{Cov}(X, Y)}{\sigma_X \sigma_Y}.$$

The sign of the covariance gives information regarding the direction of the relationship between two random variables. The covariance is positive when large values of X tend to be observed with large values of Y, and small values of X with small values of Y. Conversely, when large values of X are observed in association with small values of Y, and vice versa, the sign of the covariance is negative. The covariance is influenced by the units of measurement of the random variables. The numerical value itself does not give information about the strength of the relationship between the variables; information about the strength is given by the correlation. The correlation always lies between -1 and 1, with the values -1 and

1 indicating a perfect linear relationship between the random variables of negative and positive slope, respectively.

In the univariate case, the plot of the probability density represents a basic tool to display the main aspects of the underlying distribution. For bivariate continuous random variables, the graphical representation of the joint density can be less informative due to the difficulties of representing a three-dimensional object in two dimensions. In particular, it is hard to visualize the shape of a mode clearly. A graphical summary that avoids this problem is the contour plot. A hypothetical example is shown in Figure 3.5. The contour lines join points of equal joint density. A mode is identified as the centre of a system of nested contour lines, joining points of increasing density moving inwards. The elongation of the contours show the strength of the relationship between the random variables. Figure 3.5 shows a single mode around the point $(1, 1)$ for respectively uncorrelated (left), positively correlated (centre), and negatively correlated (right) variables.

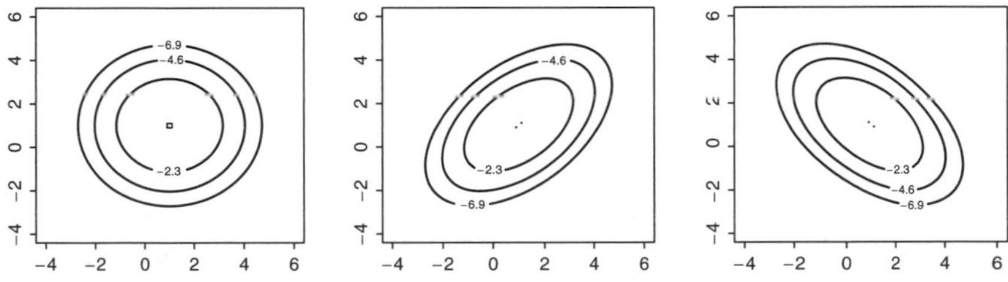

Figure 3.5 Examples of contour plots for uncorrelated (left), positively correlated (centre), and negatively correlated (right) variables.

3.2 STATISTICAL INFERENCE AND DECISION THEORY

Statistical inference represents a fundamental problem towards which statistical studies are addressed. Starting from scientific or observational results which generally consist of a set of data $x_1, \ldots, x_n$, the principal aim of statistical methods is to make statements, known as *inference*, about the process which has produced those observations and on the expected behaviour of future instances of the same process. The basic idea of statistical modelling is to consider data, $x_1, \ldots, x_n$, as the observed values of random variables $X_1, \ldots, X_n$. If the random variables are mutually independent, and each of the X_i, $i = 1, \ldots, n$, has the same probability function $f_X(x)$, then the collection $(X_1, \ldots, X_n)$ is said to be a *random sample of size* n, and the collected data are the observed value of a random sample, $(x_1, \ldots, x_n)$. The joint pdf or pmf of a set of observations from a random sample is given, using the product rule for independent observations, by

$$f_X(x_1, x_2, \ldots, x_n) = f_X(x_1) \cdot \ldots \cdot f_X(x_n) = \prod_{i=1}^{n} f_X(x_i).$$

In practice, the probability function $f_X(x)$ is unknown (in total or in part, as will be explained later). Some of its properties might be inferred by summarizing the observed data in terms of a few numbers. Any summary of a random sample, expressed as a function $t(X_1, \ldots, X_n)$, is a random variable $T = t(X_1, \ldots, X_n)$ and is called a *statistic*. The probability distribution of a statistic is called a *sampling distribution*. The basic features of a sample are its typical value (location), and a measure of how spread out the sample is (scale), that can be summarized with the sample moments (e.g. the sample mean). Good and frequently used sample summaries are the sample mean, sample variance and sample standard deviation.

The *sample mean* is the arithmetic average of the random variables, formally

$$\bar{X} = \frac{1}{n}\sum_{i=1}^{n} X_i.$$

The *sample variance* is the statistic defined by

$$S^2 = \frac{1}{n-1}\sum_{i=1}^{n}(X_i - \bar{X})^2.$$

The *sample standard deviation* is the statistic defined by $S = \sqrt{S^2}$.

As usual, the observed values of random variables are denoted with lower-case letters. So, $\bar{x} = \frac{1}{n}\sum_{i=1}^{n} x_i$, $s^2 = \frac{1}{n-1}\sum_{i=1}^{n}(x_i - \bar{x})^2$, and $s = \sqrt{s^2}$ denote the observed values of $\bar{X}$, S^2, S. The term 'sample' is used to avoid confusion with the theoretical quantities (such as mean and variance), in the sense that sample values are computed from the observed data.

Obviously, problems of inference can arise in a great variety of different circumstances. For example, consider a forensic laboratory that receives a consignment of individual items (such as tablets, or electronic data storage devices), whose characteristics may be of interest within a criminal investigation. Sampling inspection produces an answer of type 'yes' or 'no' to the question as to whether the inspected items contain something illegal. The process that gives rise to the observed data consists of sampling without replacement from a population of individual items. Each observation is assigned the value 1 if the corresponding item contains something illegal, and the value 0 otherwise. It is a realization from a binomial or hypergeometric distribution, for large and small consignments, respectively (Aitken and Taroni 2004). The aim is to find a probabilistic model that incorporates the available information and the uncertainty characterizing this information. Inferences are conditional on the adequacy of the probabilistic model: if such a model is inadequate, relevant information can be distorted or excluded. In the hypothetical scenario considered, the distribution underlying the generation of observations is assumed known (in this specific case a binomial), and is denoted $f(x \mid \theta)$. The only unknown feature is the proportion, θ, of items that possess the characteristic of interest. The object of inference is to use the available information in the sample to make statements about θ. The probabilistic model, that is the probability distribution underlying the generation of observations, is assumed to be of a known form but with a parameter θ, that is unknown. The model is called a *parametric model* and is denoted $f(x \mid \theta)$ where $\theta \in \Theta$, the parameter space.

It is worth noting that circumstances may be encountered where the process which has produced the observations has other unknown features of interest; for example, the sampling mechanism adopted may be ignored. In such cases, the probability function underlying the generation of observations may be unknown, and a model that enables estimation of the function should be used. This approach is known as *nonparametric*. A general method is the kernel density estimation (Silverman 1986; Simonoff 1996; Wand and Jones 1995). However, in this book only parametric models will be considered. It will be assumed that the observations $x_1, \ldots, x_n$ have been generated from a parametrized distribution $f(x_i \mid \theta)$, $i = 1, \ldots, n$, such that the parameter θ is unknown and the function f is known.

In this book, a decision-oriented approach to statistical inference is supported. There can be at least two motivations found for this choice (Robert 2001). First, the purpose of many inferential studies is to inform a decision (e.g. whether the defendant is the source of a crime stain, or about the necessity of performing DNA analyses), which have uncertain consequences. Consequences may be favourable or unfavourable, such as a correct or false identification in a trial, but will be unknown at the moment the decision is taken. A second reason is that in order to propose an inferential procedure, the statistician must be able to explain why they are preferable to alternative procedures. Decision theory provides fundamental tools that allow a comparison among alternative courses of action to guarantee a rational behaviour in the presence of uncertainty.

A decision problem, as described in Chapter 2, can be defined as a situation where choices are to be made among alternative courses of actions (alternatively termed decisions) with uncertain consequences. Uncertainty may depend on observable or unobservable circumstances. The proportion of items that contain something illegal is unknown unless the entire consignment is inspected. Vice versa, uncertainty may depend on circumstances which are unobservable because they refer to something that happened in the past (e.g. the defendant is the offender), or because they refer to something that will take place in the future (such as the outcome of future DNA analyses). Scientists face uncertainty that cannot be suppressed (Section 1.1): decision makers should find a way to handle it.

The basic elements of a decision problem can now be described using the more formal symbolism of probability that will be used in the rest of the book:

1. A set of possible *decisions* (also called *actions*). Decisions will be denoted d, while the set of all possible decisions will be denoted $\mathcal{D}$, the *decision space*.
2. A set of *uncertain events* (also called *states of nature*), denoted θ. The set of all possible states of nature will be denoted Θ.
3. A set $\mathcal{C}$ of *consequences* (sometimes called *rewards*), where $c(d, \theta) \in \mathcal{C}$ denotes the consequence of having taken decision d when event θ takes place.

The triplet $\{\mathcal{D}, \Theta, \mathcal{C}\}$ describes the structure of the decision problem.

***Example 3.2.1** (Forensic identification). Suppose material (a crime mark) is collected at a crime scene and an individual is apprehended. A pair of uncertain events are defined as 'The suspect is the origin of the crime mark' (θ_1), and 'Someone else is the origin of the crime mark' (θ_2). The set Θ of these states of nature is discrete, $\Theta = \{\theta_1, \theta_2\}$. Inevitably, decisions have to be made with partial knowledge of facts. The process of claiming, or 'identifying', an individual as being the source of a crime mark can be interpreted as a decision (d_1). Alternative decisions may cover statements such as 'inconclusive' (d_2) or 'exclusion' (d_3) (Biedermann et al. 2008). This decision problem is illustrated in Table 3.8. The outcomes of an 'identification' ('exclusion') statement can be favourable if the suspect is truly (is truly not) the origin of the crime mark. Consequences are then listed as 'correct identification' and 'correct exclusion' respectively. Vice versa, the outcomes of an 'identification' ('exclusion') statement can be unfavourable whenever the suspect is truly not (is truly) at the origin of the crime mark. Consequences are then listed as 'false identification' and 'false exclusion' respectively. The decision 'inconclusive' does not convey any information that tends to associate or otherwise the suspect with the issue of the source of the crime mark. Therefore, the respective consequences are listed as 'neutral'.*

Table 3.8 Decision table for an identification problem with d_i, $i = 1, 2, 3$, denoting decisions, θ_j, $j = 1, 2$, denoting states of nature and C_{ij} denoting the consequence of taking decision d_i when θ_j is the true state of nature.

	States of nature	
Decisions	θ_1 suspect is donor	θ_2 some other person is donor
d_1: identification	C_{11}: correct identification	C_{12}: false identification
d_2: inconclusive	C_{21}: neutral	C_{22}: neutral
d_3: exclusion	C_{31}: false exclusion	C_{32}: correct exclusion

The consequences are uncertain because they depend on events whose status is unknown at the time the decision is taken. It is therefore necessary to provide a criterion for evaluation which assesses the consequences of alternative courses of action, allows a comparison of different decisions and avoids irrational solutions. Decision theory investigates decision making under uncertainty and offers valuable assistance in terms of a formal model for determining the optimal course of action. The formal decision model that will be developed involves the following:

1. It is assumed that the decision maker can express preferences amongst possible consequences he might face. Preferences are measured with a function called

a *utility function* and denoted $U(\cdot)$, which specifies their desirability on some numerical scale.

2. The uncertainty about the states of nature is measured with a probability function $f(\theta \mid I, D)$ describing the plausibility of these states given the specific conditions under which the decision must be taken (e.g. previous information I, available data D). For the sake of simplicity the letters I and D will be omitted.
3. The desirability of available actions is measured by their *expected utility*

$$\bar{U}(d,f) = \begin{cases} \sum_{\Theta} U(d,\theta)f(\theta) & \text{if } \Theta \text{ discrete} \\ \int_{\Theta} U(d,\theta)f(\theta)d\theta & \text{if } \Theta \text{ continuous} \end{cases}$$

where $U(d,\theta)$ represents the utility of taking decision d when θ turns out to be the true state of nature. The best strategy is to take the action which maximizes the expected utility.

In what follows, a more detailed presentation will be given of the set of logical rules required for rational behaviour.

3.2.1 Utility theory

Consider a situation in which a decision maker is asked to make a decision which will have uncertain consequences. Consequences summarize the set $\mathcal{C}$ of outcomes resulting from a decision. They can be of various different kinds: the essential requirement is that $\mathcal{C}$ must be a well-defined set of elements, in the sense that all possible consequences are able to be defined explicitly. Note that outcomes are not necessarily monetary, nor necessarily desirable. For instance, in an identification process, consequences of an 'identification' statement might be a correct or a false identification, depending on whether the suspect is guilty or not; conversely, consequences of an 'exclusion' statement might be a false or correct exclusion, depending again on whether the suspect is guilty or not. In another statistical scenario, such as estimation, the outcome represents a measure of the distance between the estimate and the true value of the parameter; consequences might be an overestimation or underestimation of the parameter of interest.

Personal evaluations about the consequences of several actions will lead the decision maker to have preferences among consequences in any set $\mathcal{C}$. When any two consequences $c_1, c_2 \in \mathcal{C}$ are compared, the notation $c_1 \prec c_2$ indicates that c_2 is preferred to c_1; $c_1 \sim c_2$ indicates that c_1 and c_2 are equivalent (or equally preferred), while $c_1 \preceq c_2$ indicates that c_1 is not preferred to c_2 (when $c_1 \preceq c_2$ then either $c_1 \prec c_2$, or $c_1 \sim c_2$ holds). In some situations these preferences are readily recognizable: if the reward is monetary, then the greater the reward the more preferable it is. However, situations may be encountered in which the ordering of the preferences may not be immediately recognizable. In such cases, for the reasons discussed in Section 2.2.3, it is reasonable to ask the decision maker to be able to order the consequences in such a way as to satisfy the following conditions:

1. For any couple of rewards $(c_1, c_2) \in \mathcal{C}$, it must be always possible to express a preference or equivalence among them (one of the following relations must hold: $c_1 \prec c_2$, $c_2 \prec c_1$, $c_1 \sim c_2$).
2. The preference pattern is transitive, that is for any c_1, c_2 and $c_3 \in \mathcal{C}$, if $c_1 \preceq c_2$ and $c_2 \preceq c_3$, then $c_1 \preceq c_3$.

Finally, it is assumed that not all the consequences are equivalent to each other. This situation is eliminated by assuming a strict preference for at least a pair of consequences $c_1 \in \mathcal{C}$ and $c_2 \in \mathcal{C}$, that is either $c_1 \prec c_2$ or $c_2 \prec c_1$.

Let $\mathcal{P}$ be the class of all probability distributions P on the set of consequences. The decision maker's preferences among consequences will lead him to have preferences among probability distributions. As an explanation for this, let us recall that an individual cannot choose the most desirable consequence because it is uncertain. A probability distribution over consequences can be introduced, however. A question then is what gives 'value' to a particular probability distribution. The expected value of what one expects to obtain is one answer. This takes into account the desirability of the consequence, measured by the utility, and its plausibility, measured by the probability distribution. For instance, an individual may aim at a highly desirably consequence, but which may be one that is unlikely to occur: therefore, the expected value is low. The 'value' of a probability distribution P is given by the expected utility of the reward that will be received under that distribution: $E^P[U(c)]$. Thus, on the basis of his preferences among these distributions, the decision maker can specify an ordering on $\mathcal{P}$; in other words the order relation $\preceq$ is also assumed to be available on $\mathcal{P}$. What in Chapter 2 has been called a *gamble* can be thought of here as a probability distribution with a set of consequences. The completeness condition of Section 2.2.2 can now be formulated as follows.

A_1. If $P_1, P_2 \in \mathcal{P}$, then exactly one of the following three relations must hold: $P_1 \prec P_2$, $P_1 \sim P_2$, $P_2 \prec P_1$.

A_2. For any P_1, P_2 and $P_3 \in \mathcal{P}$, such that $P_1 \preceq P_2$ and $P_2 \preceq P_3$, then $P_1 \preceq P_3$.

Expected utilities are ordered in the same way as the preferences concerning the class $\mathcal{P}$. A probability distribution P_2 will thus be preferred to another one, P_1, if and only if the expected utility of the reward to be received is larger under P_2 than under the other distribution, P_1, that is $E^{P_1}[U(c)] < E^{P_2}[U(c)]$.

Two more axioms guarantee that the ordering among expected utilities satisfy the ordering among probability distributions (Berger 1988; DeGroot 1970; Robert 2001). Define a distribution P which generates a reward from P_1 with probability α, and a reward from P_2 with probability $(1-\alpha)$, $P = \alpha P_1 + (1-\alpha)P_2$. For instance, $\alpha c_1 + (1-\alpha)c_2$ is the distribution P which gives the reward c_1 with probability α, and the reward c_2 with probability $(1-\alpha)$.

A_3. If $P_1 \preceq P_2$, then there must be conservation of the order under indifferent alternatives:

$$\alpha P_1 + (1-\alpha)P \preceq \alpha P_2 + (1-\alpha)P \qquad \forall P \in \mathcal{P},\ \alpha \in (0,1).$$

Consider two situations which are identical except in the first the uncertainty is represented by $\alpha P_1 + (1-\alpha)P$ and in the second by $\alpha P_2 + (1-\alpha)P$. Then the situation involving P_2 will be preferred.

A_4. If $P_1 \preceq P_2 \preceq P_3$, there must exist α and $\beta \in (0,1)$ such that

$$\alpha P_1 + (1-\alpha)P_3 \preceq P_2 \quad \text{and} \quad P_2 \preceq \beta P_1 + (1-\beta)P_3.$$

This means that, given a favourable P_3, an unfavourable P_1, and an intermediate probability distribution P_2, it is possible to find weights such that mixing the favourable one P_3 with the unfavourable one P_1 will make the result worse (or better) that the intermediate. This axiom states that there are not infinitely desirable (or bad) consequences. In other words, no probability P_3 can be so desirable that there would be no α for which $P_2 \succeq \alpha P_1 + (1-\alpha)P_3$ Similarly, no probability P_1 can be so undesirable that there would be no β, no matter how small, for which one would take the risk of having P_1 occur (i.e. $P_2 \preceq \beta P_1 + (1-\beta)P_3$).

Consider any consequences $c, c_1, c_2 \in \mathcal{C}$ such that $c_1 \prec c_2$ and $c_1 \preceq c \preceq c_2$. Then, on the basis of Axioms A_3 and A_4, there exists a unique number $\alpha \in (0,1)$ such that

$$c \sim \alpha c_1 + (1-\alpha)c_2. \tag{3.1}$$

When the relation (3.1) is satisfied, it can be proved that

$$U(c) = \alpha U(c_1) + (1-\alpha)U(c_2). \tag{3.2}$$

See DeGroot (1970) for a proof of (3.1) and (3.2).

This utility function U is such that for any $(c_1, c_2) \in \mathcal{C}$, then $c_1 \preceq c_2$ if and only if $U(c_1) \le U(c_2)$, and the order relation in $\mathcal{C}$ is preserved. Utility functions are invariant under linear transformations. So, if $U(c)$ is a utility function, then for any $a > 0$, $aU(c) + b$ is also a utility function since it leads to the same preference pattern (Berger 1988).

There are several ways to proceed with the construction of the utility function. A simple way, as mentioned in Section 2.2.1, is to start with a pair of consequences $c_1, c_2 \in \mathcal{C}$, that are not equivalent with $c_1 \prec c_2$, and assign them a utility value. In this way, the origin and the scale of the utility function is fixed. Then, each consequence $c \in \mathcal{C}$ will be compared with c_1 and c_2, and its desirability will be measured by assigning it a value which is felt to be reasonable in comparison with the utilities assigned to c_1 and c_2. Since utility functions are invariant under linear transformations, the choice of c_1 and c_2 which are treated as benchmarks, and the choice of the scale of the utility are not relevant; it can be easily proved that different choices do not affect the optimal decision (Lindley 1985). Though any choice of c_1 and c_2 is acceptable, they are generally identified with the worst and the best consequence, respectively. It is assumed for simplicity that $U(c_1) = 0$ and $U(c_2) = 1$. The utilities of the remaining consequences are computed using results (3.1) and (3.2).

Example 3.2.2 *(Forensic identification – continued). Consider the identification problem illustrated in Table 3.8. Preferences among consequences may be evaluated with the following ranking:*

$$C_{12} \prec C_{31} \prec C_{21} \sim C_{22} \prec C_{32} \sim C_{11}$$

The construction of the utility function is started with a choice of the worst consequence (C_{12}) and the best consequence (in this case the pair C_{11} and C_{32}), and let $U(C_{11}) = U(C_{32}) = 1$ and $U(C_{12}) = 0$. The utilities of the remaining consequences, C_{21}, C_{22} and C_{31}, will be established by comparison with, respectively, the best and the worst consequences, assigning a value which seems reasonable with respect to the fixed utilities. Let us start with assigning a value to C_{21} (the 'neutral' consequence). The preference ranking outlined above states that:

$$C_{12} \prec C_{21} \prec C_{11},$$

that is, C_{11} is preferred to C_{21}, while C_{21} is preferred to C_{12}. The utility of C_{21} can be quantified using results (3.1) and (3.2). There exists a unique number α such that the consequence C_{21} is equivalent to a gamble where the worst consequence is obtained with probability α and the best consequence is obtained with probability $(1 - \alpha)$:

$$C_{21} \sim \alpha C_{12} + (1 - \alpha) C_{11}.$$

The utility of C_{21} can then be computed as:

$$U(C_{21}) = \alpha \underbrace{U(C_{12})}_{0} + (1 - \alpha) \underbrace{U(C_{11})}_{1} = 1 - \alpha.$$

Determination of such an α is the most difficult part: what would make one indifferent between a neutral consequence, and a situation in which a false identification might incur? One might agree that to be indifferent, there must *be a non-zero probability of a false identification: if such a probability were zero, then one would prefer the gamble. One might also agree that α cannot be too high: if there is an increased probability of finishing with a false identification, then the equivalence relation would become one of preference, that is the neutral consequence would be preferred. Consider $\alpha = 0.001$ is felt to be correct. Then,*

$$U(C_{21}) = \alpha U(C_{12}) + (1-\alpha)U(C_{11}) = 1 - \alpha = 0.999.$$

Let us now consider the consequence C_{31} (false exclusion). The preference ranking outlined above states that:

$$C_{12} \prec C_{31} \prec C_{11}.$$

Likewise, the utility of C_{31} will be quantified in comparison with $U(C_{12})$ and $U(C_{11})$,

$$U(C_{31}) = \alpha U(C_{12}) + (1-\alpha)U(C_{11}) = 1 - \alpha.$$

For rational behaviour, the value of α must necessarily be higher than the previous one (0.001) since the decision maker is facing, on the right side the same gamble, while on the left side a less preferred consequence ($C_{31} \prec C_{21}$). If the value of α was unchanged, then a rational decision maker would prefer the gamble. Consider $\alpha = 0.01$ is felt to be correct: $U(C_{31}) = 0.99$.

Note that the order relation in the space of consequences is preserved. Nevertheless, one might object that there is no assurance that guarantees the coherence of the quantified utility values. Stated otherwise, does this utility function reflect personal preferences? This question can be examined by comparing different combinations of consequences, such as:

$$C_{31} \prec C_{21} \prec C_{11} \quad ; \quad C_{12} \prec C_{31} \prec C_{21}$$

Consider the case on the left, for instance. According to (3.2),

$$U(C_{21}) = \alpha U(C_{31}) + (1-\alpha)U(C_{11})$$
$$0.999 = \alpha 0.99 + (1-\alpha).$$

When solving this equation one obtains $\alpha = 0.1$. Now, if this value is felt to be correct, in the sense that one is indifferent between a neutral consequence and a gamble where one might have a false exclusion with probability 0.1, *then the utility function is coherent. Otherwise, one needs to go back and check previous assessments. Likewise,*

$$U(C_{31}) = \alpha U(C_{12}) + (1-\alpha)U(C_{21})$$
$$0.99 = (1-\alpha)0.999.$$

Solving this equation yields $\alpha \simeq 0.01$. *Now, if this value is felt to be correct, in the sense that one is indifferent between a false exclusion and a gamble where one might have a false identification with probability* 0.01, *then the utility function is said to be coherent.*

3.2.2 Maximizing expected utility

In the previous section it has been shown that the plausibility of the states of nature and the value of uncertain outcomes of alternative actions can be quantified numerically, in terms of probabilities and utilities, respectively. In this section it will be shown that the desirability of each possible decision d can be measured in terms of expected utility $\bar{U}(d,f)$,

$$\bar{U}(d,f) = \sum_{\theta \in \Theta} U(d,\theta)f(\theta).$$

In the presence of uncertainty, a rational behaviour requires a person to choose an action to maximize his personal expected utility (Bernardo and Smith 2000; de Finetti 1937; Dickey 1973; Lindley 1977a, 1985; Raiffa 1968; Savage 1972):

$$\max_{d \in \mathcal{D}} \bar{U}(d,f).$$

Assume decision d is taken and θ turns out to be the true state of nature, so that the outcome is $c(d,\theta)$. Equation (3.1) can be used to show that a value α can be found such that the consequence $c(d,\theta)$ is equivalent to a hypothetical gamble where the worst consequence c_1 occurs with probability α and the best consequence c_2 occurs with probability $(1-\alpha)$

$$c(d,\theta) \sim \alpha c_1 + (1-\alpha)c_2, \qquad c_1 \preceq c(d,\theta) \preceq c_2.$$

Then, Equation (3.2) allows calculation of the utility $U(d,\theta)$ of the consequence $c(d,\theta)$:

$$U(d,\theta) = \alpha \underbrace{U(c_1)}_{0} + (1-\alpha)\underbrace{U(c_2)}_{1} = 1-\alpha.$$

Given that this hypothetical gamble can always be played, under any circumstances, this means that, for any d and any θ, taking decision d is equivalent to taking a probability $U(d,\theta)$ of obtaining the most favourable consequence or, in other words, that the conditional probability of obtaining c_2, given decision d has been taken and state of nature θ occurred, is:

$$P(c_2 \mid d,\theta) = U(d,\theta).$$

Note that, by extending the conversation (Section 2.1.2),

$$P(c_2 \mid d) = \sum_{\theta \in \Theta} P(c_2 \mid d,\theta) f(\theta). \tag{3.3}$$

Therefore, Equation (3.3) can be rewritten as

$$P(c_2 \mid d) = \sum_{\theta \in \Theta} U(d,\theta) f(\theta) = \bar{U}(d,f),$$

the expected utility, that quantifies the probability of obtaining the best consequence once decision d is taken. The strategy of taking the decision which maximizes the expected utility is optimal because it is the decision which has associated with it the highest probability of obtaining the most favourable consequence (Lindley 1985).

Example 3.2.3 *(Forensic identification – continued). The expected utilities of the alternative courses of action can be easily computed. Starting with the decision d_1,*

$$\bar{U}(d_1) = U(C_{11}) f(\theta_1) + U(C_{12}) f(\theta_2).$$

Following the utilities summarized in Table 3.9, it is readily seen that the expected utility of decision d_1 reduces to:

$$\bar{U}(d_1) = f(\theta_1).$$

Table 3.9 Illustrative values for utilities and loss in an identification scenario (θ_1: 'suspect is donor'; θ_2: 'some other person is donor').

	Uncertain events			
Decisions	θ_1	θ_2	θ_1	θ_2
	Utilities		Losses	
d_1: identification	1	0	0	1
d_2: inconclusive	0.999	0.999	0.001	0.001
d_3: exclusion	0.99	1	0.01	0

Analogously, the expected utilities of decisions 'inconclusive' (d_2) and 'exclusion' (d_3) can be found as follows

$$\begin{aligned}\bar{U}(d_2) &= U(C_{21})f(\theta_1) + U(C_{22})(1 - f(\theta_1)) \\ &= U(C_{21}) = U(C_{22}) \text{ since } U(C_{21}) \text{ and } U(C_{22}) \text{ both equal } 0.999. \\ \bar{U}(d_3) &= U(C_{31})f(\theta_1) + U(C_{32})(1 - f(\theta_1)) \\ &= U(C_{31})f(\theta_1) + (1 - f(\theta_1)).\end{aligned}$$

The optimal decision depends on the relative magnitude of $f(\theta_1)$, $U(C_{21})$, $U(C_{31})$.

Example 3.2.4 *(Bayesian networks for forensic identification). The formal relationship between the various components of a decision problem can be represented in terms of an influence diagram. An influence diagram is a Bayesian network extended by nodes for decisions and utilities, commonly represented as squares and diamonds, respectively, see Section 2.2.4 (and Cowell et al. 1999; Jensen 2001, for example).*

In forensic contexts, such influence diagrams support decision making by offering an explicit representation of the decisions under consideration and the value (utility) of the resulting outcomes, that is the states that may result from a given decision (Taroni et al. 2006a).

Figure 3.6 represents the identification problem presented in Example 3.2.1 in terms of an influence diagram. The binary variable θ encodes the proposition 'the suspect (or some other person) is the source of the crime mark'. The node D represents the available decisions d_1 ('identification'), d_2 ('inconclusive') and d_3 ('exclusion'). The node U accounts for the utility values as defined in Example 3.2.2.

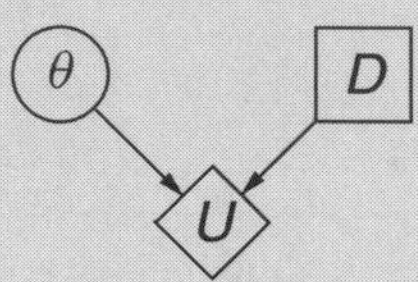

Figure 3.6 Influence diagram for the identification problem.

3.2.3 The loss function

Each pair (d,θ) gives rise to a consequence $c(d,\theta)$ whose desirability is measured by the utility function. The utility $U(d,\theta)$ represents the true gain to the decision maker when decision d is taken and θ turns out to be the true state of nature. In statistical inference, it is often convenient to work in terms of non-negative *loss functions*. Suppose that information is available about the true state of nature, say θ. Then, the decision maker will choose, in the column corresponding to the true state of nature θ, that decision to which the highest utility is associated, $\max_{d\in\mathcal{D}}\{U(d,\theta)\}$. The loss function can be simply derived from the utility function (Lindley 1985) as

$$L(d,\theta) = \max_{d\in\mathcal{D}}\{U(d,\theta)\} - U(d,\theta), \tag{3.4}$$

and measures the penalty for choosing the wrong action, that is the true amount lost by not having the most favourable situation occur. Press (1989, p. 26–27) noted that loss indicates 'opportunity loss', that is the difference between the utility of the best consequence that could have been obtained and the utility of the actual one received. Note that the loss cannot by definition be negative since $U(d,\theta)$ will be lower or at best equal to $\max_{d\in\mathcal{D}}\{U(d,\theta)\}$. The expected loss, denoted by $\bar{L}(d,f)$, measures the undesiderability of each possible action, and can be quantified as follows:

$$\bar{L}(d,f) = \begin{cases} \sum_{\Theta} L(d,\theta)f(\theta) & \text{if } \Theta \text{ is discrete} \\ \int_{\Theta} L(d,\theta)f(\theta)d\theta & \text{if } \Theta \text{ is continuous} \end{cases}$$

The best strategy, which is that of taking the decision that maximizes the expected utility, becomes that of choosing the action that minimizes the expected loss:

$$\min_{d\in\mathcal{D}} \bar{L}(d,f).$$

One might argue that the assumption of non-negativity of the loss function is too stringent. It is observed, however, that the loss function represents the error due to a bad choice. Therefore, it makes sense that even the most favourable action will induce at best a null loss.

Example 3.2.5 *(Forensic identification – continued). Table 3.9 summarizes the loss function derived from the utility function previously computed using Equation 3.4. The loss associated with each possible consequence is determined, for each possible θ, by subtracting the utility of the consequence at hand from the utility of the most favourable situation.*

3.3 THE BAYESIAN PARADIGM

Assuming the distribution underlying the generation of observations is known, the main purpose of statistical methods is to make an inference about a parameter of interest θ, starting from the collected observations, while probabilistic modelling provides the probability of any hypothetical data set before any observation is taken conditional on θ. This inverting aspect can be found in the notion of likelihood function.

Definition 3.3.1 *Let $f(x \mid \theta)$ denote the joint pdf or pmf of the sample $X = (X_1, \ldots, X_n)$. Then, given that $X = x$ is observed, the function of θ defined by*

$$l(\theta \mid x) = f(x \mid \theta)$$

is called the likelihood function.

The joint distribution is rewritten as a function of θ, conditional on the observed value x. The intuitive reason for the term 'likelihood' is that the data x for which $f(x \mid \theta)$ is large are more 'likely' with that value of θ than for a value of θ for which $f(x \mid \theta)$ is small. If we compare the likelihood function at different parameter points, θ_1 and θ_2, and find that $l(\theta_1 \mid x) > l(\theta_2 \mid x)$, then the observed sample x is more likely to have occurred if $\theta = \theta_1$ since the probability density function is greater.

A general description of the inversion of probabilities is given by Bayes' theorem, presented in Section 2.3.1. A continuous version of the Bayes' theorem states that given two random variables x and y, with conditional distribution $f(x \mid y)$ and marginal distribution $g(y)$, the conditional distribution of y given x is:

$$g(y \mid x) = \frac{f(x \mid y)g(y)}{\int f(x \mid y)g(y)\mathrm{d}y}.$$

The fundamental element of the Bayesian paradigm states that all uncertainties characterizing a problem must be described by probability distributions (Bernardo and Smith 2000). Probabilities are interpreted as a conditional measure of uncertainty associated with the occurrence of a particular event given the available information, the observed data and the accepted assumptions about the mechanism which has generated the data. They provide a measure of personal degrees of belief in the occurrence of an event in these conditions. Statistical inference about a quantity of interest is described as the modification of the uncertainty about its true value in the light of evidence, and Bayes' theorem specifies how this should be done. Hence, under the Bayesian paradigm, the uncertainty about a parameter θ is modelled through a probability distribution π on Θ, called prior distribution, that summarizes the knowledge that is available on the values of θ before the data are obtained. Notice that parameter θ is treated as a random variable in order to describe the uncertainty about its true value and not its variability (parameters are typically fixed quantities). Specifically, a Bayesian analysis treats parameters, not having been observed, as random, whereas observed data are treated as fixed. Contrary to this, frequentist analyses proceed in the opposite way: the data are treated as random even after observation whereas the parameter is considered as a fixed unknown constant to which no probability distribution can be assigned (Bolstad 2004; Kadane 1995).

A Bayesian statistical model is made by a parametric statistical model, $\{f(x \mid \theta), \theta \in \Theta\}$, and a prior distribution on the parameter, $\pi(\theta)$. All probabilities and distributions are subjective (Section 2.1): the function $f(x \mid \theta)$ measures an individual's personal degree of belief in the data taking certain values given the hypothetical information that θ takes a certain value, while the prior distribution is a measure of personal degree of belief about θ prior to observing the data. The Bayes' theorem allows the initial information on the parameter (θ) to be updated by incorporating the information contained in the observations (x). Inference is then based on the posterior distribution, $\pi(\theta \mid x)$, the distribution of θ conditional on x:

$$\pi(\theta \mid x) = \frac{f(x \mid \theta)\pi(\theta)}{\int_{\Theta} f(x \mid \theta)\pi(\theta)d\theta} = \frac{f(x \mid \theta)\pi(\theta)}{f(x)}, \tag{3.5}$$

where $f(x)$ is the marginal distribution of x. Statistical inference about the parameter θ is based on the modification of the uncertainty about its value in the light of evidence. Note that a value of θ with zero density would led to a zero posterior density. Thus, it is typically assumed that priors are strictly positive.

Available data x often take the form of a set $x = \{x_1, \ldots, x_n\}$ of 'homogeneous' observations, in the sense that only their values matter, and not the order in which they appear. Formally, this is captured by the notion of *exchangeability* (Section 2.3.4). Consider the following scenario. A laboratory receives a consignment of discrete units whose attributes may be relevant within the context of a criminal investigation. A forensic scientist is called on to conduct analyses in order to gather information that should allow one to draw an inference about, for instance, the proportion of units that are of a certain kind. The term 'positive' is used in order to refer to a unit's property that is of interest; otherwise the result is termed as 'negative'. This allows a random variable X to be defined that takes the value 1 (i.e. success) if the analyzed unit is 'positive' and 0 (i.e. failure) otherwise[1]. This is a generic scenario, which applies well to many situations such as surveys or more generally sampling procedures conducted to estimate the proportion of individuals or items in a population who share a given property or possess certain characteristics. Suppose that $n = 10$ units are analyzed, so there are $2^n = 1024$ possible outcomes. The forensic scientist should be able to assign a probability to each of the 1024 possible outcomes. The idea of exchangeability allows us to simplify greatly

[1] Experiments which lead to such events are called Bernoulli trials and the sequence of X_is a Bernoulli sequence.

this task: it is assumed that all sequences are assigned the same probability if they have the same number of 1s, successes. This is possible if it is believed that all the items are indistinguishable, and that it does not matter which particular one produced a success (a positive response) or a failure. This means that the probability assignment is invariant under changes in the order of successes and failures: if they were permuted in any way, probabilities would be unchanged. Formally, the set of observations $x = \{x_1, \ldots, x_n\}$ is said to be *exchangeable* if their joint distribution is invariant under permutation. In the hypothetical scenario considered, if the idea of exchangeability is accepted, the total number of probabilities to assign reduces from 1024 to 11. Under the assumption of exchangeability, the set of observations constitutes a random sample from some probability model $\{f(x \mid \theta), \theta \in \Theta\}$, labelled by some parameter θ, that is the joint pdf or pmf is given by

$$f(x_1, \ldots, x_n \mid \theta) = \prod_{i=1}^{n} f(x_i \mid \theta).$$

Observations are identically distributed and are independent only conditionally on the parameter value θ: exchangeability is not synonymous with independence. In fact, when any of the random variables X_is is observed, Bayesians will reconsider their opinion about the remaining future observations (Section 3.3.4). It must be underlined that in many statistical analyses data are assumed to be a random sample of conditionally independent observations from the same probability model. Since $\prod_{i=1}^{n} f(x_i \mid \theta)$ is invariant under permutation, any random sample is exchangeable, but exchangeability is a more powerful concept. An important consequence of exchangeability is that it provides an existence theorem for a probability distribution $\pi(\theta)$ on the parameter space Θ. This, and further important consequences, are discussed in Bernardo and Smith (2000), Singpurwalla (2006) and Section 2.3.4.

Example 3.3.2 *(The Binomial model). Suppose that it is of interest to study in which proportion of a given population a particular Y-chromosomal short tandem repeat (STR) haplotype occurs. An example of a haplotype sequence is presented in Table 3.10.*

Define a random variable X which takes value 1 *if the sequence of interest is observed (call this event S, success), and* 0 *otherwise (call this event F, failure). Let θ denote the probability of observing the given sequence*[2]. *A sample of dimension n is inspected and the haplotypes sequences noted: let y be the total number of observed S's (successes), and $n - y$ the total number*

[2] Notice that the estimation of population frequencies of Y-chromosomal haplotypes is problematic since haplotypes at different loci on the Y chromosome are not, owing to a lack of recombination, statistically independent. This implies that the 'Product rule' as used for autosomal loci, is not applicable. Haplotype frequencies cannot be extrapolated from allele frequencies simply by multiplication (Krawczak 2001).

Table 3.10 Example of a Y-chromosomal sequence.

Marker	Haplotype
DYS19	17
DYS3891	13
DYS389II	30
DYS390	25
DYS391	10
DYS392	11
DYS393	13
DYS385	10–14

of observed F's (failures). This experiment can be modelled as a sequence $X_1, \ldots, X_n$ of Bernoulli trials, $X_i \sim Br(\theta)$ (see Appendix A). The distribution of the total number of observed S's in n trials, $Y = \sum_{i=1}^{n} X_i$, is binomial (see Appendix A for further details about this distribution), $Y \sim Bin(n, \theta)$, with probability function

$$f(y \mid n, \theta) = \binom{n}{y} \theta^y (1-\theta)^{n-y}.$$

The parameter θ is allowed to be continuous because the fraction among n individuals that have a given sequence, lies between 0 and 1. Thus, a continuous distribution is appropriate to capture the uncertainty about the value of θ.

Let us assume the prior knowledge about θ is described by a beta distribution $Be(\alpha, \beta)$[3] (see Appendix B), with probability function

$$\pi(\theta) = \theta^{\alpha-1}(1-\theta)^{\beta-1} \frac{\Gamma(\alpha+\beta)}{\Gamma(\alpha)\Gamma(\beta)}.$$

The marginal distribution of Y can be easily computed and is given by

$$\begin{aligned} f(y) &= \int_0^1 f(y \mid n, \theta)\pi(\theta)d\theta \\ &= \binom{n}{y} \frac{\Gamma(\alpha+\beta)\Gamma(\alpha+y)\Gamma(\beta+n-y)}{\Gamma(\alpha)\Gamma(\beta)\Gamma(\alpha+\beta+n)}. \end{aligned}$$

[3]Note that in the context of the beta distribution, $\alpha > 0$, $\beta > 0$ are parameter values and not probabilities as they are in Section 3.2.1. Hopefully, the context will make clear the appropriate meaning of α and β.

The posterior distribution of θ *is:*

$$\pi(\theta \mid y) = \frac{f(y \mid n, \theta)\pi(\theta)}{f(y)}$$
$$= \frac{\Gamma(\alpha + \beta + n)}{\Gamma(\alpha + y)\Gamma(\beta + n - y)} \theta^{\alpha+y-1}(1-\theta)^{\beta+n-y-1}, \qquad (3.6)$$

which is $Be(\alpha^* = \alpha + y, \beta^* = \beta + n - y)$*. Inference is based on the posterior distribution, which gives a conditional measure of the uncertainty about* θ *given the information provided by the data* (n, y)*, the assumptions made on the mechanism which has generated the data (a random sample of* n *Bernoulli trials), and any relevant knowledge on the value of* θ *available a priori, before observations become available. The beta distribution has very convenient properties, in particular when observational data are assumed to have been generated from a Binomial model. The posterior density of* θ *has the same form as the prior: the acquisition of data has the effect of updating* α *to* $\alpha + y$*, and* β *to* $\beta + n - y$*. As the mean of the* $Be(\alpha, \beta)$ *is* $\frac{\alpha}{\alpha+\beta}$*, the posterior mean is* $\frac{\alpha+y}{\alpha+\beta+n}$*, which is roughly* $\frac{y}{n}$ *in large samples. In the same way, with the variance being equal to* $\frac{\alpha\beta}{(\alpha+\beta)^2(\alpha+\beta+1)}$*, the posterior variance is* $\frac{(\alpha+y)(\beta+n-y)}{(\alpha+\beta+n)^2(\alpha+\beta+n+1)}$ *and tends to zero as the number of trials increases to*

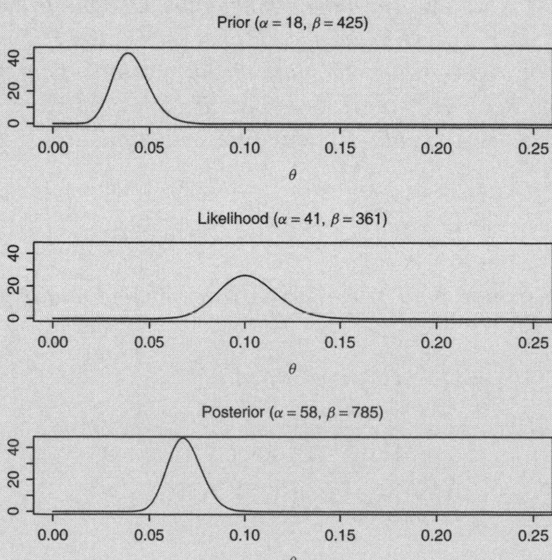

Figure 3.7 Prior, likelihood and posterior distributions for the binomial trial model (Example 3.3.2).

infinity. The binomial distribution and the beta density are said to be 'conjugate', a concept that will be discussed in Section 3.3.3.

Next, suppose that $y = 40$ haplotypes sequences have been observed in a sample of size $n = 400$. The parameters of the beta prior are chosen on the basis of available prior knowledge[4] *equal to $\alpha = 18$, $\beta = 425$. The posterior distribution is of type beta with parameters $\alpha^* = 58$, $\beta^* = 785$. The prior, the standardized likelihood*[5]*, and the posterior density are plotted in Figure 3.7. The figure shows the role of the likelihood function $l(\theta \mid x)$ in Bayes' formula. It is the function through which the data modify prior knowledge of θ.*

3.3.1 Sequential use of Bayes' theorem

Bayes' theorem allows a continual update of the uncertainty about the parameter of interest θ as more observations become available. Suppose there is an initial sample of observations, $x = (x_1, \ldots, x_n)$, then Bayes' theorem gives the posterior distribution, $\pi(\theta \mid x)$, that incorporates the initial belief about θ, the information contained in the observed data, x, and the assumptions made about the mechanism which has generated the data. Suppose a second sample of observations $w = (w_1, \ldots, w_n)$ is observed. The posterior distribution will be updated:

$$\pi(\theta \mid x, w) = \frac{f(w \mid \theta, x)\pi(\theta \mid x)}{\int_\Theta f(w \mid \theta, x)\pi(\theta \mid x)\mathrm{d}\theta} = \frac{f(w \mid \theta, x)\pi(\theta \mid x)}{f(w \mid x)}. \qquad (3.7)$$

Notice that the posterior distribution of θ given x, $\pi(\theta \mid x)$, becomes the prior distribution once new information becomes available. Bayes' theorem describes the

[4]Section 4.2.1 will illustrate in more detail how available information can be used to choose the beta parameters.

[5]The standardized likelihood is computed as

$$\frac{l(\theta \mid x)}{\int_\Theta l(\theta \mid x)\mathrm{d}\theta},$$

when the integral is finite, and the likelihood is scaled so that the area under the curve is 1. It is easy to show that in this case the standardized likelihood is a beta density with parameters $y + 1$ and $n - y + 1$, in fact

$$\begin{aligned}\frac{l(\theta \mid x)}{\int_\Theta l(\theta \mid x)\mathrm{d}\theta} &= \frac{\binom{n}{y}\theta^y(1-\theta)^{n-y}}{\int_\Theta \binom{n}{y}\theta^y(1-\theta)^{n-y}\mathrm{d}\theta} \\ &= \frac{\binom{n}{y}\theta^y(1-\theta)^{n-y}}{\frac{\Gamma(y+1)\Gamma(n-y+1)}{\Gamma(n+2)}\int_\Theta \frac{\Gamma(n+2)}{\Gamma(y+1)\Gamma(n-y+1)}\binom{n}{y}\theta^y(1-\theta)^{n-y}\mathrm{d}\theta} \\ &= \theta^y(1-\theta)^{n-y}\frac{\Gamma(n+2)}{\Gamma(y+1)\Gamma(n-y+1)}.\end{aligned}$$

process of learning from experience, and shows how beliefs about θ are continually modified as new data are observed. This two stage process is equivalent to updating directly from $\pi(\theta)$ to $\pi(\theta \mid x, w)$ by a single application of Bayes' theorem with the full data. If we rewrite in expression (3.7) the formula for $\pi(\theta \mid x)$ given by Bayes' theorem, we get:

$$\begin{aligned}\pi(\theta \mid x, w) &= \frac{f(w \mid \theta, x)}{f(w \mid x)} \times \frac{f(x \mid \theta)\pi(\theta)}{f(x)} \\ &= \frac{f(x, w \mid \theta)\pi(\theta)}{f(x, w)}.\end{aligned}$$

This is exactly what is obtained by a single application of the Bayes' theorem to the entire set of data $z = (x, w)$.

Example 3.3.3 *(The Binomial model – continued). The sequential use of Bayes' theorem is illustrated in Table 3.11. Beta parameters are first updated after observing x, which consists of $n = 400$ observations and $y = 40$ successes. The prior parameters are further updated once additional observations, w, become available ($y = 10$ successes and $n = 100$ observations). It is easily observed that they could have been directly updated by a single application of Bayes' theorem, by taking the total number of observations and successes (last column) $z = (x, w)$ with $n = 500, y = 50$.*

Table 3.11 Sequential use of Bayes' theorem. Values for prior and posterior parameters, α and β.

Parameters	Prior	Posterior		
		x $n = 400;\ y = 40$	w $n = 100;\ y = 10$	$z = (x, w)$ $n = 500;\ y = 50$
α	18	$18 + 40 = 58$	$58 + 10 = 68$	$18 + 50 = 68$
β	425	$400 - 40 + 425 = 785$	$100 - 10 + 785 = 875$	$500 - 50 + 425 = 875$

3.3.2 Principles of rational inference in statistics

Many statisticians have found it natural to adopt certain principles to justify the choice or otherwise of proposed methodologies for statistical analysis. These are principles of rational behaviour that a statistical methodology should follow in order to be accepted. Three principles are considered here. The Bayesian paradigm naturally incorporates two fundamental principles, namely the *likelihood principle* and the *sufficiency principle*, which has led some statisticians to argue that any rational analysis must correspond to some form of Bayesian analysis (Berger 1988). Alternatively, Bayesian analyses are also said to be internally consistent (Kadane 1995).

The *likelihood principle* says that in making inferences or decisions about a quantity of interest θ, all relevant experimental information is contained in the likelihood function. Moreover, if the likelihood functions for different sample points x and y are proportional, they contain identical information about θ, and lead to identical posterior distributions, so the conclusions drawn from x and y should be identical.

It is useful to note that any quantity that does not depend on θ cancels from the numerator and denominator of (3.5). If $f(x \mid \theta)$ is multiplied by an arbitrary constant, or even a function of x, that constant will cancel since $f(x \mid \theta)$ appears in both numerator and denominator of (3.5), and the same posterior density of θ will be obtained. For these reasons, the denominator of (3.5) is often omitted and the following expression is used:

$$\pi(\theta \mid x) \propto l(\theta \mid x)\pi(\theta). \tag{3.8}$$

Thus, two different experiments with likelihoods that are identical up to multiplication by an arbitrary function that does not depend on θ, contain identical information about θ. Provided the same prior is used, they lead to identical inferences. To illustrate this idea, consider the following example.

Example 3.3.4 *(Black toners – continued). Consider the population of printed documents introduced in Section 3.1. Suppose it is of interest to make an inference about the unknown proportion θ of documents printed with single-component toners (S). An experiment is conducted in which four documents are found to be printed with single-component toners and eight documents with bi-component toners. There is not a unique way to specify the probability distribution for these observations, because no information is given about the way the experiment is performed. The experiment might have consisted of a predetermined number, twelve say, of documents to be analyzed. If so, the number X of documents printed with toner of type S would be modelled as a binomial, $X \sim Bin(n = 12, \theta)$, and the likelihood function would be:*

$$l_1(\theta \mid x) = \binom{n}{x}\theta^x(1-\theta)^{n-x} = \binom{12}{4}\theta^4(1-\theta)^8. \tag{3.9}$$

But there is also a second possibility: the experiment might have consisted in the analysis of printed documents until a number equal to four documents printed with toner of type S were observed. If so, X would be modelled as a negative binomial, a random variable that counts the number of Bernoulli *trials required to get a fixed number of successes (r), $X \sim Nb(r = 4, \theta)$ (see Appendix A for further details about this distribution). The likelihood function would be:*

$$l_2(\theta \mid x) = \binom{r+x-1}{x}\theta^r(1-\theta)^x = \binom{11}{8}\theta^4(1-\theta)^8. \qquad (3.10)$$

The two likelihoods in expressions (3.9) and (3.10) are identical, except for the terms $\binom{n}{x}$ and $\binom{r+x-1}{x}$ which do not depend on θ, and will lead to identical inferences.

Note that two different experiments (with different probability models) will generally provide different information. In the previous example we have supposed to have the same number of observations and the same number of successes. Likelihoods (3.9) and (3.10) are proportional only because the numbers of observations and of successes coincide. Yet, if the actual results of different experiments yield proportional likelihood functions, provided the same prior is used, identical posterior distributions will result. The conclusion is that it does not matter whether the number of trials was fixed and the number of successes random, or vice versa, the number of trials random and the number of successes fixed: it matters only what was observed.

This is a fundamental concept of statistical inference. It is not saying that all relevant information is contained in the likelihood function, only that the experimental information is. Other relevant information can be incorporated through the prior distribution and the loss function. The likelihood principle is automatically satisfied in a Bayesian setting. It is sufficient to observe that the posterior distribution, on which the Bayesian statistical inference is based, depends on x only through $l(\theta \mid x)$: the posterior is proportional to the likelihood times the prior.

The concept of sufficiency, on the other hand, arises when part of the data does not add relevant information. Suppose that $z = (x, y)$ has been observed, and that $f(x \mid \theta, y) = f(x \mid y)$. It follows from (3.7) that

$$\pi(\theta \mid x, y) = \frac{f(x \mid \theta, y)\pi(\theta \mid y)}{f(x \mid y)} = \frac{f(x \mid y)\pi(\theta \mid y)}{f(x \mid y)} = \pi(\theta \mid y),$$

that is the value of x is irrelevant to inference about θ. Stated otherwise, it is not necessary to observe the value of x, it is sufficient to observe y. Formally, a function T of x, $y = T(x)$ is said to be sufficient if the distribution of x conditional on $T(x)$ does not depend on θ: $f(x \mid T(x), \theta) = f(x \mid T(x))$. A sufficient statistic for a parameter θ contains all the sample information concerning θ: any additional information in the sample is not relevant for inference about θ.

More formally, the *sufficiency principle* states that two sample points x_1 and x_2, with $T(x_1) = T(x_2)$, will lead to the same inference on θ. A statistic $T(x)$ may be shown to be sufficient in several ways. One can either show that $f(x \mid T(x), \theta)$ does not depend on θ, or that $\pi(\theta \mid x, T(x))$ does not depend on x, but these approaches may not be simple. A third way uses the factorization theorem, a fundamental result due to Halmos and Savage (1949). According to this theorem, a statistic $T(x)$ is sufficient for θ if the likelihood function can be factorized as:

$$l(\theta \mid x) = f(x \mid \theta) = g(T(x) \mid \theta)h(x),$$

where the function g is the density of $T(x)$ and the function h does not depend on θ.

Example 3.3.5 *Consider a random sample $X = (X_1, \ldots, X_n)$ of Bernoulli trials, with probability of success θ. Then the likelihood is given by:*

$$f(x \mid \theta) = \theta^{\sum_{i=1}^n x_i}(1-\theta)^{n-\sum_{i=1}^n x_i}. \tag{3.11}$$

According to the factorization theorem, it can be easily observed that the number of successes $y = \sum_{i=1}^n x_i$ is a sufficient statistic for θ. Formally, expression (3.11) can be rearranged as the product of two factors:

$$f(x \mid \theta) = \binom{n}{y}\theta^y(1-\theta)^{n-y} \times \frac{1}{\binom{n}{y}}.$$

The first factor is the probability of the number of successes y, while the second factor does not depend on θ. The function $y = T(x) = \sum_{i=1}^n x_i$ is then a sufficient statistic for θ that captures all the relevant information about θ. Moreover, one can observe that $f(x \mid y, \theta)$ does not depend on θ, in fact it is:

$$f(x \mid y, \theta) = \frac{\theta^y(1-\theta)^{n-y}}{\binom{n}{y}\theta^y(1-\theta)^{n-y}} = \frac{1}{\binom{n}{y}}$$

Thus, someone who observes $x = (x_1, \ldots, x_n)$ and computes $y = T(x) = \sum_{i=1}^n x_i$, has the same information about θ with respect to a second individual who is only told the number of successes $T(x) = \sum_{i=1}^n x_i$.

The likelihood principle implies that all decisions and inferences concerning θ can be made through a sufficient statistic. O'Hagan (1994) observes that in a Bayesian setting it is not necessary to check that a statistic is sufficient, or to look for a sufficient statistic. The application of Bayes' theorem will automatically take sufficiency into account, so that if a statistic $T(x)$ is sufficient, then the posterior inference will automatically depend on the data only through it. In fact, as the likelihood function is proportional to $g(T(x) \mid \theta)$, it is easily established that the posterior distribution depends on the data x only through the sufficient statistic and may be directly computed in terms of:

$$\pi(\theta \mid x) = \pi(\theta \mid T(x)) \propto g(T(x) \mid \theta)\pi(\theta). \tag{3.12}$$

The third principle is the conditionality principle, a principle that is almost as universally accepted as the sufficiency principle[6].

Two experiments on the parameter θ are available. One is selected at random. The *conditionality principle* states that the resulting inference should depend only on the experiment performed and not on the one that may have been performed. For the purpose of illustration, consider a consignment of seized tablets which may be sent to two different laboratories for analysis. Assume that there is no reason to believe that one produces more reliable responses than the other so that one may choose between them at random. When a laboratory is chosen and experimental results are received, it is generally accepted that only the experiment actually performed is relevant; conclusions should not take into account the possibility that the other laboratory could have been chosen. An example of the conditionality principle can be found in Cox (1958). The conditionality principle and the sufficiency principle together are equivalent to the likelihood principle (Birnbaum 1962).

3.3.3 Prior distributions

Bayes' theorem does not specify which prior distribution should be defined. Originally, Thomas Bayes proposed the Bayes' theorem in a context where data followed a binomial distribution with unknown probability of success θ, for which a uniform prior was chosen. This implied that all possible values of the parameter were considered equally likely. However, the theorem is far more general, and a wide variety of data distributions and prior distributions can be adopted. Generally speaking, prior distributions are often classified as 'subjective' ('personal') or 'non-informative' ('vague'), but there is considerable justification for a claim that these priors do not exist (Howson 2002; O'Hagan 2006; Press 2003). For example, the frequently termed 'non-informative' uniform distribution reflects a well-defined opinion and as such it is just as specific as the belief that the probabilities are distributed in any other perfectly specified manner (de Finetti 1993b).

Subjective priors

Subjective prior distributions represent an attempt to incorporate prior knowledge in the analysis to take advantage of this additional information about the quantity of interest. Preliminary knowledge (prior to data collection) may contribute to the formulation of a subjective prior as a probability distribution. A simple way for specifying a prior distribution consists in choosing a few summaries of the prior, such as the mean, or the most probable value (the mode), or a value such

[6]The conditionality principle must not be confused with the conditionalization principle of Chapter 2.

that θ is equally likely to lie above or below it (the median). One should use all the summaries which allow formulation of every important feature suggested by the prior knowledge (O'Hagan 1994). Generally, the prior is specified in terms of summaries of location and dispersion. Moreover, unimodality is generally assumed, unless prior knowledge suggests differently. A graph of the resulting prior distribution allows one to check that the shape looks close to prior knowledge, in the sense that the probability is reasonably spread over the range of values in which one trusts the parameter could lie. Otherwise, it will be necessary to adjust it (Bolstad 2004).

Example 3.3.6 *(Detection of nandrolone metabolites). Suppose the parameter of interest θ is the concentration of 19-norandrosterone in sportsmen (ng/ml)*[7]*. The experimenter knows the value that the parameter assumed in earlier studies, say θ^*, and can use this information as an a priori estimate of the quantity of interest. Suppose also that there is no reason to believe the uncertainty about θ is non-symmetric. Subjective prior information can be introduced into the problem by assuming for θ a Normal prior centred about θ^* (see Appendix B for further details about this distribution). Prior information will be used to decide on the spread of the distribution. For example, suppose that $\theta^* = 0.15$ and that it is believed to be extremely unlikely that either $\theta < 0.12$ or $\theta > 0.18$. These bounds, $\theta = 0.12$ and $\theta = 0.18$, can be taken as three standard deviations on either side of the mean, so that*

$$P\left(0.15 - 3\sigma < \theta^* < 0.15 + 3\sigma\right) = 0.997.$$

The standard deviation being $\sigma = 0.1$, a prior distribution for θ is $N(0.15, 0.1^2)$ and is illustrated in Figure 3.8. This example is continued in Example 6.3.3.

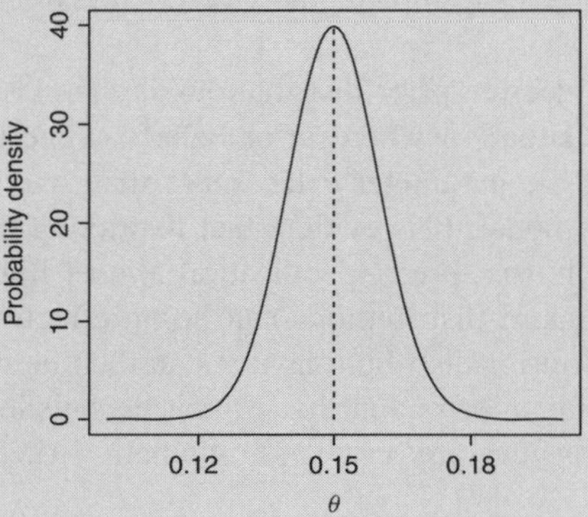

Figure 3.8 Prior density for Example 3.3.6.

[7] 19-norandrosterone is a widely used anabolic steroid in sports where strength plays an essential role.

Note that parameters indexing the prior distribution are generally called *hyperparameters*, to avoid confusion with parameters about which it is desired to make an inference. As far as the functional characteristics of the prior are concerned, one must then adopt a distribution that conforms to the formulated conditions. Any distribution that agrees with the specified summaries will be acceptable, but the choice is generally made on grounds of convenience. In fact, a complication may arise following the determination of the prior distribution. The resulting posterior may not be able to be determined in a closed form. In particular, if an arbitrary likelihood function $l(\theta \mid x)$ is combined with an arbitrary prior distribution $\pi(\theta)$, then the product might be mathematically intractable and it will not be possible to integrate out the parameter and compute the marginal distribution so as to determine a form for the posterior distribution[8].

There are several ways to tackle this problem. One is to approximate the value of the integral expression for the posterior distribution using one of several numerical techniques available. Another way is to opt for alternative, simpler forms of the likelihood function and the prior density that can be used as approximations and lead to known posterior densities instead of expressing as accurately as possible the investigator's beliefs about the process generating the data and the prior knowledge about θ. Prior distributions can be found in the *conjugate family* $\mathcal{F}$ with respect to the sampling distribution $f(x \mid \theta)$. The conjugate family is a set of prior distributions such that, for a particular sampling distribution $f(x \mid \theta)$ and for every prior in the family, the posterior distribution is also in the same family. The family $\mathcal{F}$ is also said to be closed under sampling from $f(x \mid \theta)$. An example has been presented in Section 3.3, Example 3.3.2, where observations followed a binomial distribution. A beta prior distribution was chosen so that the resulting posterior was a distribution of the same family, i.e. a beta distribution (with updated parameters). For more general purposes, some of the more common conjugate prior distribution families are listed in Table 3.12 and in Appendixes A and B (a complete list can be found in Bernardo and Smith 2000).

Another class of subjective prior distributions is the class of *mixture prior distributions*. Consider a situation where prior beliefs suggest that it is most likely that the distribution of the parameter θ has more than one mode; so any realistic prior will be at least bimodal. It is evident that restricting the choice to conjugate priors would make an honest prior specification almost impossible. A mixture of a small number of standard distributions can be an effective way to generate far more general classes of prior distributions that can then be used to adjust the prior to wide varieties of prior information that cannot be satisfied by any one of those distributions alone. Consider any convex combination (i.e. a linear combination

[8]Note that even if it is not always possible to identify the posterior distribution in closed form, various summaries (e.g. the posterior mode), depend on differentiation and do not require the integral in the denominator to be solved.

Table 3.12 Some conjugate prior distribution families. Formulae for these distributions are given in Appendixes A and B.

Probability distribution	Conjugate prior distribution
Binomial:	Beta:
$f(x \mid \theta) = \text{Bin}(n, \theta)$	$\pi(\theta) = \text{Be}(\alpha, \beta)$
Poisson:	Gamma:
$f(x \mid \lambda) = \text{Pn}(\lambda)$	$\pi(\lambda) = \text{Ga}(\alpha, \beta)$
Exponential:	Gamma:
$f(x \mid \lambda) = \text{Exp}(\lambda)$	$\pi(\lambda) = \text{Ga}(\alpha, \beta)$
Normal (known variance):	Normal:
$f(x \mid \mu, \sigma^2) = N(\mu, \sigma^2)$	$\pi(\mu) = N(\theta, \tau^2)$
Normal (known mean):	Inverted Gamma:
$f(x \mid \mu, \sigma^2) = N(\mu, \sigma^2)$	$\pi(\sigma^2) = \text{IG}(\alpha, \beta)$

where all coefficients are non-negative and sum up to 1) of $m (m > 1)$ priors π_i $(i = 1, \ldots, m)$:

$$\pi(\theta) = \sum_{i=1}^{m} \gamma_i \pi_i(\theta),$$

with $\gamma_i > 0$ $(i = 1, \ldots, m)$ and $\sum_{i=1}^{m} \gamma_i = 1$. It can then be shown that the posterior $\pi(\theta \mid x)$ can be expressed as a convex combination of the respective posteriors (Lee 2004; Press 2003), that is

$$\pi(\theta \mid x) = \sum_{i=1}^{m} \gamma_i^* \pi_i(\theta \mid x),$$

with $\gamma_i^* > 0$ $(i = 1, \ldots, m)$ and $\sum_{i=1}^{m} \gamma_i^* = 1$.

Let $x = (x_1, \ldots, x_n)$ be the observed outcomes of n Bernoulli trials with probability of success θ, and let $y = \sum_{i=1}^{n} x_i$ be the number of successes in the n trials. It has been observed that the beta family distribution for θ is convenient, but for any specification of hyperparameters α and β (> 1) it will necessarily be unimodal. Multimodal shapes can be easily generated by considering mixtures of beta densities

$$\pi(\theta) = \sum_{i=1}^{m} \frac{\gamma_i}{B(\alpha_i, \beta_i)} \theta^{\alpha_i - 1} (1 - \theta)^{\beta_i - 1},$$

with $\sum_{i=1}^{m} \gamma_i = 1$, $\gamma_i > 0$, $\alpha_i > 0$, $\beta_i > 0$ $(i = 1, ...m)$.

By Bayes' theorem, the posterior distribution for θ is also a mixture of beta densities (the mixture class of beta distributions is closed under sampling from the binomial) and takes the form

$$\pi(\theta \mid x) = \sum_{i=1}^{m} \frac{\gamma_i^*}{B(\alpha_i^*, \beta_i^*)} \theta^{\alpha_i^* - 1} (1 - \theta)^{\beta_i^* - 1},$$

where

$$\gamma_i^* = \frac{\gamma_i B(\alpha_i^*, \beta_i^*)}{\sum_{i=1}^m \gamma_i B(\alpha_i^*, \beta_i^*)},$$

$\alpha_i^* = \alpha_i + y$, $\beta_i^* = \beta_i + n - y$, and y is the number of success in a trial. Note that if each prior π_i in the mixture is a member of a conjugate family $\mathcal{F}$ with respect to a certain sampling distribution $f(x \mid \theta)$, then the larger family of mixtures is also a member of $\mathcal{F}$ (i.e. it is closed under sampling from $f(x \mid \theta)$).

Example 3.3.7 *(Proportion of items in a consignment). Consider the following beta mixture:*

$$0.6\text{Be}(3, 10) + 0.4\text{Be}(6, 2.5),$$

which corresponds to a mixture with $m = 2$, $(\gamma_1, \gamma_2) = (0.6, 0.4)$, $(\alpha_1, \alpha_2) = (3, 6)$, $(\beta_1, \beta_2) = (10, 2.5)$. *The resulting mixture prior is bimodal (Figure 3.9 (left)) and reflects, among other issues, a belief that* θ *is most likely to be around* 0.2, *but it could also lie around* 0.7. *Now, suppose that* $y = 3$ *successes have been observed in* $n = 20$ *trials. Then,* $(\gamma_1^*, \gamma_2^*) = (0.93, 0.07)$, $(\alpha_1^*, \alpha_2^*) = (6, 9)$, $(\beta_1^*, \beta_2^*) = (27, 19.5)$. *The posterior is illustrated in Figure 3.9 (right) and clearly depicts strong support for small values of* θ.

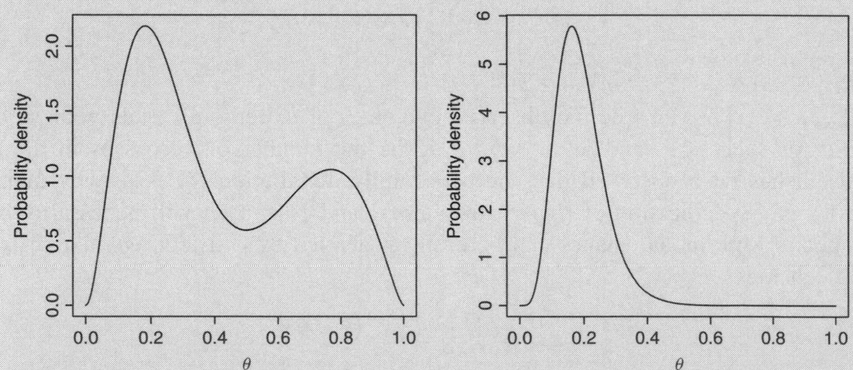

Figure 3.9 Prior (left) and posterior (right) distributions from a two-component beta mixture.

A typical situation in which a mixture prior may be useful is one that involves competing prior beliefs. The numerical example given above could refer, for instance, to settings in which a proportion (such as the proportion of *items in a consignment that contain something illegal) is thought to be either smaller than 0.5, that is about 0.2, or greater than 0.5, that is approximately around 0.7. More specifically, it is thought that the mean is centred either around the mean of a $Be(3, 10)$ or a $Be(6, 2.5)$ distribution, with more 'weight' placed on the former distribution than on the latter. That weight is actually expressed by the two values 0.6 and 0.4. It appears worth noting that such an approach to modelling a prior distribution may be of interest, for example, in a setting that involves opposing parties (e.g. prosecution and defence), which disagree with regards to their prior beliefs. A mixture prior allows the competing views to be expressed as a single distribution along with a weighting of their relative importance.*

Effect of the amount of data on the relationship between the prior distribution and the posterior distribution

There are instances in which the prior may be said to be *dominated by the likelihood*. This is typically the case when abundant data are available.

> If observations are precise, in a certain sense, relative to the prior distribution on which they bear, then the form and properties of the prior distribution have negligible influence on the posterior distribution. [...] [T]he untrammelled subjectivity of opinion about a parameter ceases to apply as soon as much data become available. More generally, two people with widely divergent prior opinions but reasonably open minds will be forced into arbitrarily close agreement about future observations by a sufficient amount of data. (Edwards *et al.* 1963, p. 527).

An illustration of this is given in Example 3.3.8.

Example 3.3.8 *(Surveillance cameras). Imagine that n distinct surveillance camera recordings are available of a male individual. Assume that it is of interest to estimate the height of that individual. Such an estimate may be needed for the purpose of comparison with other estimates obtained from recordings taken at other times and/or locations. The n experimental measurements (expressed in cm), $(x_1, \ldots, x_n)$, are assumed to be generated from a Normal distribution with known variance σ^2 equal to 2^2, that is $X \sim N(\theta, 2^2)$.*

Suppose two experts (say expert A and expert B), are asked to estimate the mean height θ. Let $\pi_A(\theta) = N(\mu_A, \tau_A^2) = N(178, 1)$ be the prior distribution expressed by expert A, and let $\pi_B(\theta) = N(\mu_B, \tau_B^2) = N(172, 2.8^2)$ be the *prior distribution expressed by expert B. These priors reflect different prior knowledge, and this difference is illustrated in Figure 3.10(a).*

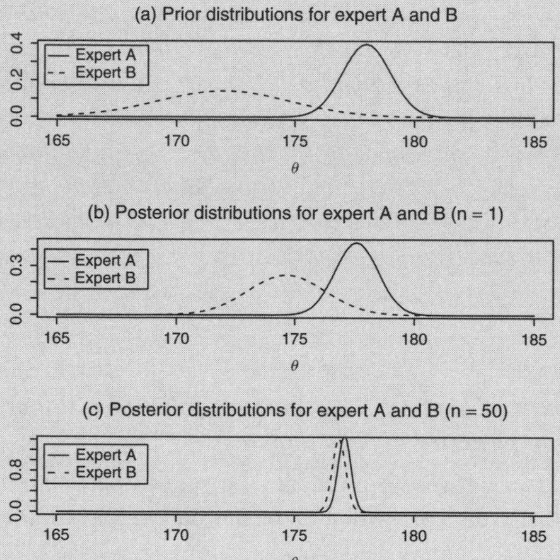

Figure 3.10 Prior and posterior distributions for expert A and expert B. Note the changes of scale in the vertical axis.

The application of Bayes' theorem allows to show how such opinions may be modified by the information contained in the data. The posterior distribution of θ given x, for each expert, is still Normal since the distributions are conjugate, with mean $\mu_p(x)$ given by

$$\mu_p(x) = \frac{\tau_p^2 \bar{x} + \mu_p \sigma^2/n}{\sigma^2/n + \tau_p^2} \qquad p = A, B$$

and variance $\tau_p^2(x)$ given by

$$\tau_p^2(x) = \frac{\tau_p^2 \sigma^2/n}{\tau_p^2 + \sigma^2/n} \qquad p = A, B.$$

The posterior mean is a weighted average of the prior mean and the observations (an example will be discussed in Section 4.4.1).

Suppose first that one observation ($n = 1$) is available, $\bar{x} = x_1 = 176$ cm. The resulting posterior distributions are plotted in Figure 3.10(b). The effects *of the prior distributions expressed by the two experts are remarkably different: in particular, the observation has much more impact on the state of uncertainty of expert B. This is not surprising because the uncertainty of expert A was considerably smaller. The effect of the likelihood on the posterior is more pronounced for expert B.*

Suppose next that fifty observations ($n = 50$) are available, and that the sample mean is equal to $\bar{x} = 177$ cm. Figure 3.10(c) shows that the posterior distributions are very similar (note also the change of scale for the vertical axis compared with case (a) and (b)), and that the two experts would now be in close agreement. Compared with the likelihood function, both priors are rather flat and have little influence in determining the posterior distribution. This example is continued in Example 4.4.4.

However, it is emphasised that in many practical situations the observed data are very poor in the sense of having high variability. An immediate consequence is that personal knowledge might be relevant in determining the posterior distribution. Notably, different priors might lead to different posteriors. This has provoked many criticisms: some people are skeptical on the use of Bayes' theorem essentially because they consider Bayesian measures to be subjective, and regard subjectivity as a liability. It is argued in Box and Tiao (1973), however, that this should not necessarily be considered a disadvantage of the Bayesian approach since it allows one to learn from one's own experience by allowing for a personal update of beliefs. All inferences are made by individuals and so they are necessarily subjective (in the sense of 'personal'). An advantage of the Bayesian approach is that it makes subjectivity explicit (Berry 1991) and allows different opinions to be expressed and evaluated formally. The sensitivity of posterior results to prior choices will be discussed in Section 4.2.1.

Subjective priors based on a limited amount of information

In some inference problems, the information available prior to data collection may be limited. In such cases, the prior distribution to be adopted should reflect that fact in some appropriate manner. In yet other situations, one may want results of scientific experiments to be reported as a general result, that is, minimally dependent on an expert's prior beliefs or dependent on a prior on which many observers can agree.

> Although a Bayesian should use her own prior, she may also need to exhibit results from a range of priors for the benefit of other statisticians, and the inclusion of one (or more) standard priors in this set makes it easier for others to judge the strength of evidence in a given body of data. (Howard 1998, p. 353).

These are situations in which so-called vague priors are sometimes recommended. It is of interest to determine the selection of a prior which provides poor information compared to the information an experiment is expected to provide.

When the unknown parameter lies in a finite interval, a uniform prior may be readily applicable. However, when at least one end point of the domain of the parameter is not finite (e.g. the mean of a Normal distribution), the uniform prior becomes *improper*. An improper prior function is defined as a non-negative function $\pi(\theta)$ such that $\int_{\Theta} \pi(\theta)\mathrm{d}\theta$ is not finite (that is, the function does not, for its admissible range, integrate or sum to 1). For example, if $\pi(\theta)$ is uniform over the entire line,

$$\pi(\theta) \propto k \qquad -\infty < \theta < \infty\,, \; k > 0,$$

then $\int_{-\infty}^{\infty} k\mathrm{d}\theta$ is not finite, and $\pi(\theta)$ is an improper prior function. This approach of assessment is acceptable only when, in spite of the fact that the prior is not proper, the application of Bayes' theorem leads to a proper posterior distribution over the entire line.

For the purpose of illustration, consider a random sample $x = (x_1, \ldots, x_n)$ from a Normal distribution $N(\theta, \sigma^2)$ with variance σ^2 known. The likelihood function of θ is

$$l(\theta \mid \sigma, x) \propto \exp\left[-\frac{1}{2}\left(\frac{\theta - \bar{x}}{\sigma/\sqrt{n}}\right)^2 \right],$$

which can be represented as a Normal distribution centred around the sample mean $\bar{x}$ with standard deviation $\sigma/\sqrt{n}$. It has been argued that all experimental information is contained in the likelihood function. In this case the experimental observation only affects the location of the likelihood. Different observations simply translate the likelihood on the θ axis, but leave the shape of the distribution unchanged, as is shown in Figure 3.11. The likelihood function is said to be 'data-translated'. The concept that only poor knowledge is available a priori (compared to the information

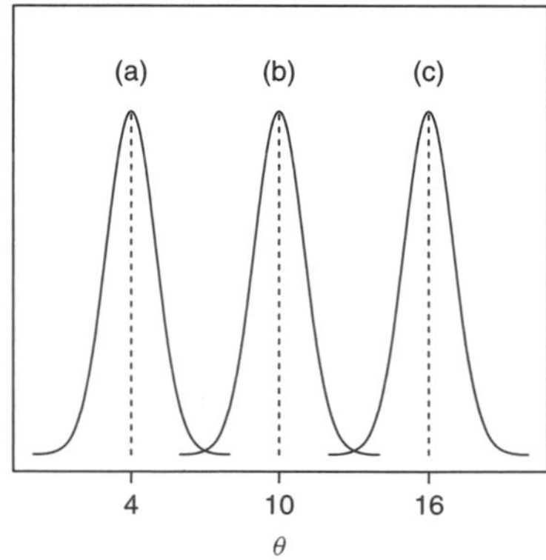

Figure 3.11 Likelihood functions resulting from different experiments of equal size giving rise to (a) $\bar{x} = 4$, (b) $\bar{x} = 10$, (c) $\bar{x} = 16$.

the experiment is expected to provide), can be expressed by assuming indifference between parameter values that lead to a likelihood which is completely determined a priori, except for its location. The state of indifference about knowledge of θ, or vagueness, may be expressed by adopting a uniform prior:

$$\pi(\theta) \propto \text{constant}, \qquad -\infty < \theta < \infty. \tag{3.13}$$

In general, the likelihood function of the parameter of interest is not necessarily data translated. Suppose it is possible to transform the unknown parameter in terms of a metric such that the likelihood function is data translated and a uniform prior can be introduced on all of its values. Take for instance the likelihood function of the unknown standard deviation σ of a Normal population, known mean θ

$$l(\sigma \mid \theta, x) \propto \sigma^{-n} \exp\left(-\frac{ns^2}{2\sigma^2}\right),$$

where $s^2 = \sum_{i=1}^{n} (x_i - \theta)^2 / n$. The likelihood is not data translated, as can be observed in Figure 3.12 (left), which depicts the likelihood curves for σ obtained from $n = 10$ observations with $s = 5$, $s = 10$, $s = 20$ (Box and Tiao 1973). However, the corresponding likelihood curves in terms of the transformation $\log \sigma$ are exactly data translated, as shown in Figure 3.12 (right). In fact, it can be noted that multiplication by the constant s^n leaves the likelihood function unchanged (data acting through s serve only to relocate the likelihood) and that the likelihood function of $\log \sigma$ can be expressed as

$$l(\log \sigma \mid \theta, x) \propto \exp\left\{-n\,(\log \sigma - \log s) - \frac{n}{2} \exp\left[-2\,(\log \sigma - \log s)\right]\right\}.$$

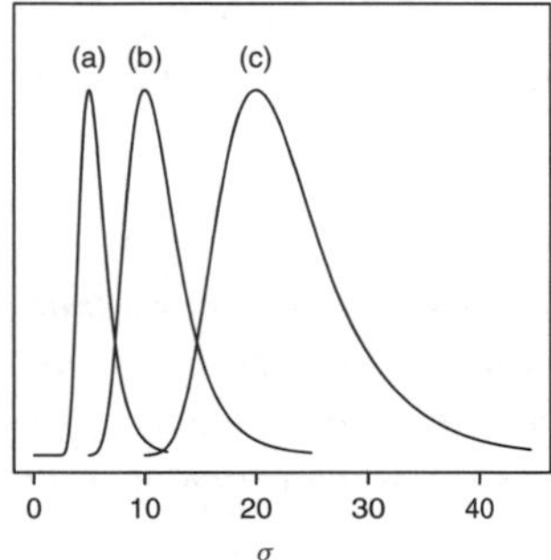

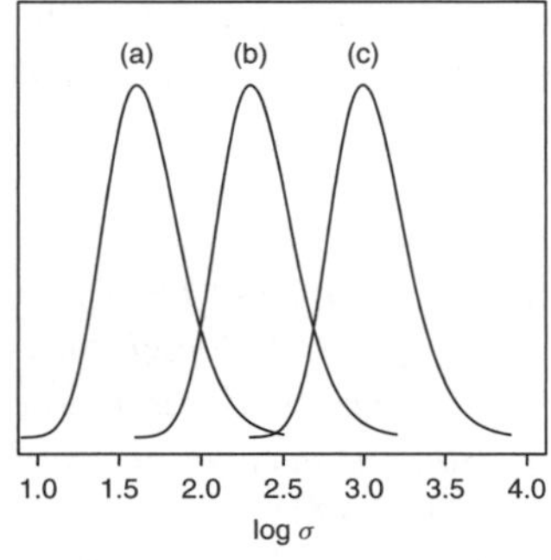

Figure 3.12 Likelihood functions resulting from different experiments giving rise to (a) $s = 5$, (b) $s = 10$, (c) $s = 20$ for the Normal standard deviation σ (left), and for $\log \sigma$ (right).

Thus, vagueness about σ may be expressed by adopting a uniform prior in $\log \sigma$:

$$\pi(\log \sigma) \propto \text{constant}, \qquad -\infty < \log \sigma < \infty. \tag{3.14}$$

Transforming variables back again gives the vague prior density:

$$\pi(\sigma) \propto \left| \frac{d \log \sigma}{d\sigma} \right| = \frac{1}{\sigma}, \qquad 0 < \sigma < \infty. \tag{3.15}$$

3.3.4 Predictive distributions

The general problem of statistical prediction may be described as that of inferring the values of unknown variables from current available information (Bernardo and Smith 2000). In fact, after having observed a random sample $x = (x_1, \ldots, x_n)$, it may be of interest to predict the value of a future observation x_{n+1}, generated by the same random mechanism, $\{f(x \mid \theta), \theta \in \Theta\}$, which has generated the data. Inference about the value of a future observation is derived from the *predictive distribution*:

$$f(x_{n+1} \mid x) = \int_{\Theta} f(x_{n+1} \mid \theta)\pi(\theta \mid x)\mathrm{d}\theta, \tag{3.16}$$

which summarizes the information concerning the value of a new observation given the prior, and the data observed so far. Parameters θ can be viewed as nuisance parameters and are integrated out from the posterior density. The predictive distribution has the form of a weighted average of all possible distributions of x with their corresponding posterior densities. It plays an important role in Bayesian analysis. The validity of the Bayesian model, that incorporates the assumptions of the sampling density $f(x \mid \theta)$ and the prior density $\pi(\theta)$ and allows inferences to be made about the parameter of interest θ, can be checked by inspecting the predictive distribution. If the observed data are consistent with the predictive density, then the model may be reasonable. Otherwise, if the observed data are in the extreme tail portion of the predictive density, then some doubts should arise on the validity of the Bayesian model, in the sense that the prior density or the sampling density may have been misspecified (Albert 2007).

Example 3.3.9 *(Errors in DNA analyses). Experience shows that errors in DNA typing do occur. Although the proportion of errors occurring in forensic DNA laboratories is perhaps low due to regular proficiency testing, for practical reasons, it may be desirable to have some estimate of error rates for a specific laboratory.*

Considering errors as rare events, their occurrence may be modelled using a Poisson distribution $Pn(\lambda)$ with parameter λ, so that $p(x \mid \lambda) = \lambda^x e^{-\lambda}/x!$,

$x = 0, 1, 2, \ldots$ (see Appendix A for further details about this distribution). In the absence of initial information about the value of λ, a vague prior may be introduced, in particular $\pi(\lambda) = \lambda^{-1/2}$. Note that the range of λ is $(0, \infty)$, and the ideas of Section 3.3.3 concerning the introduction of a prior density based on a limited amount of information for the standard deviation of a Normal distribution are applied here as $\lambda^{1/2}$ is the standard deviation of the Poisson distribution. The posterior distribution of λ given a random sample $x = (x_1, \ldots, x_n)$ is proportional to:

$$\pi(\lambda \mid x) \propto \lambda^{y} e^{-n\lambda} \lambda^{-1/2} = \lambda^{y-1/2} e^{-n\lambda},$$

where $y = \sum_{i=1}^{n} x_i$. This is the form of a gamma density with location parameter $\alpha = y + 1/2$ and scale parameter $\beta = n$ (Appendix B). The marginal distribution of x can be easily computed and is given by

$$\begin{aligned} f(x) &= \int_0^\infty f(x \mid \lambda)\pi(\lambda) d\lambda \\ &= \frac{1}{\prod_{i=1}^{n} x_i!} \frac{\Gamma(y + 1/2)}{n^{y+1/2}}. \end{aligned}$$

The posterior distribution of λ is then given by

$$\pi(\lambda \mid x) = \frac{f(x \mid \lambda)\pi(\lambda)}{f(x)} = \frac{e^{-n\lambda}\lambda^{y-1/2} n^{y+1/2}}{\Gamma(y + 1/2)}.$$

The corresponding predictive distribution is the Poisson-gamma mixture

$$\begin{aligned} f(x_{n+1} \mid x) &= \int_0^\infty \frac{e^{-\lambda}\lambda^{x_{n+1}}}{(x_{n+1})!} \frac{e^{-n\lambda}\lambda^{y-1/2} n^{y+1/2}}{\Gamma(y + 1/2)} d\lambda \\ &= \frac{n^{y+1/2}}{\Gamma(y + 1/2)} \frac{1}{(x_{n+1})!} \frac{\Gamma(y + x_{n+1} + 1/2)}{(n + 1)^{y+x_{n+1}+1/2}}, \end{aligned} \tag{3.17}$$

independent of λ as is required. Note that the sum, y, of the observations is all that is needed, not the results of each individual test.

Consider, then, a forensic DNA laboratory with $n = 15$ internal tests with no detected analytical errors (i.e. $y = 0$) and for which it may be of interest to obtain a probability for an error in a future test. This probability is obtained by Equation (3.17). In particular, $f(x_{n+1} = 0 \mid x) = 0.968$, $f(x_{n+1} = 1 \mid x) = 0.030$, $f(x_{n+1} \geq 2 \mid x) = 0.002$.

3.3.5 Markov Chain Monte Carlo methods (MCMC)

Suppose that it is of interest to find the posterior mean of a vector-valued target parameter $\boldsymbol{\theta}$, $\boldsymbol{\theta} = (\theta_1, \ldots, \theta_p)$,

$$E(\boldsymbol{\theta} \mid x) = \int_{\Theta} \boldsymbol{\theta}\pi(\boldsymbol{\theta} \mid x)\mathrm{d}\boldsymbol{\theta}.$$

Assume further that the integral does not have an analytical solution. In such cases one can implement Monte Carlo sampling techniques. The key idea of these methods is that, if one were able to sample from the posterior density, the draws could be taken to compute sample-based estimates. For example, the posterior mean could be estimated by computing the sample average of the sampled draws. Other summaries, such as the posterior quantiles could be derived from the quantiles of the sampled draws.

The task of drawing samples from the posterior density can be of modest difficulty or very complicated, depending on the problem at hand. If the prior distribution is chosen in the family of the conjugate distributions, then the posterior distribution will be known and it will be possible in general to simulate an independent sample $\boldsymbol{\theta}^{(1)}, \boldsymbol{\theta}^{(2)}, \ldots, \boldsymbol{\theta}^{(n)}$ from the exact posterior distribution. The Monte Carlo estimate of the posterior mean is given by the sample mean of the simulated draws:

$$E(\boldsymbol{\theta} \mid x) \cong \frac{1}{n}\sum_{i=1}^{n} \boldsymbol{\theta}^{(i)}.$$

As the size of the simulated draws becomes large, the Monte Carlo estimate will converge to the posterior mean. Then, it would be possible to obtain the Monte Carlo estimate of the posterior mean of any function $h(\cdot)$ of the parameter of interest, $h(\boldsymbol{\theta})$

$$E(h(\boldsymbol{\theta}) \mid x) = \int_{\Theta} h(\boldsymbol{\theta})\pi(\boldsymbol{\theta} \mid x)\mathrm{d}\theta \cong \frac{1}{n}\sum_{i=1}^{n} h(\boldsymbol{\theta}^{(i)}).$$

In other words, the computation of the integral is not necessary for parameter estimation. The estimation is determined from features of the posterior density which might be of interest (e.g. the posterior mean).

For statistical models of even moderate complexity, there may be no easy way to sample directly from the posterior distribution. The integral in Equation (3.5) may not be tractable in closed form (for example when a conjugate prior is not appropriate), and algorithms must then be used that allow a simulated sample to be obtained from the posterior density when the normalizing constant at the denominator is unknown. Rejection sampling is one of the most commonly used techniques that is available because of its generality, but it can be difficult to set up since it requires the construction of a suitable *proposal density* (Gelman *et al.* 2004; Gilks *et al.* 1996).

A set of methods which enables sampling from posterior densities is that of the so-called Markov Chain Monte Carlo methods (Gamerman and Lopes 2006; Gelman *et al.* 2004; Gilks *et al.* 1996). Their key feature is to produce draws by recursively simulating a chain that converges to the posterior density $\pi(\boldsymbol{\theta} \mid x)$. In general, the simulated values of the parameter of interest $\boldsymbol{\theta}$ obtained at the beginning of an MCMC run do not have the desired posterior distribution, because of the effect of the starting values, and they are discarded. However, after some number, say n_b, of iterations have been performed (the so-called *burn in period*), the effect of the initial values wears off, and simulated values beyond the first n_b iterations

$$\boldsymbol{\theta}^{(n_b+1)}, \boldsymbol{\theta}^{(n_b+2)}, \ldots, \boldsymbol{\theta}^{(n_b+n)}$$

can be taken as draws from the posterior density. Under general conditions, sample-based estimates from the generated chain converge to the quantities of interest as the number of iterations becomes sufficiently large. Good practice consists in monitoring the convergence, that is to check whether the simulated sample provides a reasonable approximation for the posterior density (Albert 2007; Gamerman and Lopes 2006; Gelman *et al.* 2004). At first, if the number of iterations is not long enough, simulations may be not representative of the target distribution. A second problem which might arise is the correlation between draws: simulated values of the parameter of interest at the $(j+1)^{th}$ iteration are dependent on the simulated values at the jth iteration. Serial correlation is not itself a problem for convergence, but can cause serious inefficiencies in simulations. In fact, a strong correlation between successive iterations may prevent the MCMC algorithm from exploring the entire parameter space. Moreover, a strong correlation between successive values will produce inefficient simulations because an immediate neighbouring value will provide little additional information about the posterior distribution. Convergence can be assessed by graphical and numerical (posterior summaries) diagnostics. Fundamental graphical diagnostics are the *trace* and the *autocorrelation* plots. The trace plot is obtained by plotting simulated draws against the iteration number: a plot exhibiting the same behaviour through iterations after an initial period is an indication of convergence. The autocorrelation plot is obtained by plotting the autocorrelation between sets of values $\{\boldsymbol{\theta}^j\}$ and $\{\boldsymbol{\theta}^{j+L}\}$ against L, where L is the lag, or the number of iterations separating these sets of values. If the chain is mixing adequately, the values of the autocorrelation will decrease to zero as the lag value (distance between successive sets of iterations) is increased.

Two well-known methods to construct a chain with these features are known as the Metropolis–Hastings (M-H) algorithm and the Gibbs sampling algorithm, a special case of the M-H algorithm.

The Metropolis–Hastings algoritm

Call $\boldsymbol{\theta}^{(0)}$ the starting value of the chain, and suppose that the draws $\boldsymbol{\theta}^{(1)}, \boldsymbol{\theta}^{(2)}, \ldots, \boldsymbol{\theta}^{(j-1)}$ have been obtained. The next item of the chain $\boldsymbol{\theta}^{(j)}$ is obtained by a two-step process:

1. A proposal value, called $\boldsymbol{\theta}^{\text{prop}}$, is drawn from a density $q\left(\boldsymbol{\theta}^{(j-1)}, \boldsymbol{\theta}^{\text{prop}}\right)$, called the transition density or candidate-generating density;
2. The proposal value is accepted with probability

$$\alpha(\boldsymbol{\theta}^{(j-1)}, \boldsymbol{\theta}^{\text{prop}}) = \min\left\{1, \frac{\pi(\boldsymbol{\theta}^{\text{prop}})}{\pi(\boldsymbol{\theta}^{(j-1)})}\frac{q(\boldsymbol{\theta}^{\text{prop}}, \boldsymbol{\theta}^{(j-1)})}{q(\boldsymbol{\theta}^{(j-1)}, \boldsymbol{\theta}^{\text{prop}})}\right\}. \tag{3.18}$$

Then,

$$\boldsymbol{\theta}^{(j)} = \begin{cases} \boldsymbol{\theta}^{\text{prop}} & \text{in case of acceptance} \\ \boldsymbol{\theta}^{(j-1)} & \text{otherwise.} \end{cases}$$

It is evident that the Markov chain generated in this way can repeat the current value $\boldsymbol{\theta}^{(j-1)}$ for several iterations. The transition density must be chosen to avoid the event that the chain stays at the same point for many iterations and instead moves efficiently to explore the support of the target density. If the candidate generating density is built to have the mean equal to the current value of the parameter, the algorithm is also called a *random-walk* algorithm. The candidate generating density is generally a Normal proposal density, with mean equal to the current value of the parameter, and a variance that must be carefully set as it defines the level of acceptance rate on which the success of the method depends. If the variance is too large, an extremely large proportion of iterations will be rejected and the algorithm will therefore be inefficient. Conversely, if the variance is too small, the random walk will accept nearly all proposed values but will explore the distribution inadequately and it will take many iterations to converge. Many authors generally agree that a good acceptance rate must be between 20% and 50% (Gamerman 2006). Other families of candidate-generating densities are possible, see Chib and Greenberg (1995). Note that if the candidate-generating density is symmetric, the probability of acceptance (3.18) reduces to

$$\alpha(\boldsymbol{\theta}^{(j-1)}, \boldsymbol{\theta}^{\text{prop}}) = \min\left\{1, \frac{\pi(\boldsymbol{\theta}^{\text{prop}})}{\pi(\boldsymbol{\theta}^{(j-1)})}\right\}.$$

Multiple-block M-H algorithm

In some cases it is useful to split the parameter space into smaller blocks and construct a Markov chain in each of these blocks, since it can be difficult to choose a transition density on the full parameter space that allows generation of a chain that is rapidly converging to the posterior distribution. Suppose the parameter $\boldsymbol{\theta}$ is split into two blocks (but the procedure is the same for any number of blocks),

$\boldsymbol{\theta} = (\boldsymbol{\theta}_1, \boldsymbol{\theta}_2)$. For each block a candidate-generating density $q_k(\boldsymbol{\theta}_k^{(j-1)}, \boldsymbol{\theta}_k^{\text{prop}})$, $k = 1, 2$ will be chosen with a proposed value $\boldsymbol{\theta}_k^{\text{prop}}$ conditioned on the previous value of the block and on the current value of the other block, and the computed probability of acceptance is given by

$$\alpha\left(\boldsymbol{\theta}_k^{(j-1)}, \boldsymbol{\theta}_k^{\text{prop}}\right) = \min\left\{1, \frac{\pi(\boldsymbol{\theta}_k^{\text{prop}} \mid \boldsymbol{\theta}_{-k})\, q_k(\boldsymbol{\theta}_k^{\text{prop}}, \boldsymbol{\theta}_k^{(j-1)})}{\pi(\boldsymbol{\theta}_k^{(j-1)} \mid \boldsymbol{\theta}_{-k})\, q_k(\boldsymbol{\theta}_k^{(j-1)}, \boldsymbol{\theta}_k^{\text{prop}})}\right\},$$

where $\boldsymbol{\theta}_{-k}$ denotes the parameters of the remaining blocks, and $\pi(\boldsymbol{\theta}_k^{(\cdot)} \mid \boldsymbol{\theta}_{-k})$ is called *full conditional density* for both situations prop and $(j-1)$ denoted $(\cdot)$ here for brevity. The algorithm is summarized as follows:

1. Specify an initial value $\boldsymbol{\theta} = (\boldsymbol{\theta}_1^{(0)}, \boldsymbol{\theta}_2^{(0)})$;
2. Repeat for $j = 1, \ldots, n_b + n$ iterations
   Repeat for each block $(k = 1, 2)$
   (a) Propose a value for the kth block
   $$\boldsymbol{\theta}_k^{\text{prop}} \sim q_k(\boldsymbol{\theta}_k^{(j-1)}, \boldsymbol{\theta}_k^{\text{prop}});$$
   (b) Calculate the acceptance probability $\alpha(\boldsymbol{\theta}_k^{(j-1)}, \boldsymbol{\theta}_k^{\text{prop}})$;
   (c) Update the kth block
   $$\boldsymbol{\theta}_k^{(j)} = \begin{cases} \boldsymbol{\theta}_k^{\text{prop}} & \text{in case of acceptance} \\ \boldsymbol{\theta}_k^{(j-1)} & \text{otherwise.} \end{cases}$$
3. Return the values $\{(\boldsymbol{\theta}^{(n_b+1)}, \boldsymbol{\theta}^{(n_b+2)}, \ldots, \boldsymbol{\theta}^{(n_b+n)})\}$.

3.4 BAYESIAN DECISION THEORY

Bayesian statistical inference can be seen as a decision problem, where the class of available decisions $\mathcal{D}$ is given by the possible conditional probability distributions of θ given the data

$$\mathcal{D} = \left\{\pi(\theta \mid x);\ \pi(\theta \mid x) > 0 \text{ and } \int_\Theta \pi(\theta \mid x) d\theta = 1\right\}.$$

A decision maker is asked to choose among actions which correspond to reporting a probability distribution for some unknown quantity of interest.

A fundamental basis of Bayesian decision theory is that statistical inference requires a rigorous determination of:

1. the probabilistic model underlying the observed data[9];

[9]In this book only parametric models are considered, $\{f(x \mid \theta); \theta \in \Theta\}$.

2. the prior distribution $\pi(\theta)$ which measures the uncertainty about the unknown parameter θ;
3. the loss function $L(d, \theta)$ which specifies the undesirability of available decisions.

The statistical model then involves three spaces (sets of all possible values of the appropriate concepts): the observation space $\mathcal{X}$, the parameter space Θ, the decision space $\mathcal{D}$. A decision means an action. For this reason, the decision space is sometimes known as the action space.

This approach is also known as a *fully Bayesian approach*:

> A fully Bayesian approach can be distinguished from partial Bayesian approaches, without meaning to imply that less than fully Bayesian is less than good. A fully Bayesian approach is decision theoretic and posterior probabilities are based on all available evidence, including that separate from the trial at hand. There are at least two ways to be less than fully Bayesian. First, one can calculate posterior distributions as data summaries without incorporating them into a decision analysis. Second, one can calculate posterior distributions using canonical prior distributions rather than prior distributions based on the available evidence. (Berry and Stangl 1996, p. 8).

3.4.1 Optimal decisions

In decision problems that involve uncertainty, the actual incurred loss can never be known with certainty. The best way of proceeding in contexts in which the decision maker has to make the decision in the face of uncertain outcomes of experiments is to consider the expected loss of making a decision, and choose an optimal decision with respect to this expected loss (Savage 1972). The Bayesian expected loss of a decision, denoted by $\bar{L}(d, \pi^*)$, averages the error (i.e. the loss) according to the probability distribution π^* of θ at the time of decision making:

$$\begin{aligned}\bar{L}(d, \pi^*) &= E^{\pi^*}[L(d, \theta)] \\ &= \int_{\Theta} L(d, \theta)\pi^*(\theta)\mathrm{d}\theta.\end{aligned}$$

An optimal decision, also called *Bayes decision* and denoted by d^{π^*}, is a decision $d \in \mathcal{D}$ that minimizes the Bayesian expected loss:

$$d^{\pi^*} = \min_{d \in \mathcal{D}} \bar{L}(d, \pi^*).$$

Assume that $X = x$ has been observed. Then, π^* can be replaced with $\pi(\theta \mid x)$. The Bayesian (posterior) expected loss averages the error according to the posterior distribution of the parameter θ, conditionally on the observed value x:

$$\begin{aligned}\bar{L}\,(d, \pi(\theta \mid x)) &= E^{\pi(\theta \mid x)}[L(d, \theta)] \\ &= \int_{\Theta} L(d, \theta)\pi(\theta \mid x)\mathrm{d}\theta.\end{aligned}$$

The criterion of choice presented above, and generally adopted throughout this book, assumes that the consequences of several actions can be identified with probability one, once decision d is taken and θ turns out to be the true state of nature.

The forensic identification setting discussed in Example 3.2.1 illustrates a decision problem of this type: if the conclusion (i.e. the decision) is an identification and the suspect is truly the source of the crime stain, the suspect is correctly associated with the crime stain with probability one.

Decision problems might be more complicated, however, since there may be situations where each decision d can be described by a set of probability distributions over consequences given each possible parameter value: $p_d(c \mid \theta)$. A simple example of this kind is presented and discussed, for example, in Parmigiani (2002). Assume that the decision maker must choose between two mutually exclusive treatments, treatment A (d_1) and treatment B (d_2). The unknown state of nature is the genotype, which might be 'responsive' (θ_1), or 'unresponsive' (θ_2). Specifically, the treatment A is effective in patients with a responsive genotype, but not effective with others, while the treatment B is moderately effective in all patients. For each combination of treatments and genotype, there is a potentially different probability distribution for recovery times. Both states of nature and consequences are unknown: both are assigned a probability distribution. The Bayesian expected loss is then obtained by integrating out both parameters and consequences:

$$\bar{L}(d, \pi^*) = \int_C \int_\Theta L(d, \theta) p_d(c \mid \theta) \pi^*(\theta) \mathrm{d}\theta \mathrm{d}c,$$

where the expression $p_d(c \mid \theta)$ is to be integrated over both consequences c and parameters θ. This line of reasoning is not pursued further in this book.

3.4.2 Standard loss functions

The form of a loss function has only briefly been discussed so far for a discrete case in Section 3.2.3. The aim is to specify loss functions that adequately reflect a decision maker's personal preferences. In this section, standard loss functions will be illustrated which are mathematically tractable and well documented. The first loss function considered is the *squared-error (or quadratic) loss*:

$$L(d, \theta) = k(d - \theta)^2, \tag{3.19}$$

where k denotes a constant (Press 2003). The loss associated with making a decision d when the true state of nature is θ increases by the square of the difference between d and θ. For example, if the difference doubles the loss quadruples (increases by a factor of 4). The choice of a quadratic loss may be acceptable because it is conservative in the sense that it assigns low losses to values close to the true state of nature, and high losses to values far from the true state of nature. Notice that the shape of this function (Figure 3.13) is such that large deviations from the true value of the parameter θ are more strongly penalized.

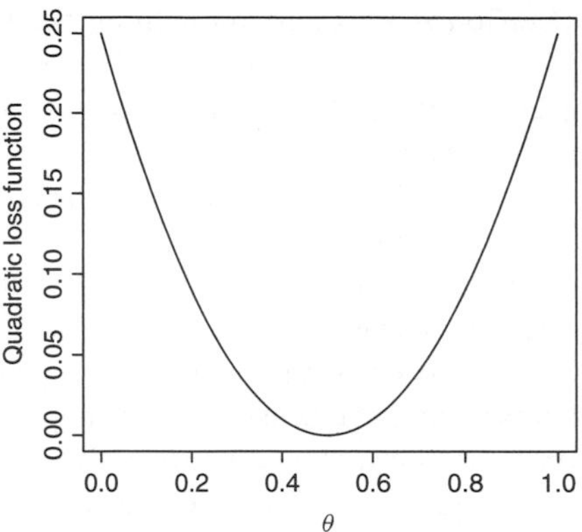

Figure 3.13 Quadratic loss function with $d = 0.5$ and $k = 1$.

One might also encounter statistical problems that call for a loss symmetric in $(d - \theta)$, while the exact functional form of the loss is not crucial. In these situations, the loss function may be approximated appropriately by the quadratic loss (Berger 1988). Sometimes, the function is also chosen for reasons of simplicity since it gives intuitive Bayesian solutions. Then, the Bayesian posterior expected loss is given by:

$$\bar{L}(d, \pi(\theta \mid x)) = \int_{\Theta} k(d - \theta)^2 \pi(\theta \mid x) \mathrm{d}\theta.$$

To find the minimum, we differentiate with respect to d and set the result equal to 0:

$$\begin{aligned}
&\frac{\partial}{\partial d}\left[\int_{\Theta} k(d - \theta)^2 \pi(\theta \mid x) \mathrm{d}\theta\right] = \\
&= \frac{\partial}{\partial d}\left[kd^2 \int_{\Theta} \pi(\theta \mid x) \mathrm{d}\theta + k \int_{\Theta} \theta^2 \pi(\theta \mid x) \mathrm{d}\theta - 2kd \int_{\Theta} \theta \pi(\theta \mid x) \mathrm{d}\theta\right] \\
&= 2kd - 2kE^{\pi(\theta \mid x)}(\theta) = 0.
\end{aligned}$$

Solving for d gives

$$d^{\pi^*} = E^{\pi(\theta \mid x)}(\theta).$$

Then, the Bayes decision under quadratic loss is the mean of the posterior distribution of θ given x, which is intuitively satisfying. This result, the posterior mean, indicates that this is the best estimator, rather than the posterior median or any other statistic. Note, however, that this is not the only loss function which provides this result. Other losses – called *proper losses* – share this property[10] (Lindley 1985).

[10]A proper loss function is one for which a Bayesian's best strategy is to tell the truth. Such a property seems reasonable. To say that the Bayesian's best strategy is to tell the truth is to say that the best estimator of the parameter is the Bayesian's best assessment of the probability of its occurrence (Hwang *et al.* 1992). A complete characterization of proper losses is given in Schervish (1989).

An example for a situation in which a symmetric loss could be used is one in which, for the purpose of identification, the height of an individual needs to be estimated from data extracted from surveillance camera images. The true state of nature θ is the height of the individual in the image. It particular, it may be accepted by a decision maker that under- and overestimation of the individual's height incurrs equal losses (see Example 4.4.4 in Section 4.4.3).

For a multidimensional model parameter $\boldsymbol{\theta} = (\theta_1, \ldots, \theta_k)$, the quadratic loss generalizes to the quadratic form:

$$L(\mathbf{d}, \boldsymbol{\theta}) = (\mathbf{d} - \boldsymbol{\theta})' A \, (\mathbf{d} - \boldsymbol{\theta}) \,,$$

where A is a non-negative definite matrix (Lütkepohl 1996), and $\mathbf{d} = (d_1, \ldots, d_k)$. It can be shown (O'Hagan 1994) that the Bayes decision (i.e. the decision that minimizes the expected loss) is the mean vector:

$$\mathbf{d}^{\pi^*} = E^{\pi(\boldsymbol{\theta}|x)}(\boldsymbol{\theta}).$$

Another loss function is the *piecewise linear loss function* which is given, for $k_0 > 0$, and $k_1 > 0$, by:

$$L(d, \theta) = \begin{cases} k_0(\theta - d) & \text{if } \theta - d \geq 0, \\ k_1(d - \theta) & \text{if } \theta - d \leq 0. \end{cases} \tag{3.20}$$

The constants k_0 and k_1 can be chosen so as to reflect the relative importance of underestimation and overestimation. Figure 3.14 shows both of these situations. The figure on the left represents a linear function that penalizes underestimation ($k_0 > k_1$) more heavily. On the right, a linear function is shown that penalizes overestimation ($k_0 < k_1$) more heavily. The linear loss function increases more

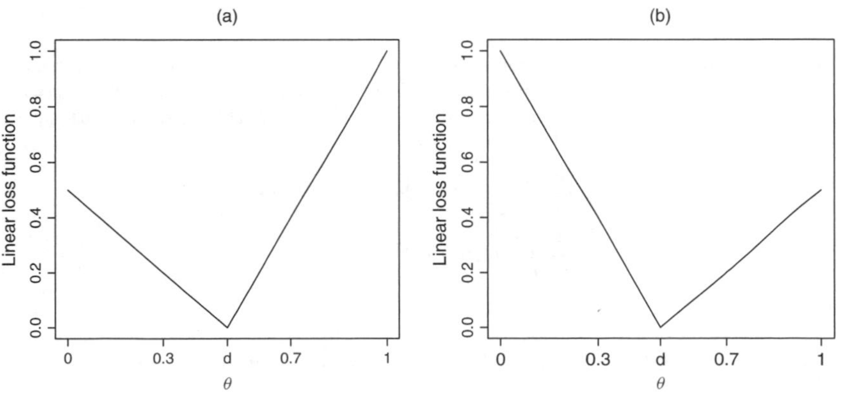

Figure 3.14 Linear loss function $L(d,\theta) = k_0(\theta - d)$, $(\theta - d) \geq 0$, $L(d,\theta) = k_1(d - \theta)$, $(\theta - d) \leq 0$, with $d = 0.5$. (a) $k_0 > k_1$, (b) $k_0 < k_1$.

slowly than a quadratic loss and does not overpenalize large but unlikely errors. If the constants k_0 and k_1 are equal, the loss function in (3.20) can be written as

$$L(d,\theta) = k \mid d - \theta \mid,$$

and is called *absolute error loss*.

The Bayesian posterior expected loss for the piecewise linear loss function, Equation (3.20), is given by:

$$\bar{L}(d, \pi(\theta \mid x)) = \int_d^{\infty} k_0(\theta - d)\pi(\theta \mid x)\mathrm{d}\theta + \int_{-\infty}^{d} k_1(d - \theta)\pi(\theta \mid x)\mathrm{d}\theta.$$

To find the minimum, we differentiate the posterior expected loss with respect to d and set the result equal to 0[11]:

$$\frac{\partial}{\partial d}\left[\int_d^{\infty} k_0(\theta - d)\pi(\theta \mid x)\mathrm{d}\theta + \int_{-\infty}^{d} k_1(d - \theta)\pi(\theta \mid x)\mathrm{d}\theta\right] = -k_0 + k_0 P_{\pi(\theta|x)}(\theta \leq d) + k_1 P_{\pi(\theta|x)}(\theta \leq d).$$

Solving for $P_{\pi(\theta|x)}(\theta \leq d)$ gives

$$P_{\pi(\theta|x)}(\theta \leq d) = \frac{k_0}{k_0 + k_1}.$$

11

$$\begin{aligned}\bar{L}(d, \pi(\theta \mid x)) &= \int_d^{\infty} k_0(\theta - d)\pi(\theta \mid x)\mathrm{d}\theta + \int_{-\infty}^{d} k_1(d - \theta)\pi(\theta \mid x)\mathrm{d}\theta \\ &= k_0 \int_d^{\infty} \theta\pi(\theta \mid x)\mathrm{d}\theta - k_0 d \int_d^{\infty} \pi(\theta \mid x)\mathrm{d}\theta \\ &- k_1 \int_{-\infty}^{d} \theta\pi(\theta \mid x)\mathrm{d}\theta + k_1 d \int_{-\infty}^{d} \pi(\theta \mid x)\mathrm{d}\theta.\end{aligned}$$

Note that the second term can be rewritten simply as $k_0 d\left(1 - P_{\pi(\theta|x)}(\theta \leq d)\right)$, and the fourth equals to $k_1 d P_{\pi(\theta|x)}(\theta \leq d)$. To find the minimum we differentiate with respect to d and then set the result equal to 0. Note that the range of the integrals depends on the unknown variable d, and that the application of the Leibniz's rule (Casella and Berger 2002) gives

$$\begin{aligned}\frac{\partial}{\partial d}\int_d^{\infty} \theta\pi(\theta \mid x)\mathrm{d}\theta &= -d\pi(d \mid x) \\ \frac{\partial}{\partial d}\int_{-\infty}^{d} \theta\pi(\theta \mid x)\mathrm{d}\theta &= d\pi(d \mid x)\end{aligned}$$

Then

$$\begin{aligned}&\frac{\partial}{\partial d}\left[\int_d^{\infty} k_0(\theta - d)\pi(\theta \mid x)\mathrm{d}\theta + \int_{-\infty}^{d} k_1(d - \theta)\pi(\theta \mid x)\mathrm{d}\theta\right] = -k_0 d\pi(d \mid x) - k_0 \\ &\quad + k_0 d\pi(d \mid x) + k_0 P_{\pi(\theta|x)}(\theta \leq d) - k_1 d\pi(d \mid x) + k_1 P_{\pi(\theta|x)}(\theta \leq d) + k_1 d\pi(d \mid x) \\ &\quad = -k_0 + k_0 P_{\pi(\theta|x)}(\theta \leq d) + k_1 P_{\pi(\theta|x)}(\theta \leq d) = 0.\end{aligned}$$

Then, for any linear loss function (3.20), the Bayes decision is the $\frac{k_0}{k_0+k_1}$ quantile of the distribution of θ at the time of decision making. In particular, if $k_0 = k_1$ (i.e. in the case of absolute error loss), the Bayes decision is the posterior median.

Consider the problem of estimating blood alcohol concentration through the analysis of breath. Here, a decision maker may prefer to penalize the underestimation of the true alcohol concentration more than the overestimation. This may be so because falsely concluding a low alcohol concentration in an individual with increased blood alcohol is regarded as a more serious error (because such an individual may represent a serious danger in traffic) than falsely assigning a high blood alcohol concentration to an individual which has actually a low concentration level (see Example 4.4.1 in Section 4.4.1).

There are also generalizations of the absolute error loss function for vector-valued parameters, $\boldsymbol{\theta} = (\theta_1, \ldots, \theta_k)$. The problem is that, while there is a natural definition of the mean of a vector random variable, there is no natural definition of the median. Consider instead the following loss function:

$$L(\mathbf{d}, \boldsymbol{\theta}) - \sum_{j=1}^{k} k_j \mid d_j - \theta_j \mid .$$

The expected loss is also a sum whose components are minimized by letting each decision d_j be the median of the marginal distribution of θ_j (O'Hagan 1994).

A third loss function is the *'0–1' loss function* which may be considered in relation with a two-action decision problem: $\mathcal{D} = \{d_0, d_1\}$. Let action d_0 be correct if $\theta \in \Theta_0$ and action d_1 be correct if $\theta \in \Theta_1$. This corresponds to testing $H_0 : \theta \in \Theta_0$ versus $H_1 : \theta \in \Theta_1$. The *'0–1' loss function* is then defined as:

$$L(d_i, \theta) = \begin{cases} 0 & \text{if } \theta \in \Theta_i \\ 1 & \text{if } \theta \in \Theta_j \ (j \neq i). \end{cases} \tag{3.21}$$

The loss is zero if a correct decision is made, and one if an incorrect decision is made. The optimal decision is found by minimizing the posterior expected loss, that is equal to

$$\begin{aligned} \bar{L}(d_0, \pi(\theta \mid x) &= \int_{\Theta} L(d_0, \theta)\pi(\theta \mid x)\mathrm{d}\theta \\ &= \int_{\Theta_1} \pi(\theta \mid x)\mathrm{d}\theta = P\,(\theta \in \Theta_1 \mid x)\,, \end{aligned}$$

for decision d_0, and to

$$\bar{L}(d_1, \pi(\theta \mid x) = P\,(\theta \in \Theta_0 \mid x)\,,$$

for decision d_1. Decision d_0 should be taken when

$$\bar{L}(d_0, \pi(\theta \mid x) < \bar{L}(d_1, \pi(\theta \mid x),$$

or equivalently when $P(\theta \in \Theta_0 \mid x) > P(\theta \in \Theta_1 \mid x)$. See O'Hagan (1994) for a generalization to the multidimensional case.

The *'0–1' loss* may not be a good approximation of the true loss, however. More realistic losses may be:

$$L(d_i,\theta) = \begin{cases} 0 & \text{if } \theta \in \Theta_i \\ k_i & \text{if } \theta \in \Theta_j \ (j \neq i). \end{cases} \tag{3.22}$$

and

$$L(d_i,\theta) = \begin{cases} 0 & \text{if } \theta \in \Theta_i \\ f_i(\theta) & \text{if } \theta \in \Theta_j \ (j \neq i). \end{cases} \tag{3.23}$$

The loss in (3.23) is a function (which could be linear, quadratic or something else) of the distance between the decision and the true value of θ: the loss depends on the severity of the mistake. This may be of interest, for example, in the context of estimating proportions in sampling surveys, where one may need to decide whether or not a proportion is greater than a certain threshold.

Considering the loss function in Equation (3.22), the Bayesian expected losses for decisions d_0 and d_1 are readily computed as

$$\bar{L}(d_0, \pi(\theta \mid x)) = k_0 P(\theta \in \Theta_1 \mid x)$$
$$\bar{L}(d_1, \pi(\theta \mid x)) = k_1 P(\theta \in \Theta_0 \mid x).$$

So, decision d_0 should be taken when

$$k_0 P(\theta \in \Theta_1 \mid x) < k_1 P(\theta \in \Theta_0 \mid x),$$

or equivalently when

$$\frac{k_0}{k_1} < \frac{P(\theta \in \Theta_0 \mid x)}{P(\theta \in \Theta_1 \mid x)}.$$

3.5 R CODE

A symbol '$*$', '+', ',' and so on at the end of a line indicates that the command continuous to the following line. The absence of such a symbol indicates the end of a command.

Example 3.3.2

Data and prior parameters

```
y=40
n=400
alpha=18
beta=425
```

Posterior parameters

```
alphap=alpha+y
betap=beta+n-y
print(paste('alpha*=',alphap))
print(paste('beta*=',betap))
```

Prior, likelihood and posterior distributions

```
par(mfrow=c(3,1))
plot(function(x) dbeta(x,alpha,beta),0,0.25,ylim=c(0,45),
ylab='',xlab=expression(paste(theta)),
main=expression(paste("Prior (")*paste(alpha==paste(18, )*
paste(", ")*paste(beta==paste(425)))*paste(")")))

plot(function(x) dbeta(x,y+1,n-y+1),0,0.25,ylim=c(0,45),
ylab='',xlab=expression(paste(theta)),
main=expression(paste("Likelihood (")*paste(alpha==paste(41,)*
paste(", ")*paste(beta==paste(361)))*paste(")")))

plot(function(x) dbeta(x,alphap,betap),0,0.25,ylim=c(0,45),
ylab='',xlab=expression(paste(theta)),
main=expression(paste("Posterior (")*paste(alpha==paste(58,)*
paste(", ")*paste(beta==paste(785)))*paste(")")))
```

Example 3.3.7

Data, prior parameters and prior distribution

```
n=20
y=3
alpha1=3
beta1=10
alpha2=6
beta2=2.5
g1=0.6
g2=1-g1
plot(function(x) g1*dbeta(x,alpha1,beta1)+(1-g1)*
dbeta(x,alpha2,beta2),0,1,
xlab=expression(paste(theta)),ylab='Probability density')
```

Posterior parameters and posterior distribution

```
alphap1=alpha1+y
betap1=beta1+n-y
alphap2=alpha2+y
betap2=beta2+n-y
gp1=round((g1*beta(alphap1,betap1))/(g1*beta(alphap1,betap1)
+g2*beta(alphap2,betap2)),2)
gp2=1-gp1
plot(function(x) gp1*dbeta(x,alphap1,betap1)+(1-gp1)*
dbeta(x,alphap2,betap2),0,1,
xlab=expression(paste(theta)),ylab='Probability density')
```

Example 3.3.8

Data, prior parameters

```
mu=c(178,172)
tau2=c(1,2.8^2)
n=1
xbar=176
sigma2=4
```

Posterior parameters

```
mux=c(0,0)
taux2=c(0,0)
for (i in 1:length(mu)){
mux[i]=(mu[i]*sigma2/n+tau2[i]*xbar)/(sigma2/n+tau2[i])
taux2[i]=sqrt((tau2[i]*sigma2/n)/(tau2[i]+sigma2/n))
}
```

For a sequential use of the Bayes theorem, assign

```
mu=mux
tau2=taux2
```

and rerun the codes above to compute posterior parameters.

Prior and posterior distributions

```
par(mfrow=c(2,1))
plot(function(x) dnorm(x,mu[1],sqrt(tau2)[1]), 165, 185,
main = 'Prior distributions for expert A and B',lty=1,
xlab=expression(paste(theta)),ylab='')
plot(function(x) dnorm(x,mu[2],sqrt(tau2)[2]), 165, 185,
main = '',lty=2,add=TRUE)
legend(164.5,0.4,c('Expert A','Expert B'),lty=c(1,2))

plot(function(x) dnorm(x,mux[1],sqrt(taux2)[1]), 165, 185,
main = paste('Posterior distributions for expert A and B (n=',
n,paste(')')),lty=1,xlab=expression(paste(theta)),ylab='')
plot(function(x) dnorm(x,mux[2],sqrt(taux2)[2]), 165, 185,
main = "",lty=2,add=TRUE)
legend(164.5,0.4,c("Expert A","Expert B"),lty=c(1,2))
```

Example 3.3.9

Data

```
n=15
y=0
```

Probability of future observations

```
x=c(0,1)
(n^(y+1/2))/(gamma(y+1/2)*factorial(x))*(gamma(x+y+1/2)/(n+1)^
(x+y+1/2))
```

Part II
Forensic Data Analysis

4

Point Estimation

4.1 INTRODUCTION

Point estimation, also called *parameter estimation*, essentially refers to the process of using sample data to estimate the value of a population parameter, such as the mean, the variance or the proportion[1]. As for many scientists in various disciplines, parameter estimation is also of interest to forensic scientists. The idea of this chapter is to illustrate how forensic scientists can 'learn' from past experience to draw inferences about the value of a population parameter. Particular attention will be drawn on how one may learn about, for example, a population proportion or a population mean. The topic of multi-parameter (vector-valued parameter) learning will also be covered.

Sample surveys, that is a means for collecting sample data, commonly serve as a basis for estimating, for example, the proportion of items or individuals in a given population that possess a certain characteristic. Suppose, for instance, that the population is represented by a collection of individuals which can be divided into categories according to the characteristic actually possessed by each member of the population. For the purpose of illustration, a forensic scientist might be interested in estimating the proportion of individuals whose mitochondrial DNA sequence is of a certain kind. In such a case, the population can be divided into two categories: the individuals that have the target DNA sequence (first category), and the individuals which have another DNA sequence, different from the first (second category). Another scenario in which estimation may play an important role could

[1]Note that, generally speaking, 'point estimation' is taken as some process of arriving at an estimate without regard to its precision. Chapter 5 will focus on 'interval estimation' where the precision of the estimate is to some extent taken into account.

Data Analysis in Forensic Science: A Bayesian Decision Perspective Franco Taroni, Silvia Bozza, Alex Biedermann, Paolo Garbolino and Colin Aitken

be one in which each member of a population of interest (i.e. ecstasy pills) has an associated continuous measurement, such as a weight. In such a setting, the scientist may be interested in estimating the mean weight of the seized pills.

Let $(x_1, x_2, \ldots, x_n)$ be the available data from a survey, which are assumed to have been generated by a parametric statistical model $\{f(x \mid \theta), \theta \in \Theta\}$, where θ denotes the unknown model parameter. In other words, the available data are the observed values of a random sample $(X_1, \ldots, X_n)$ from a probability distribution $f(x \mid \theta)$. For the time being, only singular-valued parameters (scalars), θ, will be considered. In the case of multi-parameter learning, the unknown parameter will be vector-valued and will be denoted $\boldsymbol{\theta} = (\theta_1, \ldots, \theta_p)$.

Estimation of θ is a natural requirement since knowledge of this parameter provides knowledge of the entire population. The aim is to find an accurate proxy for the actual but unknown value. A *point estimator* of the model parameter θ can be defined as a statistic, or random variable, $T = t(X_1, \ldots, X_n)$, that is any function of the random sample that does not depend on θ. When the observed data are $x_1, \ldots, x_n$, one can calculate the *point estimate* of θ, denoted $\hat{\theta}$, as $t = t(x_1, \ldots, x_n)$. The latter is a realized value of the point estimator, or random variable, and is a number.

A Bayesian statistical model is specified with the choice of a prior distribution $\pi(\theta)$ that allows the scientist to express his initial beliefs about the target parameter. Assuming that the probability model is correct, all available information about the value of the parameter θ, after observing the data, is contained in the posterior distribution $\pi(\theta \mid x)$. In other words, the posterior distribution encapsulates all that is known about θ.

An intuitive summary of the main conclusions which may be possibly drawn from knowledge of the parameter gained from the data can be obtained by plotting the posterior probability density function; interesting features of the posterior distribution (e.g. bimodality) may, in this manner, be simply revealed. Generally, however, scientists seek to summarize the posterior distribution by a single 'typical' number. It is often found convenient to summarize the information contained in the posterior distribution by providing values of the quantity of interest which are likely to be good Bayesian estimates for its true value. Common Bayesian estimates are the *posterior mode* (that is the value $\hat{\theta}$ at which the posterior probability density function takes its maximum value), the *posterior mean* or the *posterior median*.

A summary value is usually accompanied by a dispersion measure that may be interpreted as an indicator of its accuracy, that is the weakness or the strength of the posterior information (O'Hagan 1994). While a customary Bayesian measure of the accuracy of an estimator is the posterior variance of the estimator (Berger 1988), the question remains of how a scientist is to choose the best (in some sense) Bayesian estimator. The question of interest may be formulated as follows:

> Should the estimate chosen be the value with the highest probability (the mode), a value such that the odds are one-to-one of being above or below it (the median), a middle value in a center-of-gravity sense (the mean), or something else? (Antelman 1997, p. 356)

This issue can be dealt with in the decision approach to Bayesian statistical inference. From a decision-theoretic point of view, the estimation of a parameter θ is seen as a decision problem where the decision space $\mathcal{D}$ corresponds to the set of all possible values of θ, the parameter space Θ. The decision maker chooses a point estimate $\hat{\theta}$ and decides to act as though $\hat{\theta}$ were θ, that is deciding $d = \hat{\theta}$. The decision problem requires one to specify a loss function that represents the penalty of acting as if the true value of the quantity of interest were d, when the true value is actually θ, that is $d \neq \theta$. The simplicity of the Bayesian approach follows from the fact that the optimal decision, called *Bayesian decision*, and so the Bayesian estimate $\hat{\theta}$, is defined as that function of the observations which minimizes the posterior expected loss (see Section 3.4). This criterion is the only one a Bayesian needs for point estimation. In fact, knowing θ, one would choose $\hat{\theta} = \theta$ and one's loss would be zero. According to this, any possible parameter summary may turn out to be a Bayesian decision: it all depends on the loss function, as shown earlier in Section 3.4.2. As has been demonstrated, if a conventional loss function is used, the derivation of the Bayesian estimator is greatly simplified.

4.2 BAYESIAN DECISION FOR A PROPORTION

It is a common need in forensic practice to obtain an idea of the proportion of individuals or items that share a given characteristic (e.g. the proportion of red woollen fibres in a given population, the proportion of pills that contain something illegal, or of pirated CDs in a large seizure). In such settings, the outcome of an experiment that reveals the presence (absence) of the characteristic of interest is typically termed *positive* (*negative*).

Such questions may also be of interest in contexts involving DNA evidence. For example, forensic DNA analysts may encounter particular sequences of a mt-DNA region from both a recovered and a control sample, which may need to be compared. If the sequences are unequivocally different, then the samples are regarded as originating from different sources. Conversely, if the sequences cannot be distinguished, then one cannot exclude the possibility that the recovered and control samples potentially come from the same source[2]. When there is no notable difference between the two samples, it is desirable to convey some information about the number of times a particular sequence (or haplotype) is observed in a relevant population. This problem can be approached with the beta-binomial model earlier introduced in Section 3.3, notably throughout Example 3.3.2. Still within this context, a particular complication that may occur is that there may be zero occurrences of the sequence of interest in a given database (representing the relevant population) at hand. The estimation of the proportion in such a situation may be somewhat troubling, essentially with regard to traditional maximum likelihood estimators, which can be poor in such circumstances, even though a

[2]Notice that issues such as nucleotide differences, mutations or heteroplasmy are outside the scope of what is considered in the remainder of this discussion.

confidence interval giving an upper bound for the parameter may be provided. At such junctures, Bayesian inference can show a viable way ahead.

Sections 4.2.1 and 4.2.2 address this topic through a Bayesian decision-analytic approach with particular consideration being given to different prior distributions and loss functions as well as the robustness of the proposed procedure.

A second issue, addressed in Section 4.2.4, is that of complications that arise in counting experiments that are conducted under real-world circumstances. As observed by D'Agostini (2004a), scientists may sometimes miss relevant items in counting, whereas on other occasions they might be confused by other items that do not belong to the classes that are actually being searched (presence of background).

A further topic considered at the end of this section is that of parameter learning for a multinomial variable which represents a generalization to k outcomes of the binomial case. The principal ideas are illustrated in terms of an example that relates to the forensic examination of physical characteristics of questioned documents. At times Bayesian networks will be used with the aim of providing further insight in the proposed procedures.

4.2.1 Estimation when there are zero occurrences in a sample

Besides DNA-related matters, settings with zero occurrences in a sample may also be encountered in relation to other types of scientific evidence. Stoney (1992), for example, considered the following scenario involving glass evidence. The scientist found a match in several physical properties between glass from a broken crime scene window and glass fragments found in connection with a suspect. The scientist, who is interested in estimating the proportion of glasses having such physical properties, surveyed a sample of $n = 64$ glass objects and found that none of these agreed with the glass found in their case. In other words, there is no match with the crime-related items in a sample of size $n = 64$. How can they estimate the proportion of interest, if 0 occurrences have been observed in the sample? From a practical point of view, this can appear troubling; the sample is, however, of modest size and intuitively, it may be felt that the sample is of insufficient size to enable one to say that the glass of the crime scene window was unique.

Consider the following general scenario. Let θ denote the unknown proportion of individuals or items in a given population having a target characteristic. A sample of size n is inspected and no observation of a positive outcome is noted. An estimate of the proportion θ can be obtained starting from the Bayesian statistical model presented earlier in Example 3.3.2. The number of positive outcomes y in n trials (e.g. the number of glass fragments with some physical properties) is then modelled through a binomial distribution, that is $f(y \mid n, \theta) = \text{Bin}(n, \theta)$. Since the proportion θ is a continuous parameter, it will be necessary to construct a prior density $\pi(\theta)$ on the interval $(0, 1)$ that represents the initial beliefs. Typically, the prior density for θ will be chosen in the conjugate family for the binomial likelihood, that is the beta family (Section 3.3.3). Next, suppose that one has no preference for any of the possible values of the unknown proportion, so that one will fix the hyperparameters α and β equal to 1, which corresponds to a uniform prior. The term *uniform* means that the beta density is constant over all values of θ between 0 and 1. An implication of this is that any interval of values between 0 and 1 has the same probability as any other interval of the same width.

Consider, for the problem at hand, a piecewise linear loss function for the estimate d of θ (Equation (3.20)):

$$L(d,\theta) = \begin{cases} k_0(\theta - d) & \text{if } \theta - d \geq 0 \\ k_1(d - \theta) & \text{if } \theta - d \leq 0. \end{cases}$$

The constants k_0 and k_1 are fixed such that $k_0 \neq k_1$, to allow the decision maker to have different penalties for under- and overestimation of the parameter of interest (i.e. the loss function is *asymmetric*). In particular, k_0 is chosen to be greater than k_1 to indicate that underestimation is more severely penalized (Section 3.4.2). This may appear reasonable to a forensic decision maker because a very rare event (e.g. in case of an observation of zero occurrences of some characteristic of interest in a sample) tends to strengthen evidence against a suspect who possesses this characteristic (e.g. a glass fragment with a given characteristic). Figure 4.1 illustrates two alternative linear loss functions, computed at $d = 0.1$, for different values of k_0 ($k_0 = 2$ and $k_0 = 4$, respectively), while the value of k_1 is unchanged ($k_1 = 1$). Note that the greater the value of k_0, the more one penalizes an underestimation of the parameter θ. Figure 4.1 also shows the loss when $d = 0.1$, a particular

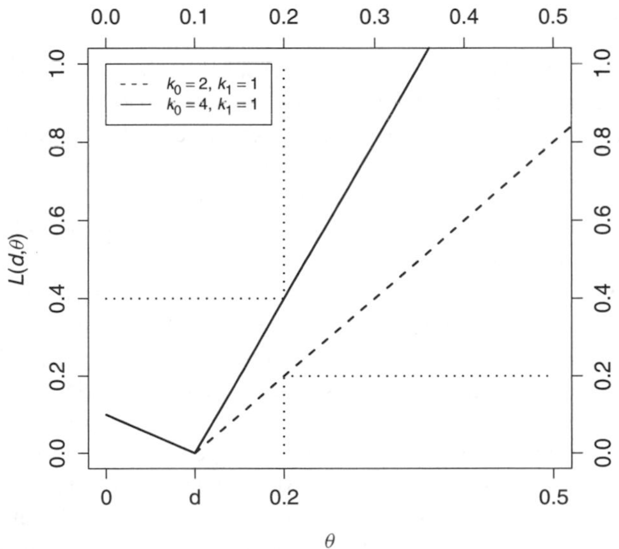

Figure 4.1 Effect of different values of k_0 on the incurred loss (for $d = 0.1$). $L(d,\theta) = 0.2$ for $\theta = 0.2$, $k_0 = 2$ and $k_1 = 1$; $L(d,\theta) = 0.4$ for $\theta = 0.2$, $k_0 = 4$ and $k_1 = 1$.

case in which θ is actually 0.2. The decision maker underestimates the value of θ, a choice of k_0 equal to 2 leads to an incurred loss $L(d, \theta) = 0.2$; if $k_0 = 4$ were chosen, then $L(d, \theta) = 0.4$. This example shows that the effect of the values for k_0 on $L(d, \theta)$ are substantial: the incurred loss doubles. Note that the two loss functions penalize equally an overestimation ($d > \theta$) of the parameter θ.

Once the decision maker has rigorously specified both a prior density that summarizes their prior beliefs about θ, and a loss function that quantifies the undesirability of the error of estimation, all elements necessary for the use of Bayesian decision theory for inference about θ are defined. The Bayesian decision in presence of a linear loss function is the $k_0/(k_0 + k_1)$ quantile of the posterior distribution (Section 3.4.2), which is a beta density with updated parameters, $\text{Be}(\alpha + y, \beta + n - y)$.

Example 4.2.1 *(Glass fragments). A sample of $n = 64$ objects is surveyed and no positive outcomes are observed ($y = 0$). The beta prior density is chosen to be uniform, that is $\alpha = \beta = 1$, so that the posterior density is a beta density with parameters $\alpha = 1$ and $\beta = 1 + 64$. Assuming a linear loss function with constants $k_0 = 2$ and $k_1 = 1$, the Bayes decision is the $2/(2 + 1) = 2/3$ quantile of the Be(1, 65) distribution, which is equal to 0.017. Alternatively, assuming $k_0 = 4$ and $k_1 = 1$, the Bayes decision is the 4/5 quantile of the same distribution, that is equal to 0.024. This result underlines the importance of the choice of the values for k_0 and k_1: the greater the value of k_0, the more an underestimation of the parameter of interest is penalized.*

Often, however, the scientist may reasonably be assumed to be more knowledgeable about the model parameter, in the sense that he may wish the prior distribution to be based on available knowledge. For example, when θ is the probability of occurrence of a 'rare' event, a prior distribution with mass concentrated on small values of θ may be more reasonable. For example, a triangular prior, that is a $\text{Be}(1, 2), \pi(\theta) = 2(1 - \theta)$, may be felt to be more appropriate than a rectangular prior, $\text{Be}(1, 1), \pi(\theta) = 1$.

Suppose thus that a substantial amount of past experience (e.g. due to previous experiments) is available, and advantage is taken of this to specify a few summaries of the prior distribution. Generally, this will be a location summary (such as the mean) and a dispersion summary (e.g. the variance). The problem then is to choose a beta prior whose shape matches these prior beliefs about the mean and the variance. Let μ and σ^2 denote, respectively, the prior mean and the prior variance for the proportion of interest as suggested by the available knowledge. Recalling that the mean of $\text{Be}(\alpha, \beta)$ is $E^{\pi}[\theta] = \alpha/(\alpha + \beta)$, and the variance is $Var^{\pi}[\theta] = \alpha\beta/((\alpha + \beta)^2(\alpha + \beta + 1))$, parameters α and β can be chosen by solving the following system:

$$\mu = \frac{\alpha}{\alpha + \beta},$$
$$\sigma^2 = \frac{\alpha\beta}{(\alpha + \beta)^2(\alpha + \beta + 1)}.$$

Solution of these two equations gives

$$\alpha = \mu \left[\frac{\mu(1 - \mu)}{\sigma^2} - 1 \right], \tag{4.1}$$
$$\beta = (1 - \mu) \left[\frac{\mu(1 - \mu)}{\sigma^2} - 1 \right], \tag{4.2}$$

and hence the exact form of the prior density is determined. An example will be given in Section 4.2.4.

Next, suppose that the available knowledge comes from a previous experiment in the form of an observation of a random sample of size n from a Bernoulli process with probability of success θ and a total number $y = \sum_{i=1}^{n} x_i$ of positive outcomes (successes). A common estimator of the parameter θ is the sample proportion[3] $P = \frac{1}{n}\sum_{i=1}^{n} X_i$, which has a mean equal to θ and variance equal to $\{\theta(1 - \theta)/n\}$.[4] Parameters α and β can be fixed by equating the mean and the variance of the sample proportion to the prior mean μ and the prior variance σ^2 at $p = \frac{1}{n}\sum_{i=1}^{n} x_i$, the estimate of θ. Substituting $\mu = p$ and $\sigma^2 = p(1 - p)/n$ in Equation (4.1) and (4.2) gives

$$\alpha = p \left[\frac{p(1 - p)n}{p(1 - p)} - 1 \right] = p(n - 1), \tag{4.3}$$
$$\beta = (1 - p) \left[\frac{np(1 - p)}{p(1 - p)} - 1 \right] = (1 - p)(n - 1). \tag{4.4}$$

One can readily see that the parameters obtained in this way are such that $n = \alpha + \beta + 1$, which means that one obtains a prior distribution that provides an

[3]This estimator is obtained by the maximum likelihood approach which consists of estimating the unknown parameter θ by a number $\hat{\theta}$ which maximizes the likelihood function: $\hat{\theta}$ is the value of the parameter for which likelihood is maximized.

[4]Let $(X_1, \ldots, X_n)$ be a random sample from a population with mean θ and variance $\theta(1 - \theta)$. Then, the mean of the sample proportion is

$$E(P) = E\left(\frac{1}{n}\sum_{i=1}^{n} X_i\right) = \frac{1}{n}\sum_{i=1}^{n}(E(X_i)) = \frac{1}{n}\sum_{i=1}^{n}\theta = \frac{1}{n}n\theta = \theta,$$

and the variance

$$Var(P) = Var\left(\frac{1}{n}\sum_{i=1}^{n} X_i\right) = \frac{1}{n^2}\sum_{i=1}^{n}(Var(X_i)) = \frac{1}{n^2}\sum_{i=1}^{n}\theta(1 - \theta) = \frac{1}{n^2}n\theta(1 - \theta) = \frac{\theta(1 - \theta)}{n}.$$

amount of information about θ equivalent to the amount that comes from a random sample of size n. This suggests an important caveat when the choice is made of the form of a beta density from a given location summary (such as a prior mean) and a given dispersion summary (such as a prior variance), that is computing α and β directly from Equations (4.1) and (4.2) rather than from the number of positive outcomes occurred in previous experiments. If the choice is made from location and dispersion statistics, Bolstad (2004) proposes that, once α and β have been obtained, the so-called *equivalent sample size* $n_{eq} = \alpha + \beta + 1$ is computed and a check is made as to whether the available knowledge is realistically comparable with the knowledge that would have been obtained from an experiment of that size. If not, one should increase the prior variance incrementally in order to avoid the use of a prior density that reflects more knowledge than provided by the 'prior data'.

***Example 4.2.2** (Glass fragments – continued). Recall the previously discussed scenario (Example 4.2.1) where $n = 64$ glass objects are surveyed and no positive outcomes are found ($y = 0$). Imagine that the prior knowledge is given by a sample of 40 observations among which there is 1 positive result. Then, $p = 0.025$ and the application of Equations (4.3) and (4.4) gives $\alpha = 0.97$ and $\beta = 38.03$, approximated to $\alpha = 1$ and $\beta = 38$ respectively, that is a $Be(1, 38)$ prior density. The posterior distribution will be a beta distribution with parameters $\alpha = 1$ and $\beta = 102$. The effect of such an informed prior is illustrated in Figure 4.2. Given the same linear loss function of Example 4.2.1 ($k_0 = 2$, $k_1 = 1$), the 2/3 quantile moves from 0.017 if a $Be(1,1)$ is adopted, to 0.011 if the parameters α and β are chosen on the basis of the informed prior knowledge, $Be(1,38)$.*

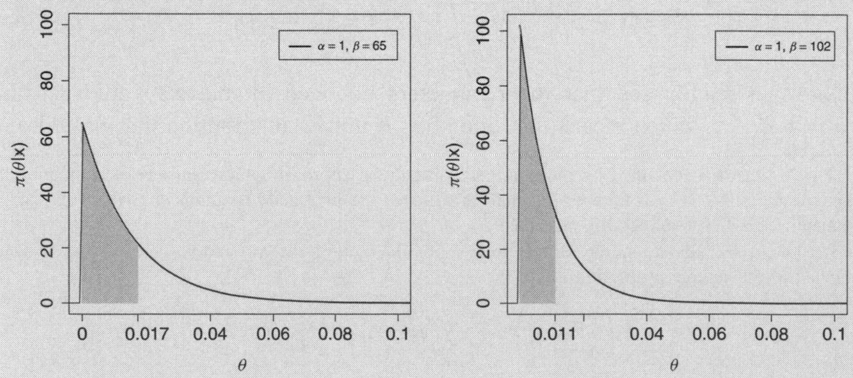

Figure 4.2 Effect of different beta priors on the Bayes decision in Example 4.2.2.

4.2.2 Prior probabilities

At this point, it appears useful to take a closer look at the relation between a posterior inference and assumptions about prior probabilities. As shown in Figure 4.2, differences in prior opinions might affect Bayesian estimates and this may be relevant within forensic contexts. A recurrent question thus is how one is to choose an appropriate prior distribution. A general and common answer to this is to say that appropriate prior distributions are simply those that reflect one's beliefs about the subject matter, conditioned as these may be by one's background evidence (Howson 2002). From a practical point of view, however, there is a tendency to seek criteria for determining in some sense 'objective' prior distributions, even though such endeavours are viewed cautiously. Fienberg (2006, p. 431) notes that '(...) the search for the objective prior is like the search of the Holy Grail in Christian myth. The goal is elusive, the exercise is fruitless, and the enterprise is ultimately diverting our energies from *doing* quality statistics'. Other authors take the position that there is no 'objective' prior, only an illusion of objectivity (Berger and Berry 1988).

A frequently invoked recommendation says that when an unknown parameter lies in a finite interval, then the uniform distribution would serve one well, in an 'objective' way. It is generally useful, however, to see where recommendations lead to in practice and to examine carefully if they reasonably reflect one's prior knowledge, as is illustrated by the following:

> Suppose you are asked to assess a prior distribution for the density (weight per unit volume) of potassium in your bloodstream, θ. Unless you have a strong background in biology, at first you may throw up your hands and say 'I am totally uninformed about θ', and then you may take all values of θ as equally likely. But when you think about it a bit you realize that there are really some things you know about θ. For example, since you know that a potassium density can't be negative, you know $\theta > 0$. (Press 1989, pp. 48–49)

That is to say, so-called 'subjective' (i.e. personal) prior distributions represent an attempt to bring prior knowledge about the phenomenon under study into the problem. One takes full advantage of additional information. This is not so by using a uniform prior density by default. Actually, it seems doubtful to consider that an individual 'has no idea at all' about the possible values for a parameter of interest. In forensic science contexts, there usually is substantial past experience and literature that can offer a thorough background and a wealth of knowledge on a wide range of specific subject matters.

An insightful discussion on this point is due to de Finetti (1993b) who argued that situations rarely, if ever, arise, in which there is no knowledge of a priori probabilities. De Finetti insists on being consistent about the implications of understanding probabilities in their very subjective sense, that is, if they are expressions of one's own belief, they cannot be unknown to oneself. An awareness is required not to reduce probabilities to some abstract property but as a standpoint intimately associated with an individual:

> The belief that the *a priori* probabilities are distributed uniformly is a well defined opinion and is just as specific as the belief that these probabilities are distributed in any other perfectly specified manner. Accordingly, there is no reason why the absence of opinion on the distribution of the *a priori* probabilities should be taken as equivalent to the opinion that their distribution is of this or that specified form. To be misinformed of the prevailing temperature is not the same as to believe that it is zero degrees. (de Finetti 1993b, p. 382)

With regards to Example 4.2.1, it may thus be emphasized that the use of a uniform prior as an a priori distribution for the parameter of interest, θ, should be a choice of an opinion, as in all other cases, and not necessarily a consequence of supposed absence of any opinion. To quote de Finetti again:

> This opinion is therefore a subjective one, as is every other, and not one specifically endowed with an objective significance. (de Finetti 1993b, p. 383)

A further issue to consider, for example in case of a beta density, is that the choice of values for the prior parameters will not matter too much when there is a large amount of data. For y successes in n trials, the posterior mean, $(\alpha + y)/(\alpha + \beta + n)$, is roughly y/n when y and n are large relative to α and β, so that prior parameters α and β may have very little or even no effect on the posterior. As underlined by Raiffa and Schlaifer (1961), this holds for both y and $n - y$ large: when the number of both positive (y) and negative ($n - y$) outcomes is large, then even very different prior distributions lead to similar posterior distributions. Conversely, if the number of positive outcomes is zero, as in the current example, it can be shown that the posterior inference is sensitive to the choice of the prior, even though the sample size n is large. An illustration of this is given in Figure 4.3.

In forensic contexts, where scientists have access to considerable sources of information (such as literature or past experience), efforts to use informed, subjective prior distributions can be seen as an attempt to make the best use of the full potential of the Bayesian method. However, such so-called *prior elicitation*, the translation of appropriate knowledge into a probability distribution, is a distinct subject in its own right (O'Hagan *et al.* 2006). Whenever one commits oneself to a particular prior distribution, it may be that another one sufficiently close could be equally acceptable. If a conflict exists between several experts that hold far different prior specifications, then it could be a matter for the customer of the inferential process (e.g. a judge) to ignore the opinion of some or all of these experts, to ask them to reformulate their priors in view of the conflicting opinions, to ask them to consider a mixture of the prior specifications, or, finally, to decide that additional data are to be collected in order to lower the influence of, and therefore possible conflict amongst, the differing opinions.

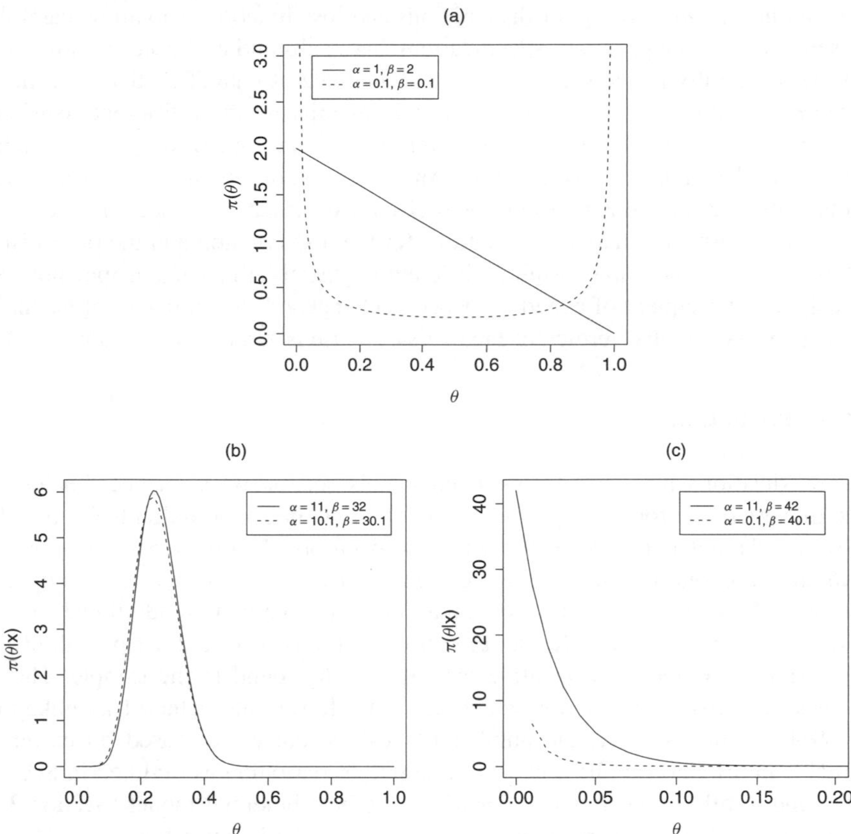

Figure 4.3 Comparison of posterior distributions. (a) Two different prior distributions, Be$(1, 2)$ and Be$(0.1, 0.1)$. (b) The resultant posterior distributions given $y = 10$ and $n = 40$, Be$(11, 32)$ and Be$(10.1, 30.1)$ with 2/3 quantiles equal to 0.2813 and 0.2773, respectively. (c) The resultant posterior distributions given $y = 0$ and $n = 40$, Be$(1, 42)$ and Be$(0.1, 40.1)$ with 2/3 quantiles equal to 0.0258 and 0.0003, respectively.

Besides prior probabilities, discussion on Bayesian decision-theoretic procedures should also draw attention to the choice of the loss function. Bayesian estimates are sensitive to the way in which losses are specified; for example, different values for the constants k_0 and k_1 result in different posterior quantiles, and, consequently, different parameter estimates. This should not be considered, however, as an inconvenience because, as pointed out by O'Hagan (1994), there are situations in which it may be worse to underestimate a parameter than to overestimate it, and conversely, so that it is rather natural that the resulting estimates will be dependent on the context from which they were obtained.

In summary thus, both prior distributions and loss functions should be carefully chosen because inappropriate specifications may substantially affect an inference process, especially in the presence of small amounts of data. This does not mean, however, that different Bayesian estimates necessarily result in different decisions. As argued by Box and Tiao (1973), different Bayesian estimates may be acceptable because the decision remains the same. An illustration of this may be considered in terms of the example of zero occurrences discussed earlier in this section where one may well imagine alternative propositions for the loss function and the prior distribution. Although these may result in different Bayesian estimates, this may not necessarily lead a recipient of expert evidence to change a decision from, for example, a guilty to a non-guilty verdict, or vice versa. A case is discussed in Example 4.4.2.

4.2.3 Prediction

So far, situations have been considered in this section where a decision maker was interested in, for example, the proportion of items or individuals, θ, who possess a characteristic of interest within a given population. A particular setting of interest was one in which zero occurrences of a positive outcome are noted in a sample of size n ($y = 0$). In such situations it may be that an additional sample of size m will be collected. It may then become relevant to find the probability of there being a number z of positive outcomes being found in the sample. This is the task of statistical prediction which aims at inferring the value of an unknown observable value, such as the number of positive outcomes, based on currently available information. Following the same line of reasoning as outlined in Section 3.3.4, and recalling the posterior density in the beta-binomial model (Section 3.3) the predictive density of the number of positive results is given by:

$$P(z \mid m,n,y) = \int_{\Theta} P(z \mid m,\theta)\pi(\theta \mid y,n)\mathrm{d}\theta \tag{4.5}$$

$$= \int_{\Theta} \binom{m}{z} \theta^{z}(1-\theta)^{m-z} \frac{\Gamma(\alpha+\beta+n)}{\Gamma(\alpha+y)\Gamma(\beta+n-y)} \theta^{\alpha+y-1}(1-\theta)^{\beta+n-y-1}\mathrm{d}\theta$$

$$= \binom{m}{z} \frac{\Gamma(\alpha+\beta+n)\Gamma(\alpha+y+z)\Gamma(\beta+m+n-y-z)}{\Gamma(\alpha+y)\Gamma(\beta+n-y)\Gamma(\alpha+\beta+m+n)},$$

where Equation (4.5) is the probability mass function of a beta-binomial distribution (see Appendix A for further details about this distribution).

Example 4.2.3 *(Glass fragments – continued). Given the prior density in Example 4.2.2, Be(1, 38), where no positive outcomes ($y = 0$) were found among $n = 64$ trials, the probability of finding 0 objects fragments sharing* *the characteristic of interest among* 100 *trials is given by*[5]*:*

$$P(0 \mid 100, 64, 0) = \binom{100}{0} \frac{\Gamma(103)\Gamma(1)\Gamma(202)}{\Gamma(1)\Gamma(102)\Gamma(203)} = \frac{102}{202} = 0.505.$$

In the same way, one can compute:

$$P(1 \mid 100, 64, 0) = 100 \times \frac{102}{202 \times 201} = 0.251,$$

$$P(2 \mid 100, 64, 0) = 4950 \times \frac{2 \times 102}{202 \times 201 \times 200} = 0.124.$$

Any inferential statement about the number of positive outcomes observable in a future sample is contained in the posterior predictive density. As no parameters are involved, there are no further developments needed. Decision problems relating to future observations will not be addressed here, but refer to Section 2.4.2 for the notion of the 'expected value of information' and Section 7.4 for sampling decision. It is solely noted that one would need to specify a utility function $u(d, z)$ to measure the gain in taking decision d when z turns out to be the observed value. The optimal decision follows from maximizing the expected utility:

$$\bar{u}(d \mid y) = \int_Z u(d, z) P(z \mid m, n, y) \mathrm{d}z.$$

Note that the posterior predictive distribution must be used and not the posterior distribution as usual (Bernardo and Smith 2000).

4.2.4 Inference for θ in the presence of background data on the number of successes

The scenarios set forth so far considered experiments that aimed at the estimation of a proportion θ of elements within a given population that possesses a characteristic of interest. An important assumption that has tacitly been admitted throughout these settings was that the counting process is free of error. Unfortunately, however, experiments in real-world circumstances might be affected by inefficiencies because of objects which might be lost in counting, or because of the presence of a background, that is objects that are observationally indistinguishable from the object of interest. This might alter the observed number of successes and/or the observed number of trials (D'Agostini 2004a).

[5]Note that the gamma function $\Gamma(\cdot)$ satisfies the relation $\Gamma(n) = (n-1)!$, for any integer $n > 0$, $0! = 1$.

For the purpose of illustration, consider a scenario in which the number of observed successes, y, could contain an unknown number of background events. An inference on the proportion of interest θ must then account for the fact that the observed number of successes is due to the sum of two individually unobservable contributions. On the one hand, there is the true but unobservable number of inspected items that actually possess the characteristic of interest, y_s. On the other hand, there is a number of positive outcomes due to background, y_b. The binomial distribution is a widely applied discrete distribution for modelling experiments which consist in counting objects. It can serve the purpose of modelling the first component, that is a binomial distribution with parameters n and θ, written $Y_s \sim \text{Bin}(n,\theta)$ for short. The background component can be modelled by recurring to a Poisson distribution as an approximation for the binomial. Actually, for large n and small θ, the binomial distribution can be approximated by a Poisson distribution with parameter $\lambda = n\theta$. The quantity of background items being unknown, the number of occurrences y_b can then be modelled with a Poisson distribution with parameter λ that is always dimensionless, $Y_b \sim \text{Pn}(\lambda)$. Suppose, for the time being, that the expected value λ of the background component is known from previous experiments, while the uncertainty about θ is modelled through a beta prior density with parameters α and β, $\theta \sim \text{Be}(\alpha,\beta)$.

The distribution of the total number of successes, $Y = Y_s + Y_b$, can be obtained by applying methods for finding the distribution of a sum of random variables (Casella and Berger 2002). It can be shown that:

$$P_Y(y \mid n,\theta,\lambda) = \sum_{y_b=0}^{y} \binom{n}{y-y_b} \theta^{y-y_b}(1-\theta)^{n-y+y_b}\frac{e^{-\lambda}\lambda^{y_b}}{y_b!}. \tag{4.6}$$

The posterior distribution, up to the normalizing constant, is given by:

$$\pi(\theta \mid n,y,\lambda) \propto P_Y(y \mid n,\theta,\lambda)\pi(\theta) =$$
$$= \sum_{y_b=0}^{y} \binom{n}{y-y_b} \theta^{y-y_b}(1-\theta)^{n-y+y_b}\frac{e^{-\lambda}\lambda^{y_b}}{y_b!}\frac{\Gamma(\alpha+\beta)}{\Gamma(\alpha)\Gamma(\beta)}\theta^{\alpha-1}(1-\theta)^{\beta-1}. \tag{4.7}$$

The integral in the denominator of the Bayes' theorem, Equation (3.5), is not tractable analytically, therefore the posterior distribution cannot be obtained in closed form. The unnormalized posterior, (4.7), can be obtained at a list of θ values $(0.000, 0.001, \ldots, 1)$ by multiplying the prior density $\pi(\theta)$ and the likelihood function $P_Y(y \mid n,\theta,\lambda)$ at each point. Samples from the posterior distribution can be obtained by normalizing the distribution on the discrete grid of θ values by its sum (Gelman *et al.* 2004).

Example 4.2.4 *(Textile fibres). Textile fibres found on a given surface, such as a car seat, may need to be counted. This is typically done in so-called fibre* *population studies that focus on the occurrence of fibres, classified according to fibre type and colour, for example, on various surfaces (such as T-shirts, cinema seats etc.). For the purpose of the current example, it is assumed that the scientist is interested in counting red-woollen fibres. The target surface may however have been in contact with different sources of red-woollen fibres so that the number of successes (that is the counts) may be affected by the background. Previous knowledge on the subject matter may suggest a background incidence of successes (that is observationally indistinguishable fibres of another category of classification) of about 2 per unit of surface. The expected value* λ *of the background component could thus be fixed equal to* 2.5 *in order to have the mode equal to* 2 *and the probability mass concentrated around this value. With regard to the mean of the beta prior density, it is assumed that existing research suggests a value equal to* 0.05*. In addition, it is believed that proportions lower than 1.5% or greater than* 10% *are very unlikely, so the prior variance* σ^2 *is taken to be* 0.0005*. Equations (4.1) and (4.2) allow one to determine the hyperparameters,* $\alpha = 4.7$ *and* $\beta = 89.3$*, such that the prior mean is equal to* 0.05 *(the location summary suggested by available knowledge) and* $P\ (0.015 \leq \theta \leq 0.1) \approx 0.95$*, where* θ *is the proportion of positive items. The prior density is depicted in Figure 4.4 (left). Suppose that a unit of surface is examined and* $n = 100$ *textile fibres are counted from which a total of 25 are noted as 'red-wool'. Figure 4.4 (right) shows the approximated posterior density obtained on a discrete grid of* 1000 *values. Given an asymmetric linear loss function with constants* $k_0 = 2$ *and* $k_1 = 1$*, the Bayes decision is the 2/3 quantile of the posterior distribution, which can be computed numerically using the grid of values and is equal to 0.1405.*

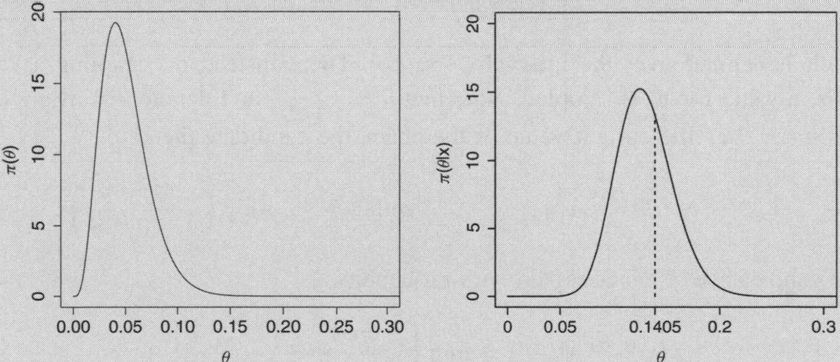

Figure 4.4 Prior density (left) and approximated posterior density (right) for Example 4.2.4. The dotted line indicates the Bayes estimate (2/3 quantile) for an asymmetric linear loss function with constants $k_0 = 2$ and $k_1 = 1$; see Section 3.4.2.

An additional complication that may need to be accounted for is that the expected value of background events is actually unknown. Uncertainty about λ can, however, be quantified by a gamma density with parameters a and b, $\lambda \sim \text{Ga}(a,b)$ (see Appendix B for details about this distribution). The joint posterior distribution, up to the normalizing constant, is given by:

$$\begin{aligned}\pi(\theta,\lambda \mid n,y) &\propto P_Y(y \mid n,\theta,\lambda)\pi(\theta)\pi(\lambda)\\ &\propto \sum_{y_b=0}^{y}\binom{n}{y-y_b}\theta^{y-y_b}(1-\theta)^{n-y+y_b}\frac{e^{-\lambda}\lambda^{y_b}}{y_b!}\theta^{\alpha-1}(1-\theta)^{\beta-1}\lambda^{a-1}e^{-b\lambda}. \quad (4.8)\end{aligned}$$

The marginal posterior of θ, $\pi(\theta \mid n,y)$, cannot be obtained analytically, but a sample from the posterior can be drawn using the MCMC techniques illustrated in Section 3.3.5, in particular the multiple-block M-H algorithm (with $k = 2$ univariate blocks, θ and λ). The candidate-generating density, for each block, is taken to be Normal with mean equal to the current value of the parameter, and variance τ_k^2 ($k = 1,2$) fixed to obtain a good acceptance rate (Gamerman 2006).

Start by considering the first block, that is the parameter θ. The full conditional density which is necessary to compute the probability of acceptance can be easily derived from the joint posterior density, Equation (4.8), and gives

$$\pi(\theta \mid \lambda,n,y) \propto \sum_{y_b=0}^{y}\binom{n}{y-y_b}\theta^{y-y_b}(1-\theta)^{n-y+y_b}\frac{\lambda^{y_b}}{y_b!}\theta^{\alpha-1}(1-\theta)^{\beta-1}. \quad (4.9)$$

The candidate value for θ cannot be sampled directly from a Normal density, since $0 < \theta < 1$. The solution is straightforward. Consider the new parameter:

$$\psi = \log\left(\frac{\theta}{1-\theta}\right), \qquad 0 < \theta < 1,$$

which is defined over the interval $(-\infty,\infty)$. Then, instead of sampling a value for θ, a value for ψ is sampled. Note that $\theta = \frac{e^{\psi}}{(1+e^{\psi})}$, and denote this as $g^{-1}(\psi)$. Given $\theta^{(j-1)}$ as the current value of the chain, the candidate then is

$$\psi^{prop} \sim N\left(\psi^{(j-1)},\tau_1^2\right), \qquad \text{with } \psi^{(j-1)} = \log\left(\frac{\theta^{(j-1)}}{1-\theta^{(j-1)}}\right).$$

The candidate ψ^{prop} is accepted with probability

$$\alpha\left(\psi^{(j-1)},\psi^{prop}\right) = \min\left\{1, \frac{\pi(\psi^{prop} \mid \lambda^{(j-1)})}{\pi(\psi^{(j-1)} \mid \lambda^{(j-1)})}\right\},$$

where $\pi(\psi \mid \lambda)$ is the reparametrized full conditional density of θ and is given by

$$\pi(\psi \mid \lambda) = \left|\frac{d}{d(\psi)}\left(g^{-1}(\psi)\right)\right|\pi\left(g^{-1}(\psi) \mid \lambda\right) = \frac{e^{\psi}}{\left(1+e^{\psi}\right)^2}\pi\left(g^{-1}(\psi) \mid \lambda\right).$$

If the candidate is accepted, then $\theta^{prop} = \frac{e^{\psi^{prop}}}{1+e^{\psi^{prop}}}$ becomes the current value $\theta^{(j)}$ of the chain.

Next, consider the second block, that is the parameter λ. The full conditional density of the parameter λ is obtained from Equation (4.8) and gives

$$\pi(\lambda \mid \theta, n, y) \propto \sum_{y_b=0}^{y} \binom{n}{y - y_b} \theta^{y-y_b}(1-\theta)^{n-y+y_b} \frac{e^{-\lambda}\lambda^{y_b}}{y_b!} \lambda^{a-1} e^{-b\lambda}. \quad (4.10)$$

As before, the candidate value for λ cannot be sampled directly from a Normal density since $\lambda > 0$. Consider the new parameter

$$\phi = \log(\lambda), \qquad \lambda > 0,$$

which is defined over the interval $(-\infty, \infty)$. Then, instead of sampling a value for λ, a value is sampled for ϕ. Note that $\lambda = e^{\phi}$ and denote this as $g^{-1}(\phi)$. Given $\lambda^{(j-1)}$ as the current value of the chain, the candidate then is

$$\phi^{prop} \sim N\left(\phi^{(j-1)}, \tau_2^2\right), \qquad \text{with } \phi^{(j-1)} = \log\left(\lambda^{(j-1)}\right).$$

The candidate ϕ^{prop} is accepted with probability

$$\alpha\left(\phi^{(j-1)}, \phi^{prop}\right) = \min\left\{1, \frac{\pi(\phi^{prop} \mid \theta^{(j)})}{\pi(\phi^{(j-1)} \mid \theta^{(j)})}\right\},$$

where $\pi(\phi \mid \theta)$ is the reparametrized full conditional density of λ and is given by

$$\pi(\phi \mid \theta) = \left|\frac{d}{d(\phi)}\left(g^{-1}(\phi)\right)\right| \pi\left(g^{-1}(\phi) \mid \theta\right) = e^{\phi}\pi\left(g^{-1}(\phi) \mid \theta\right).$$

If the candidate is accepted, then $\lambda^{prop} = e^{\phi^{prop}}$ becomes the current value $\lambda^{(j)}$ of the chain.

The algorithm can be summarized as follows:

1. Specify an initial value $(\theta^{(0)}, \lambda^{(0)})$;
2. Repeat for $j = 1, \ldots, n_b + n$ iterations:

   (a) Propose a value for the rescaled θ:

   $$\psi^{prop} \sim N(\psi^{(j-1)}, \tau_1^2).$$

   Calculate the probability of acceptance

   $$\alpha\left(\psi^{(j-1)}, \psi^{prop}\right) = \min\left\{1, \frac{\pi(\psi^{prop} \mid \lambda^{(j-1)})}{\pi(\psi^{(j-1)} \mid \lambda^{(j-1)})}\right\}.$$

   Update

   $$\theta^{(j)} = \begin{cases} \frac{e^{\psi^{prop}}}{1+e^{\psi^{prop}}} & \text{in case of acceptance.} \\ \theta^{(j-1)} & \text{otherwise.} \end{cases}$$

(b) Propose a value for the rescaled λ:

$$\phi^{prop} \sim N(\phi^{(j-1)}, \tau_2^2).$$

Calculate the probability of acceptance:

$$\alpha\left(\phi^{(j-1)}, \phi^{prop}\right) = \min\left\{1, \frac{\pi(\phi^{prop} \mid \theta^{(j)})}{\pi(\phi^{(j-1)} \mid \theta^{(j)})}\right\}.$$

Update

$$\lambda^{(j)} = \begin{cases} e^{\phi^{prop}} & \text{in case of acceptance} \\ \lambda^{(j-1)} & \text{otherwise.} \end{cases}$$

3. Return $\{\theta^{(n_b+1)}, \lambda^{(n_b+1)}, \ldots, \theta^{(n_b+n)}, \lambda^{(n_b+n)}\}$.

Example 4.2.5 *(Textile fibres – continued). Uncertainty about λ is modelled through a gamma distribution with parameters $a = 3$ and $b = 2$. A sample*

$$(\theta^{(n_b+1)}, \lambda^{(n_b+1)}), \ldots, (\theta^{(n_b+n)}, \lambda^{(n_b+n)})$$

of size $n = 15000$ (with a burn-in n_b of 5000 iterations) is taken from the posterior distribution, following the algorithm outlined above. Initial values $\theta^{(0)}$ and $\lambda^{(0)}$ are set equal to 0.5 and 10, respectively, while parameters τ_1^2 and τ_2^2 are set equal to 0.8^2 and 1.8^2, respectively (the resulting acceptance rate is around 34%). Trace plots (Figure 4.5 top) exhibit the same behaviour through iterations and do not show dependence on the starting values (the reader can verify taking different starting values). The autocorrelation plots (Figure 4.5 bottom) suggest a good performance of the algorithm since values of the autocorrelation decrease to zero as the lag value is increased. Monte Carlo estimates of the posterior quantities of interest can be obtained from the simulated draws. The estimated posterior mean is equal to 0.1315, while the 2/3 posterior quantile (the Bayes decision with linear loss function and $k_0 = 2$, $k_1 = 1$), which is of interest here, is equal to 0.1427.

An algorithm using the R package is given at the end of the chapter. Interested readers may also consider the BUGS Project (Bayesian inference Using Gibbs Sampling[6]), a package specifically designed to carry out MCMC computations for a wide variety of complex statistical models. This tool is frequently referenced in specialized statistical literature on the topic.

[6] `http://www.mrc-bsu.cam.ac.uk/bugs.`

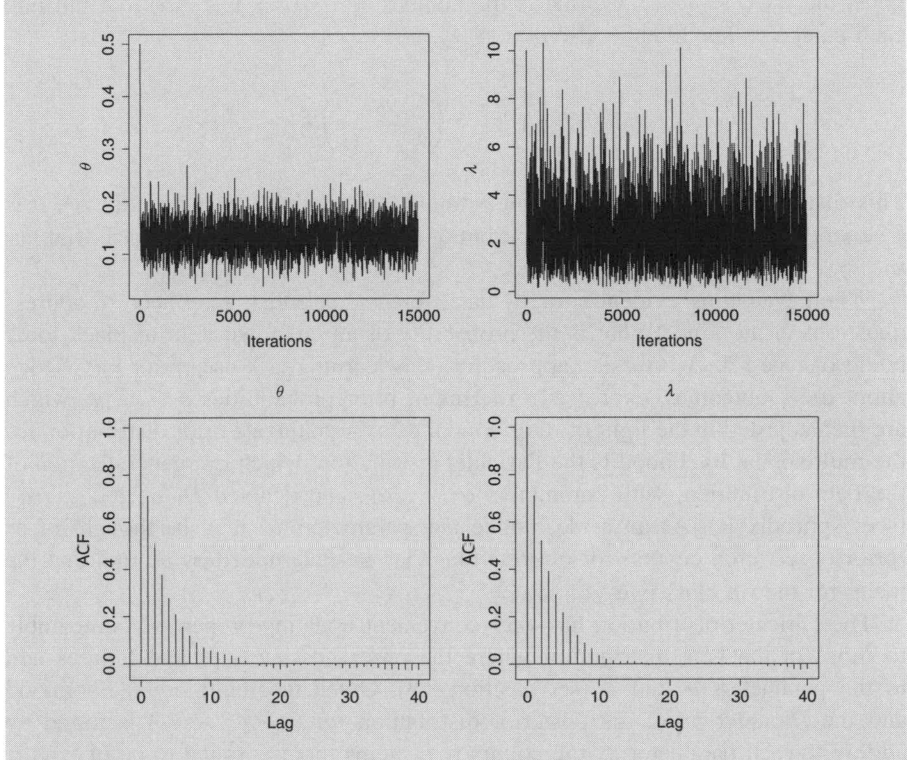

Figure 4.5 MCMC diagnostic: Trace plots of simulated draws of θ (top left) and λ (top right), autocorrelation plots of simulated draws of θ (bottom left) and λ (bottom right).

4.2.5 Multinomial variables

The kind of samples discussed so far are particular in the sense that they consist of sequences of independent trials in which the target characteristic will take exactly one of two possible mutually exclusive outcomes.

A different situation is one that has been encountered earlier with Example 3.1.5, where a population of printed documents was considered. In particular, attention was being drawn to a descriptor that may assume one of more than two possible outcomes ($k > 2$). In the case of black toner that may be found on printed documents, forensic scientists commonly analyze resins by means of Fourier Transform Infrared Spectroscopy (FTIR), the results of which (so-called IR data) may be classified in one of several mutually exclusive categories.

Let $x_1, \ldots, x_n$ be the observed values of n random variables, each of which can take one of k possible values with probability $\theta_1, \ldots, \theta_k$. Suppose that one only notes the y_js ($j = 1, \ldots, k$), that is the number of x_is ($i = 1, \ldots, n$) that fall into each category. The likelihood function is

$$f(y_1, \ldots, y_k \mid \theta_1, \ldots, \theta_k) = \binom{n}{y_1 \cdots y_k} \theta_1^{y_1} \cdot \ldots \cdot \theta_k^{y_k}.$$

This distribution is known as the multinomial distribution (see Appendix A). It is a generalization of the binomial distribution to the situation in which each trial has k ($k > 2$) distinct possible outcomes.

When evaluating evidence for IR data, forensic scientists may need to address questions of the kind 'What is the probability of an unknown item of black toner being of type j?'. A Bayesian approach to this learning task uses prior knowledge about the k outcomes, expressed in terms of prior probabilities $\theta_1, \ldots, \theta_k$, which are then adjusted in the light of multinomial data. A conjugate prior distribution for the multinomial likelihood is the Dirichlet distribution, which is a generalization of the beta distribution, with parameters $\alpha_1, \ldots, \alpha_k$, and denoted $D_k(\alpha_1, \alpha_2, \ldots, \alpha_k)$ (see Appendix B). As an aside, notice that parameters α_j may be thought of as 'prior observation counts' for observations that are determined by θ_j, and that the mean for the jth class is α_j/α_0, $\alpha_0 = \sum_{j=1}^{k} \alpha_j; j = 1, \ldots, k$.

The Dirichlet distribution has very convenient updating properties, comparable to those of the beta distribution where the observed successes and failures add to the parameters α and β (see Section 3.3). Given the multinomial likelihood and the Dirichlet prior, the posterior distribution for $(\theta_1, \theta_2, \ldots, \theta_k)$ is found by adding to each parameter α_j the counts y_j of items that are found to be of type j: $D_k(\alpha_1 + y_1, \alpha_2 + y_2, \ldots, \alpha_k + y_k)$.

The model parameter $\boldsymbol{\theta} = (\theta_1, \ldots, \theta_k)$ being vector-valued, an appropriate generalization of the loss function to the multivariate case needs to be introduced. Several choices are possible, as discussed in Section 3.4.2. If the loss is believed to be quadratic, the optimal Bayes decision $\mathbf{d} = (d_1, \ldots, d_k)$ is equal to the mean of the posterior distribution. No more information is needed. For the Dirichlet-multinomial model, posterior means are easily derived since the posterior distribution is still in the same family as the prior, with updated parameters, so that

$$E(\theta_j \mid y_1, \ldots, y_k) = \frac{\alpha_j + y_j}{\alpha_0 + n}, \qquad j = 1, \ldots, k.$$

Example 4.2.6 *(Black toners). Consider a sample of $n = 100$ printed documents, for each of which the toner's resin is analyzed. There are $k = 7$ resin groups in total, so each variable x_i ($i = 1, \ldots, n$) can assume one of seven possible values. The total number of analyzed documents falling into each group, y_j ($j = 1, \ldots, k$), are noted. Data are illustrated in Table 4.1.*

Table 4.1 Total number of observations (counts) for each of seven resin groups for black printer toner. The sample covers a total of 100 items (documents with black toner), each of which falls in exactly one category.

Resin group (number (j) – name)	counts (y_j)
1 – Styrene-co-acrylate	83
2 – Epoxy A	11
3 – Epoxy B	2
4 – Epoxy C	1
5 – Epoxy D	1
6 – Polystyrene	1
7 – Unknown	1

For the seven classes represented in Table 4.1 the prior distribution may be written as $D_k(\alpha_1, \alpha_2, \ldots, \alpha_7)$. Assume $\alpha_j = 1$ for all k categories. This is equivalent to assuming a uniform prior. Given the observed data, Table 4.1, the updating procedure leading to posterior distribution is outlined in Table 4.2.

Given a quadratic loss function, the optimal Bayesian estimates of the unknown parameters, θ_j, are given by the posterior means which are shown in the column on the far right-hand side of Table 4.2.

Table 4.2 Example for updating a Dirichlet distribution with parameters α_j. The data are taken from Table 4.1 (represented here in column four) and consist of the numbers y_j of observations (counts) of items found to be of type j (the sample size is 100). Column three represents the prior means, column five the revised parameters and column six the posterior means.

Resin group j	α_j	α_j/α_0	y_j	$\alpha_j + y_j$	$\frac{\alpha_j+y_j}{\alpha_0+n}$
1	1	1/7	83	84	0.78505
2	1	1/7	11	12	0.11215
3	1	1/7	2	3	0.02804
4	1	1/7	1	2	0.01869
5	1	1/7	1	2	0.01869
6	1	1/7	1	2	0.01869
7	1	1/7	1	2	0.01869
Sums:	$\alpha_0 = 7$	1	$n = 100$	$\alpha_0 + n = 107$	1

Another example of an application of such a procedure can be found in Foreman *et al.* (1997) where a genetic parameter of correlation, F_{ST}, used in forensic DNA evidence evaluation, is estimated.

4.3 BAYESIAN DECISION FOR A POISSON MEAN

Some forensic science applications focus on the number of occurrences of certain events that occur, at a constant rate, randomly through time (or space). Examples for such events are corresponding matching striations in the comparative examination of toolmarks or the numbers of gunshot residue (GSR) particles collected on the surface of the hands of individuals suspected to be involved in the discharge of a firearm.

A commonly used statistical model in such situations assumes that counts follow a Poisson distribution with parameter λ, $f(y \mid \lambda) = \text{Pn}(\lambda)$. The parameter λ may be estimated on the basis of data collected in surveys or during experiments conducted under controlled situations. As has been outlined already, since knowledge of the parametric model yields knowledge of the entire population, it is natural to try to estimate the value of the parameter. A Bayesian decision-theoretic approach is developed for this scenario within perspectives that account for, respectively, the presence and absence of background information that may affect a count. A third part of this section will focus on the use of Poisson distributed variables for forensic inference about the selected propositions of interest using graphical models. An additional complication that will be addressed at that point is that a given count may, in part, also originate from other sources unrelated to the alleged incident.

4.3.1 Inference about the Poisson parameter in the absence of background events

Consider a random sample $(y_1, \ldots, y_n)$ from a Poisson distribution with parameter λ. Then, the likelihood is given by:

$$l(y_1, \ldots, y_n \mid \lambda) = \prod_{i=1}^{n} \frac{e^{-\lambda}\lambda^{y_i}}{y_i!} = \frac{\lambda^{\sum_{i=1}^{n} y_i} e^{-n\lambda}}{\prod_{i=1}^{n} y_i!}. \tag{4.11}$$

At first, a prior density reflecting one's initial beliefs must be specified. Typically, the choice falls in the family of conjugate gamma densities $\text{Ga}(\alpha, \beta)$:

$$\pi(\lambda) = \frac{\beta^{\alpha}}{\Gamma(\alpha)} \lambda^{\alpha-1} e^{-\beta\lambda}. \tag{4.12}$$

The shape of the posterior distribution is obtained as follows:

$$\pi(\lambda \mid y) \propto \pi(\lambda) l(y_1, \ldots, y_n \mid \lambda)$$
$$= \frac{\beta^\alpha \lambda^{\alpha-1} e^{-\beta\lambda}}{\Gamma(\alpha)} \frac{\lambda^{\sum_{i=1}^n y_i} e^{-n\lambda}}{\prod_{i=1}^n y_i!}$$
$$\propto \lambda^{\alpha+\sum_{i=1}^n y_i - 1} e^{-(\beta+n)\lambda}.$$

The final expression can be recognized to be in the form of a gamma density with shape parameter $\alpha^* = \alpha + \sum_{i=1}^n y_i$, and scale parameter $\beta^* = \beta + n$.

In order to obtain a posterior distribution using Bayes' theorem, a gamma prior distribution needs to be specified for the Poisson parameter λ. There are several possible ways to do this. Assuming no relevant prior knowledge is available, the scientist can take a positive uniform prior density $\pi(\lambda) = 1$, for $\lambda > 0$, as in Example 4.3.1.

Conversely, imagine that the scientist has relevant experience in the field (e.g. from previous experiments conducted under controlled conditions) and is willing to summarize his prior beliefs in terms of the prior mean μ and the prior variance σ^2. Then, following the same line of reasoning illustrated in Section 4.2.1 for the elicitation of the beta parameters, and recalling that the mean of a Ga(α, β) is $E^\pi[\lambda] = \alpha/\beta$, and the variance is $\mathrm{Var}^\pi[\lambda] = \alpha/\beta^2$, the shape parameter α and the scale parameter β can be chosen by solving the following system:

$$\mu = \frac{\alpha}{\beta}$$
$$\sigma^2 = \frac{\alpha}{\beta^2}.$$

Solution of these two equations gives

$$\alpha = \frac{\mu^2}{\sigma^2}$$
$$\beta = \frac{\mu}{\sigma^2}.$$

A practical case will be discussed in Example 4.3.2.

Example 4.3.1 *(Gunshot residue particles). The search, detection and identification of gunshot residue particles (GSR) is regularly conducted on individuals (or items belonging to individuals) suspected of involvement in a shooting incident. In order to help their clients to interpret GSR evidence appropriately,* *forensic scientists may need a clear idea of the occurrence of such evidence on both (i) individuals known to be associated with the firing of a gun and (ii) individuals unrelated to a shooting incident. The latter setting is also known in the context as 'accidental contamination' or 'presence by chance'.*

Consider, for example, a situation in which a scientist intends to compute probabilities for the number y of GSR particles (where $y = 0, 1, 2, \ldots$) on the hands of an individual unrelated to a shooting incident, based on data obtained from a new experiment. The GSR count y is assumed to follow a Poisson distribution with mean λ. Some knowledge about the value of λ is available from literature, believed to be appropriate as a source of prior information (because of, for example, similar experimental conditions). For instance, there may be a survey in which a total of two GSR particles were found after analysing samples taken from the surfaces of the hands of $n = 50$ individuals. A random sample $(y_1, \ldots, y_n)$ thus is available with the reported numbers of GSR particles for each individual. Assume a positive uniform prior density $\pi(\lambda) = 1$, for $\lambda > 0$. Then, the posterior distribution is gamma with parameters $\alpha = 3$ and $\beta = 50$, Ga(3, 50). Notice that this distribution is proper even though the positive uniform density $\pi(\lambda) = 1$ was not.

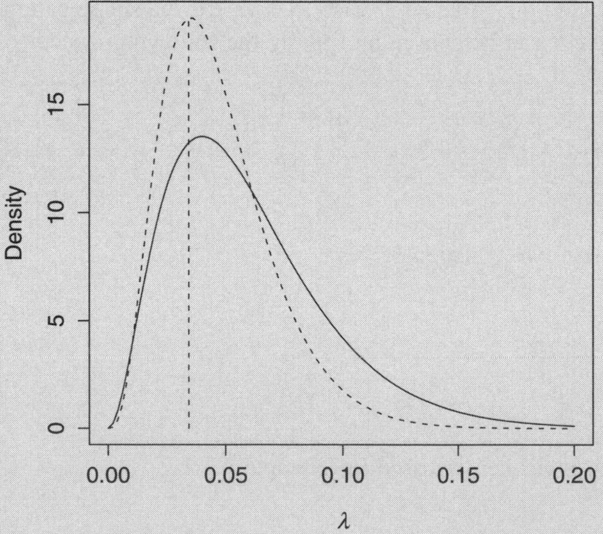

Figure 4.6 Ga(3,50) prior distribution (solid line) updated by Poisson distributed data. The dashed line shows the Ga(4,85) posterior distribution. The vertical line represents the $1/3$ posterior quantile, $\lambda = 0.034$.

Next, consider an additional experiment where the counts of GSR particles are observed after analysing a further sample of $m = 35$ *individuals. In total, one additional particle is found. The resulting gamma posterior distribution for* λ *has parameters* $\alpha^* = \alpha + 1 = 4$ *and* $\beta^* = \beta + m = 85$. *An asymmetric linear loss function is introduced with constants* $k_0 = 1$ *and* $k_1 = 2$. *This choice of constants favours an underestimate of the true number of GSR particles. Therefore, the Bayesian decision is the* 1/3 *quantile of the* Ga(4, 85) *distribution, which is equal to* 0.034. *The posterior distributions obtained sequentially are depicted in Figure 4.6. The vertical line represents the* 1/3 *posterior quantile for* Ga(4, 85). *The value 0.034 is the estimate of the rate of finding GSR particles per individual in the general population.*

Example 4.3.2 *(Gunshot residue particles – continued). Consider a setting in which a scientist intends to estimate the mean number of gunshot particles that may be detected on samples taken from the surfaces of the hands of individuals that discharged a firearm some specified time ago (e.g. 5 hours).*

Imagine that the scientist assumes – based on similar past cases – a prior mean of 6.5 *and a prior standard deviation of* 3. *The gamma conjugate prior that matches these two prior moments thus has the parameters* $\alpha = 6.5^2/3^2 = 4.69$ *and* $\beta = 6.5/3^2 = 0.72$.

When the scientist wishes to update this prior using data from a new experiment in which a total particle count $\sum_{i=1}^{m} y_i = 26$ *is noted after analysing samples taken from* $m = 6$ *individuals (5 hours after the discharge of a firearm) – where* y_i *is the particle count for individual i – , then the posterior gamma distribution can be found as follows:*

$$\alpha^* = \alpha + \sum_{i=1}^{m} y_i, \ \beta^* = \beta + m.$$

Note that only $\sum_{i=1}^{m} y_i$ *and m matter for the final inference and the individual* y_i*'s do not matter. The posterior distribution is* Ga(30.69, 6.72). *Given the same linear asymmetric loss function of Example 4.3.1 (*$k_0 = 1$, $k_1 = 2$*), the Bayesian deision is the* 1/3 *posterior quantile that is about* 4.2 (4.174). *Note that the mean number of gunshot particles that may be detected on samples taken from the surfaces of the hands of individuals that discharged a firearm some specified time ago is about 4.6: the specifc choice of the loss function penalizes an overestimate of the number of particles more than an underestimation.*

4.3.2 Inference about the Poisson parameter in the presence of background events

Consider a scenario in which the number of occurrences of a given event (e.g. the number of GSR particles) could contain an unknown number of background events. The total number of counts is then given by two individually unobservable contributions. On the one hand, there are the unknown number of occurrences truly related to the event of interest (e.g. the transferred particles), y_s. On the other hand, there are a number of occurrences due to the background, y_b, independent of y_s, (e.g. particles already present on the hands of individuals). Assume for both components a Poisson distribution, that is $Y_s \sim \text{Pn}(\lambda_s)$ and $Y_b \sim \text{Pn}(\lambda_b)$. The distribution of the total number of counts, $Y = Y_s + Y_b$, is still Poisson, $P(y \mid \lambda) = \text{Pn}(\lambda)$, with parameter $\lambda = \lambda_s + \lambda_b$. Suppose that the expected value λ_b of the background component is known from previous experiments, while the uncertainty about λ_s is modelled through a gamma prior density with parameters α and β, $\lambda_s \sim \text{Ga}(\alpha, \beta)$. The posterior distribution, up to the normalizing constant, is given by

$$\pi(\lambda_s \mid y_1, \ldots, y_n, \lambda_b) \propto l(y_1, \ldots, y_n \mid \lambda_s, \lambda_b)\pi(\lambda_s)$$

$$= \frac{e^{-n(\lambda_s+\lambda_b)}(\lambda_s + \lambda_b)^{\sum_{i=1}^{n} y_i}}{\prod_{i=1}^{n} y_i!} \frac{\beta^\alpha}{\Gamma(\alpha)} \lambda_s^{\alpha-1} e^{-\beta\lambda_s}. \quad (4.13)$$

The integral in the denominator of the Bayes' theorem, Equation (3.5), is not tractable analytically in this example, therefore the posterior distribution cannot be obtained in closed form. The unnormalized posterior, (4.13), can be obtained as a list of λ_s values, as illustrated in Section 4.2.4.

Example 4.3.3 *(Gunshot residue particles – continued). Consider the scenario presented in Example 4.3.2 where the uncertainty about the mean of the number of particles on an individual is represented by a gamma prior distribution with hyperparameters* $\alpha = 4.69$ *and* $\beta = 0.72$. *The prior distribution,* $\lambda_s \sim \text{Ga}(4.69, 0.72)$, *is shown in Figure 4.7 (left). A new experiment reveals 26 particles on samples taken from* $m = 6$ *individuals. Knowing from previous studies that people at random in the relevant population carry a tiny number of background particles on hands, parameter* λ_b *is fixed equal to* 0.05 *and is substituted in Equation (4.13). The posterior distribution for the unknown parameter* λ_s *is presented in Figure 4.7 (right). The* 1/3 *posterior quantile moves from* 4.174 *(the value given in Example 4.3.2) to* 4.125.

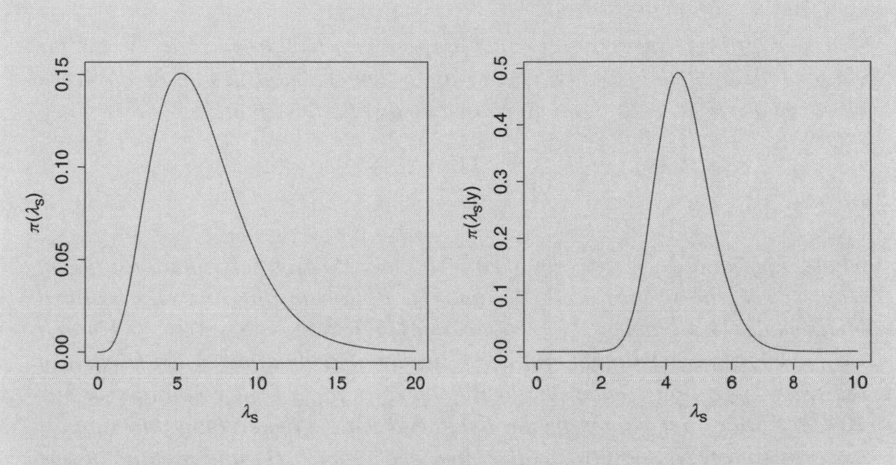

Figure 4.7 Prior density (left) and approximated posterior density (right) for Example 4.3.3.

4.3.3 Forensic inference using graphical models

Forensic scientists commonly require probabilities – assuming varying circumstantial settings (presented, for example, in the form of competing propositions forwarded by the prosecution and the defence) – for scientific evidence that consists of a count (such as a GSR particle count). The probabilities thus obtained are compared against each other in order to obtain an indication of how well the evidence, in the form of a count, allows one to distinguish between competing scenarios (Aitken and Taroni 2004).

In the context, discrimination among propositions is typically referred to as forensic inference. Examples 4.3.4 and 4.3.5 consider this topic, using graphical models, for situations in which the evidential counts are treated as Poisson distributed variables.

Example 4.3.4 *(Forensic inference based on a Poisson distributed variable using Bayesian networks). Imagine a scenario in which three GSR particles are detected on an individual suspected of involvement in a shooting incident. A question of interest may then be the level of discrimination provided by that evidence between the pair of propositions H_p (the suspect was recently exposed to the discharge of a firearm, e.g. 5 hours ago) and H_d (the suspect is not associated with a recent discharge of a firearm).*

A standard measure used for this purpose is a likelihood ratio[7] *V (V for value of the evidence), which is the ratio of the probabilities of the observed GSR count* $y = 0, 1, 2, \ldots$ *as given by two distinct distributions:*

$$V = \frac{f(y \mid \lambda_{H_p})}{f(y \mid \lambda_{H_d})},$$

where λ_{H_p} *and* λ_{H_d} *are the means of the Poisson distributions given, respectively, the propositions* H_p *and* H_d*. Following the analyses outlined in Examples 4.3.1 and 4.3.2, let* $\lambda_{H_p} = 31/7$ *(exact value* $30.69/6.72$*) and* $\lambda_{H_d} = 4/85$*;* i.e., *the means of the posterior distributions of* λ *for each of the two propositions. Then, the likelihood ratio for a GSR count of* $y = 3$ *is* 10423. *Notice that values of the likelihood ratio greater than one support the proposition* H_p*, values smaller than one support* H_p *and a value of one favours neither of the two propositions (Aitken and Taroni 2004).*

Figure 4.8 proposes a simple Bayesian network useable for calculating Poisson probabilities for GSR counts $y = 0, 1, 2, \ldots$*. The node* H *is Boolean whose states, true and false, represent, respectively,* H_p *and* H_d*. The nodes* $a_{H_p}, b_{H_p}, a_{H_d}, b_{H_d}$ *have numeric states that correspond to the value of the parameters of the gamma distributed variables* λ_{H_p} *and* λ_{H_d}*.* Y_{H_p} *and* Y_{H_d} *are numeric nodes that model the GSR particle count for the settings assuming the truth of, respectively,* H_p *and* H_d*. The tables of* Y_{H_p} *and* Y_{H_d} *can be completed using the expressions* `Poisson(aHp/bHp)` *and* `Poisson(aHd/bHd)`, *respectively (* HUGIN *syntax*[8]*). The node* Y *models the GSR count* $y = 0, 1, 2, \ldots$ *as a function of the truthstate of* H*, that is* $Y = Y_{H_p}$ *if* $H = H_p$ *and* $Y = Y_{H_d}$ *if* $H = H_d$*. Thus, the node table of* Y *can be completed, referring again to* HUGIN *syntax, using the expression* `if(Hp, YHp, YHd)`.

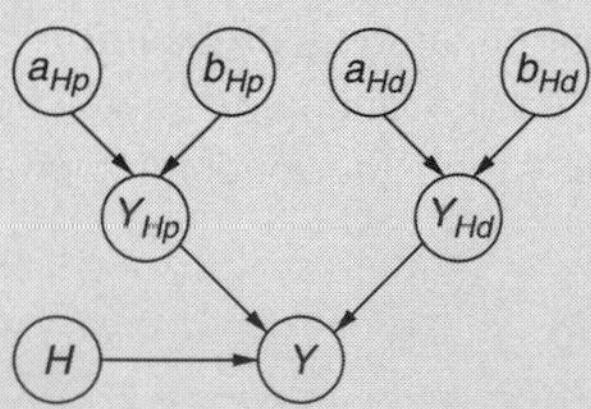

Figure 4.8 Bayesian network for evaluating GSR particle evidence. The node descriptors are given in Example 4.3.4.

[7] The concept of a likelihood ratio is introduced in Chapter 2 and more extensively discussed in Chapter 6.

[8] The program Hugin is available at www.hugin.com.

Figure 4.9 provides examples of calculations that the proposed Bayesian network can offer. The node Y in the figure on the left displays the probabilities for observing $y = 0, 1, 2, \ldots$ particle counts whereas the figure on the right shows the inference about the main proposition of interest in a scenario in which $y = 3$ particles are observed.

The model considered in Example 4.3.4 is a general one essentially because it focuses solely on the presence of the overall GSR count after a given time, irrespective of fact that this count may include particles unrelated to the shooting incident of interest. In fact, GSR particles may not only originate from the discharge of a firearm, but also from a contaminated collection kit. Stated otherwise, there may be GSR particles present on the lifting device prior to the sampling of the hands of a suspect. The overall GSR count thus is, in a more strict sense, the sum of the number of particles taken from the surface of a suspect's hands and the number of particles already present – in the sense of a contamination – on the lifting device. The probability of the number of contaminant particles on the collection kit, written, for example, $c = 0, 1, 2, \ldots$, may thus be modelled separately using distributions with parameters varying according to the degree of contamination D. To keep the development manageable, the degree of contamination may be allowed to cover, for instance, the discrete possibilities '*none*', '*low*', '*medium*' and '*high*'. Notwithstanding, the complexity of the approach may increase in case D cannot be assumed to be known and a probability distribution needs to be specified over its possible states. Example 4.3.5 illustrates that Bayesian networks can support the hierarchical modelling of such issues.

Example 4.3.5 *(Forensic inference based on a Poisson-distributed variable using Bayesian networks – continued). Consider again the Bayesian network shown in Figure 4.8 and let there be an additional node Y' that models the overall GSR particle count. This variable differs from the previously defined node Y because it includes particles that may originate from a contaminated swab. The node Y' is incorporated in the Bayesian network as a descendant of Y and C. The latter node, C, with states $c = 0, 1, 2, \ldots$, accounts for the GSR particles present on the lifting device prior to sampling. Both nodes, Y and C, determine the actual state of the node Y', the node table of which is completed as follows:*

$$P(Y' = y' \mid Y = y, C = c) = \begin{cases} 1, & y' = y + c, \\ 0, & y' \neq y + c. \end{cases} \tag{4.14}$$

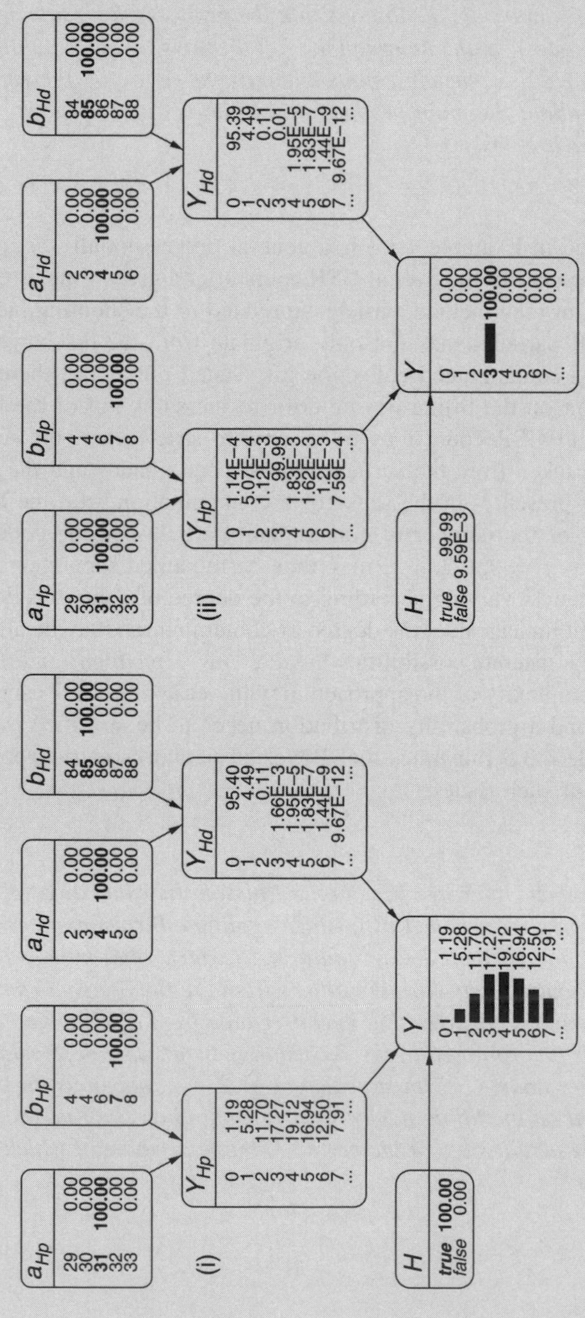

Figure 4.9 Bayesian network for evaluating GSR particle evidence with expanded nodes from Example 4.3.4. Instantiations are shown in bold and probabilities are shown in %. Figure (i) shows Poisson probabilities $f(y \mid \lambda_{H_p} = 31/7)$ ($a_{H_p} = 31, b_{H_p} = 7$) for GSR particle counts $y = 0, 1, 2, \ldots$ in the nodes Y_{H_p} and Y, whereas the node Y_{H_d} ($a_{H_d} = 4, b_{H_d} = 85$) displays Poisson probabilities $f(y \mid \lambda_{H_d} = 4/85)$. Figure (ii) shows an inference about the truthstate of H in a setting in which $y = 3$ GSR particles are observed. The distributive assumptions for the GSR particle count given H are as explained in the text.

The node table of Y' may be readily completed using expression Y+C *(* HUGIN *syntax).*

For brevity, the states of the node D are restricted to 'clean' and 'not clean'. The number of GSR particles that may be present on the surface of the collection kit (node C) is assumed to be Poisson distributed and to depend on whether the kit is or is not contaminated (node D). For the purpose of the current discussion, a Poisson distribution with parameter $\lambda_C = 0.1$ is defined for $P(C = c \mid D = \text{not clean})$. Obviously, no particles are present on a clean stub, so $P(C = 0 \mid D = \text{clean}) = 1$ and $P(C \neq 0 \mid D = \text{clean}) = 0$. The overall structure of the proposed Bayesian network is shown in Figure 4.10.

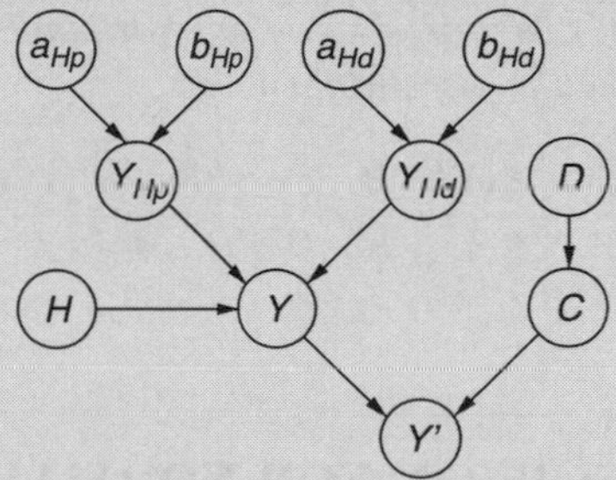

Figure 4.10 Extended Bayesian network for evaluating GSR particle evidence. Node descriptors are given in Examples 4.3.4 and 4.3.5.

Consider, then, a case in which two GSR particles are observed, while the sample collection kit is assumed to be one that was not clean. In such a setting, the likelihood ratio writes:

$$V = \frac{P(Y' = 2 \mid H_p, D = \textit{not clean})}{P(Y' = 2 \mid H_d, D = \textit{not clean})}.$$

For the Bayesian network shown in Figure 4.10 it can be shown that the numerator of the likelihood ratio is a Poisson distribution with a parameter that is given by the sum of the Poisson parameters λ_{H_p} and λ_C,

$$(Y' \mid H_p, D = \textit{not clean}) \sim \text{Pn}(\lambda_{H_p} + \lambda_C).$$

Following the previously defined parametric assumptions one obtains:

$$P(Y' = 2 \mid H_p, D = \textit{not clean}) = \frac{e^{-(31/7+0.1)}(31/7 + 0.1)^2}{2!} = 0.1107.$$

Analogously, the denominator is given by:

$$(Y' \mid H_d, D = \text{not clean}) \sim Pn(\lambda_{H_d} + \lambda_C),$$

which can be found to give:

$$P(Y' = 2 \mid H_p, D = \text{not clean}) = \frac{e^{-(4/85+0.1)}(4/85 + 0.1)^2}{2!} = 0.0093.$$

Notice that the proposed Bayesian network also efficiently supports calculations for situations in which the node D cannot assumed to be known. In such a scenario the evidence would be propagated in the network while leaving the node D uninstantiated (with probabilities assigned to the states 'clean' and 'not clean'). Writing D and $\bar{D}$ shorthand for 'D = clean' and 'D = not clean' the likelihood ratio then is:

$$V = \frac{P(Y' \mid H_p)}{P(Y' \mid H_d)} = \frac{P(Y' \mid H_p, D)P(D) + P(Y' \mid H_p, \bar{D})P(\bar{D})}{P(Y' \mid H_d, D)P(D) + P(Y' \mid H_d, \bar{D})P(\bar{D})}.$$

4.4 BAYESIAN DECISION FOR NORMAL MEAN

Quantitative data may be either counts – known as discrete data, since the counts take discrete, integer values – or measurements – referred to as continuous data, since the measurements may take any value on a continuous interval. Previous sections dealt with discrete data to estimate a proportion, for example the proportion of individuals sharing the same characteristics of interest (such as a given DNA profile). The refractive index of glass fragments, the alcohol concentration in blood, the widths of landmarks on firearm bullets or the weights of ecstasy pills are examples of continuous measurements. This section presents the Bayesian procedure for learning about population means of Normal variables.

Imagine, for example, a scientist who is interested in determining the alcohol concentration on the basis of a series of measurements taken from a given individual arrested by traffic police, or a firearm examiner who seeks to estimate the width of grooves on a bullet fired through the barrel of a suspect's gun. Section 4.4.1 deals with the special case of Bayesian inference about an unknown mean for a known variance. Section 4.4.2 develops the methodology for the inference about an unknown mean for an unknown variance. The possible presence of background data is considered in Section 4.4.3, while Section 5.2 addresses the problem of prediction.

4.4.1 Case with known variance

Suppose the data from an experiment follow a Normal distribution with unknown mean θ and known variance σ^2. The assumption of a known variance is a plausible approximation in many cases, such as those involving an analytical instrument used to measure some physical, chemical or biological quantity. As noted by Howson and Urbach (1996, p. 354), such an '[...] instrument would, as a rule, deliver a spread of results if used repeatedly under similar conditions, and experience shows that this variability often follows a Normal curve, with an approximately constant standard deviation. Making measurements with such an instrument would then be practically equivalent to drawing a random sample of observations from a Normal population of possible observations, whose mean is the unknown quantity and whose standard deviation has been established from previous calibrations'. Note that the precision of a distribution is often defined as the reciprocal of its variance. The more precise an analytical device (that is, the lower the variance of the resulting measurements), the higher the precision of the respective posterior distribution of the quantity being measured, for given prior precision.

To begin with, suppose that there is very little information and that all values of θ seem reasonably equally likely. Accordingly, let θ have a locally uniform prior (one in which the probability density function for θ is a constant over any finite interval),

$$\pi(\theta) \propto \text{constant}. \tag{4.15}$$

Given a random sample of n observations, $x_1, \ldots, x_n$ from a Normal distribution, $X \sim N(\theta, \sigma^2)$, the likelihood function can be written in terms of the sample mean $\bar{x}$, which is a sufficient statistic (see Section 3.3.2). In fact,

$$l(\theta \mid x_1, \ldots, x_n) =$$
$$\prod_{i=1}^{n} \frac{1}{\sqrt{2\pi}\sigma} \exp\left[-\frac{1}{2\sigma^2}(x_i - \theta)^2\right] \propto \exp\left[-\frac{1}{2\sigma^2}\sum_{i=1}^{n}[(x_i - \bar{x}) + (\bar{x} - \theta)]^2\right].$$

Completing the square gives

$$l(\theta \mid x_1, \ldots, x_n) \propto \exp\left[-\frac{n}{2\sigma^2}(\bar{x} - \theta)^2\right].$$

The sample mean is Normally distributed, $\bar{X} \sim N\left(\theta, \frac{\sigma^2}{n}\right)$. Applying Bayes' theorem and incorporating the uniform prior, Equation (4.15), the posterior distribution is Normal with mean $\bar{x}$, and variance σ^2/n, in fact

$$\begin{aligned}\pi(\theta \mid x_1, \ldots, x_n) &\propto l(\theta \mid x_1, \ldots, x_n) \qquad \text{since } \pi(\theta) \text{ is a constant}\\ &\propto \exp\left[-\frac{n}{2\sigma^2}(\bar{x} - \theta)^2\right].\end{aligned}$$

Note that the posterior distribution is proper and Normal, although the prior was improper. Very little information has been provided by the prior: the posterior distribution is centred at the mean, which is also the maximum likelihood estimator.

Next, suppose that some background information is available so that a proper prior may be used. A conjugate prior distribution for the mean is the Normal distribution, therefore it is assumed that the parameter θ is Normally distributed with mean μ and variance τ^2, $\theta \sim N(\mu, \tau^2)$, with both μ and τ known. The probability density of θ is thus given by

$$\pi(\theta) \propto \exp\left[-\frac{1}{2\tau^2}(\theta-\mu)^2\right].$$

Then,

$$\begin{aligned}\pi(\theta \mid x_1,\ldots,x_n) &\propto l(\theta \mid x_1,\ldots,x_n)\pi(\theta)\\ &\propto \exp\left[-\frac{1}{2\sigma^2}\sum_{i=1}^{n}(x_i-\theta)^2\right]\exp\left[-\frac{1}{2\tau^2}(\theta-\mu)^2\right]=\\ &= \exp\left[-\frac{1}{2\sigma^2}\sum_{i=1}^{n}\{(x_i-\bar{x})+(\bar{x}-\theta)\}^2-\frac{1}{2\tau^2}(\theta-\mu)^2\right].\end{aligned}$$

Completing the squares, it follows with a small effort (Berger 1988; Press 2003, e.g) that

$$\pi(\theta \mid x_1,\ldots,x_n) \propto \exp\left[-\frac{1}{2}\frac{\frac{\sigma^2}{n}+\tau^2}{\frac{\sigma^2}{n}\tau^2}\left(\theta-\left(\frac{\frac{\sigma^2}{n}}{\frac{\sigma^2}{n}+\tau^2}\mu+\frac{\tau^2}{\frac{\sigma^2}{n}+\tau^2}\bar{x}\right)\right)^2\right].$$

The posterior density is Normal $N(\mu(x), \tau^2(x))$ with mean

$$\mu(x) = \frac{\frac{\sigma^2}{n}}{\frac{\sigma^2}{n}+\tau^2}\mu + \frac{\tau^2}{\frac{\sigma^2}{n}+\tau^2}\bar{x}, \tag{4.16}$$

and variance

$$\tau^2(x) = \frac{\frac{\sigma^2}{n}\tau^2}{\frac{\sigma^2}{n}+\tau^2}, \tag{4.17}$$

or precision

$$\frac{1}{\tau^2(x)} = \frac{1}{\tau^2} + \frac{1}{\sigma^2/n}. \tag{4.18}$$

The posterior mean is a weighted average of the prior mean μ and the sample mean $\bar{x}$, with weights proportional to the variances corresponding to the prior distribution and the sampling distribution. The mean of the distribution with lower variance (higher precision) receives greater weight. The posterior precision is the sum of the precisions of the prior and the likelihood.

Example 4.4.1 *(Alcohol concentration in blood). Consider a case in which a person is stopped because of suspicion of driving under the influence of alcohol. A sample taken from that individual is submitted to a forensic laboratory. It is common practice in many forensic laboratories to estimate the concentration of alcohol in blood by performing two independent analytical procedures. Let them be denoted HS and ID. Two measurements are obtained from each procedure. For HS, these are* 0.6066 *and* 0.5778 *g/kg. For ID, these are* 0.4997 *and* 0.5197 *g/kg. Let X be the Normally distributed random variable of measurements of blood alcohol for a particular person, with x denoting the value of a particular measurement,* $X \sim N(\theta, \sigma_p^2)$*, where* θ *is the same for both procedures and the variance* σ_p^2 *has different values for* $p = HS, ID$*. The variances* σ_p^2 *are assumed known as they have been estimated from many previous experiments conducted by both procedures. In particular, it is assumed that the standard deviations are constants, independent of* θ*, and both equal to* 0.0229 *(g/kg) for procedure HS, and equal to* 0.0463 *(g/kg) for procedure ID. Available knowledge (typically, circumstantial information, such as the fact that the person has been stopped while driving at midnight, exceeding the speed limit, etc.) suggests, for example, a prior mean* $\mu = 1$*. The prior variance is chosen according to the procedure described in Example 3.3.6. Values for* $\theta < 0.1$ *and* > 1.9 *are believed to be extremely unlikely, so that a value of* $\tau = 0.3$ *is chosen for the standard deviation. The prior distribution for* θ *is thus* $N(1, 0.3^2)$*. The readings for the first experiment are* $x_1 = 0.6066$ *g/kg and* $x_2 = 0.5778$ *g/kg. The sample mean of the measurements recorded by the first procedure is* $\bar{x} = (x_1 + x_2)/2 = (0.6066 + 0.5778)/2 = 0.5922$ *which is above the legal limit of* 0.5 *(g/kg). The posterior distribution is still Normal with mean*

$$\mu(x) = \frac{(0.0229^2/2)1 + 0.09(0.5922)}{0.0229^2/2 + 0.09} = 0.593,$$

and variance

$$\tau^2(x) = \frac{(0.0229^2/2)0.09}{0.0229^2/2 + 0.09} = 0.00026.$$

It can be observed that, after procedure HS, the probability of being over the legal limit is

$$P\,(\theta > 0.5 \mid \theta \sim N(0.593, 0.00026)) \approx 1.$$

Bayes' theorem allows a continual update of the uncertainty about the alcohol concentration (Section 3.3.1) . Typically, this may be the case when results of the second series of analyses (that is, by the ID method) become available. The readings for the second experiment are $x_1 = 0.4997$ *g/kg and* $x_2 = 0.5197$ *g/kg. The new sample mean for the quantity of alcohol is* $\bar{x} = (x_1 + x_2)/2 = (0.4997 + 0.5197)/2 = 0.5097$. *Note that the posterior density of* θ *from the first set of data becomes the prior density for the new set of data. In the case at hand, the posterior belief for* θ *becomes* $N(0.577, 0.00021)$*; see row 2 of Table 4.3. Then, after procedure ID, the probability of being over the legal limit is*

$$P\,(\theta > 0.5 \mid \theta \sim N(0.577, 0.00021)) \approx 1.$$

The sequential use of Bayes' theorem is illustrated in Figure 4.11 where the posterior distributions are plotted sequentially.

The loss function is assumed, as mentioned earlier in Section 3.4.2, to be piecewise linear and asymmetric, with $k_0 = 2$ *and* $k_1 = 1$*, since an underestimation of the alcohol concentration is regarded as a more serious error. The Bayesian optimal decision following procedures HS and ID is the* $2/(2+1)$ *quantile of the posterior distribution, which is equal to* 0.583*; see row 1 of Table 4.3.*

An approach to inference based on hypothesis testing is illustrated in Example 6.3.4

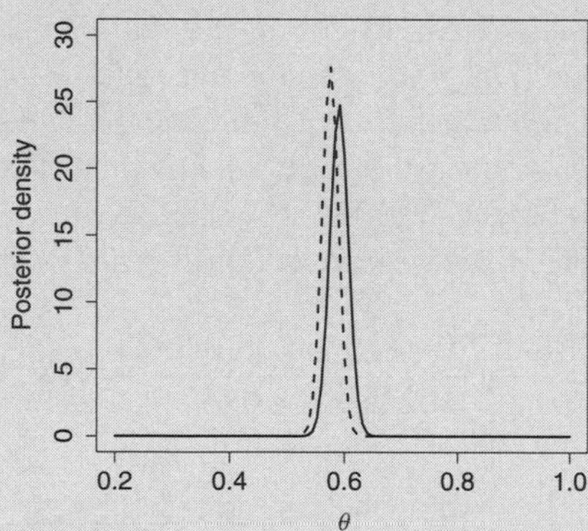

Figure 4.11 Sequential use of Bayes' theorem as outlined in Example 4.4.1. Posterior distributions obtained with the HS data (solid line), subsequently updated by the ID data (dotted line).

Example 4.4.2 *(Alcohol concentration in blood – continued). In Section 4.2.2, it has been argued that the choice of prior distributions and loss* *functions may alter a decision. This is correct within a 'point estimation' perspective. But, even if such choices may lead to different Bayesian estimates, this may not necessarily influence the decision to be taken by a recipient of expert evidence. In a legal context, this may be, for instance, a change from a guilty to a non-guilty verdict, or vice versa. Table 4.3 illustrates this point using different prior distributions and loss functions.*

Neither a change in the prior distribution nor in the loss fuction influences the decision according to which the parameter of interest is greater than the (legally defined) threshold of 0.5.

Table 4.3 Sensitivity of the final decision on alcohol level (g/kg) to different choices of the prior distribution and the loss function, using measurements for HS (0.5922) and for ID (0.5097) from Example 4.4.1.

Prior distribution	Loss function	Point estimate	Decision
$N(1, 0.3^2)$	$k_0 = 2; k_1 = 1$	0.583	Over legal threshold
$N(1, 0.3^2)$	$k_0 = 1; k_1 = 1$	0.577	Over legal threshold
$N(0.5, 0.13^2)$	$k_0 = 2; k_1 = 1$	0.581	Over legal threshold
$N(0.5, 0.13^2)$	$k_0 = 1; k_1 = 1$	0.575	Over legal threshold

4.4.2 Case with unknown variance

Consider now a situation in which the data X are still Normally distributed, $X \sim N(\theta, \sigma^2)$, but with both the mean θ and the variance σ^2 unknown. It is then necessary to consider a prior distribution on both parameters. If very little prior information is available, a non-informative prior distribution on both parameters can be defined (e.g., Bernardo and Smith 2000; Box and Tiao 1973; Robert 2001), and is given by

$$\pi(\theta, \sigma) = \frac{1}{\sigma}.$$

If $x_1, \ldots, x_n$ are observed, the posterior distribution of (θ, σ) associated to this prior is:

$$\theta \mid \sigma, \bar{x}, s^2 \sim N(\bar{x}, \sigma^2/n),$$
$$\sigma^2 \mid \bar{x}, s^2 \sim IG\left(\frac{n-1}{2}, \frac{(n-1)s^2}{2}\right),$$

where IG denotes an inverse gamma distribution (see Appendix B for more details about this distribution), and $s^2 = \frac{1}{n-1}\sum_{i=1}^{n}(x_i - \bar{x})^2$. The marginal posterior distribution of the prior mean θ which is of interest here is

$$\theta \mid \bar{x}, s^2 \sim \mathrm{St}\left((n-1), \bar{x}, \frac{s^2}{n}\right),$$

i.e. θ has a *Student t* distribution (see Appendix B) centred about $\bar{x}$ and with $(n-1)$ degrees of freedom. An example can be found in Section 8.4.4.

Suppose now that some knowledge is available. Conjugate prior distributions for this context are characterized by the fact that parameters are not independent. Consider then

$$\pi(\theta, \sigma^2) = \pi(\theta \mid \sigma^2)\pi(\sigma^2),$$

where $\pi(\theta \mid \sigma^2)$ is a Normal distribution with mean μ and variance σ^2/n_0 for some fixed n_0, $(\theta \mid \sigma^2) \sim N(\mu, \sigma^2/n_0)$, and $\pi(\sigma^2)$ is an inverse gamma distribution with parameters α and β, $\sigma^2 \sim \mathrm{IG}(\alpha, \beta)$. Therefore, the prior distribution on the mean depends on the precision associated with the mean. The appearance of σ^2 in the conditional distribution of the mean θ indicates, for example, that if σ^2 is large, a high variance on θ is induced. This dependency must be justified, considering that conjugate prior distributions are chosen largely for convenience and that alternative choices are possible. Although this dependency cannot hold in general for every problem of estimation, it is argued in Robert (2001) that when the prior distribution is built from previous observations, it makes sense that σ^2 is conditionally involved in the prior variance of θ, in the sense that prior beliefs about the mean are calibrated by the scale of measurements of the observations (Gelman *et al.* 1997).

The hyperparameter n_0 expresses the strength of prior belief about the chosen location μ for the mean of the data (Congdon 2001). In general, n_0/n characterizes the relative precision of the determination of the prior distribution as compared with the precision of the observations. It can be observed that if the ratio is very small, the posterior distribution approaches the posterior distribution that is obtained with a uniform prior.

These prior distributions are conjugate, since, given a random sample $x = (x_1, \ldots, x_n)$, the posterior distribution of the variance $\pi(\sigma^2 \mid x)$ is still an inverse gamma with updated parameters

$$\alpha^* = \alpha + \frac{n}{2},$$
$$\beta^* = \beta + \frac{1}{2}\left[\sum_{i=1}^{n}(x_i - \bar{x})^2 + \frac{(\bar{x}-\mu)^2}{\frac{1}{n_0} + \frac{1}{n}}\right],$$

while the posterior distribution of the mean $\pi(\theta \mid \sigma^2, x)$ is Normal with mean

$$E^{\pi(\theta|\sigma^2,x)}(\theta) = \frac{n}{n+n_0}\bar{x} + \frac{n_0}{n+n_0}\mu, \tag{4.19}$$

and variance

$$Var^{\pi(\theta|\sigma^2,x)}(\theta) = \frac{\sigma^2}{(n+n_0)}.$$

The posterior marginal distribution for the mean $\pi(\theta \mid x)$ is a *Student t* distribution with $2\alpha^*$ degrees of freedom centred at $E^{\pi(\theta|\sigma^2,x)}(\theta)$.

Example 4.4.3 *(Widths of landmarks on a firearm bullet). Suppose that a bullet is recovered on a crime scene. Initially, the forensic scientist may measure the width of the landmarks on the bullet. Let X be the Normally distributed random variable of measurements on the width of landmarks, $X \sim N(\theta, \sigma^2)$. Both parameters are unknown. A database of measurements on the bullets is available. The prior location measure μ is fixed equal to* 1.6965 *(mm) for a population of caliber* 38 *guns (the relevant population).*

Considering the great variability that characterizes the available database, the precision of the determination of the prior distribution is rather poor and the parameter n_0 is fixed equal to 0.1. Note that the elicitation of the inverse gamma distribution for the variance is not necessary in this context. A quadratic loss function is considered appropriate because the decision maker accepts that under- and overestimation of the width incur equal losses. Therefore, what is needed is the posterior mean of the marginal distribution of θ, that is a Student t *distribution centred at the posterior mean of* $(\theta \mid \sigma^2, x)$.

A sample of $n = 36$ observations is available, with a mean of $\bar{x} = 1.9306$. The Bayesian optimal decision is the posterior mean which is given by Equation (4.19) and is equal to

$$\frac{36}{36.1}1.9306 + \frac{0.1}{36.1}1.6965 = 1.9299.$$

4.4.3 Estimation of the mean in the presence of background data

Consider again the scenario outlined in Example 3.3.8, where the aim was to estimate the height of an individual from the image obtained from a surveillance camera. The true height of the individual is denoted θ. The measured height Y of the individual is assumed to have a Normal distribution. This measured height is biased because of errors introduced by the context in which the image was taken, such as the posture of the individual, the presence or absence of headgear and the angle of the camera relative to the individual (Taroni *et al.* 2006b). Denote the bias by ξ. There are two sources of variance and these are assumed independent. The first is the precision of the measurement device, denote its variance by σ^2. The second is the variation associated with the context and is denoted δ^2. Hence, $Y \sim N(\theta + \xi, \sigma^2 + \delta^2)$.

The uncertainty about θ is modelled through a Normal distribution with mean μ and variance τ^2, $\theta \sim N(\mu, \tau^2)$, while hyperparameters ξ and δ^2 are assumed to be known from the scenario of interest. This appears to be admissible because the assumption is made that the case at hand is basic in the sense that several images of an individual are available from video recording of one specific surveillance camera. That is to say parameters ξ and δ^2 could be obtained through an ad hoc reconstruction. Whenever recordings are available from different surveillance cameras on different locations, then the scenario is one in which it would be necessary to model the prior mean and variance.

Following the same steps as outlined in Section 4.4.1, it can be shown that the posterior distribution is still a Normal density with mean

$$\mu(y) = \frac{\tau^2(\bar{y} - \xi) + \mu\left(\frac{\sigma^2+\delta^2}{n}\right)}{\tau^2 + \frac{\sigma^2+\delta^2}{n}},$$

and variance

$$\tau^2(y) = \frac{\tau^2\left(\frac{\sigma^2+\delta^2}{n}\right)}{\tau^2 + \frac{\sigma^2+\delta^2}{n}}.$$

The variance is clearly higher with respect to an error-free setting, while more weight is given to the prior mean μ, according to the amplitude of the error component. This is illustrated in Example 4.4.4.

Example 4.4.4 *(Surveillance cameras – continued from Example 3.3.8). Imagine a video recording is made by a surveillance camera during a bank robbery. The recordings depict an individual appearing in* $n = 10$ *images. Measurements on the available recordings yield* $\bar{y} = 178$*(cm). It is of interest to infer the mean height of that individual. The precision of the measurement procedure (independent on the complexity of the scenario) is known, and it is set to* $\sigma^2 = 0.1$*. Parameters* μ *and* τ^2 *are chosen following the same line of argument illustrated in Example 3.3.6. In particular, there is eyewitnesses evidence based on which the mean* μ *is fixed equal to* 175*. Values less than* 170 *and greater than* 180 *are considered extremely unlikely, therefore the variance* τ^2 *is fixed equal to* 2.67*. Finally, repeated* *measurements are obtained in experiments under controlled conditions (i.e. a reconstruction), which allows to choose values for the hyperparameters of the Normal distribution of the error. These values are taken to be* $\xi = 1$ *and* $\delta^2 = 1$*, respectively. Then,*

$$\mu(y) = \frac{2.67(178 - 1) + 175\left(\frac{0.1+1}{10}\right)}{2.67 + \frac{0.1+1}{10}} = 176.92cm.$$

Assuming a quadratic loss (since it may be accepted by a decision maker that under- and overestimation of the individual's height incur equal losses), this is the optimal Bayesian estimate.

The posterior density is depicted in Figure 4.12 (solid line). The posterior density obtained in the absence of a source of distortion is also shown.

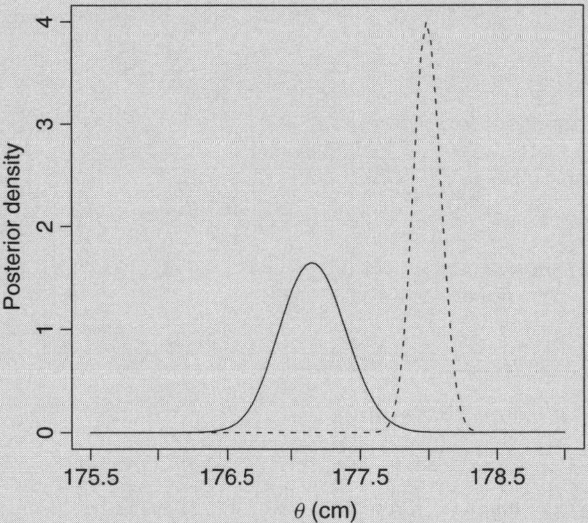

Figure 4.12 Posterior density for Example 4.4.4 obtained in presence (solid line), and in absence (dashed line) of a source of distortion.

4.5 R CODE

A symbol '$*$', '+', ',' and so on at the end of a line indicates that the command continuous to the following line. The absence of such a symbol indicates the end of a command.

Example 4.2.1

Data, prior parameters, loss values

```
n=64
y=0
alpha=1
beta=1
k0=2
k1=1
```

Posterior distribution and Bayes decision

```
alphap=alpha+y
betap=beta+n-y
plot(function(x) dbeta(x,alphap,betap),0,0.1,ylim=c(0,100),
xlab=expression(paste(theta)),ylab=expression(paste(pi)*
paste("(")*paste(theta)*paste('|') *paste(x) * paste(")")))
d=round(qbeta(k0/(k0+k1),alphap,betap),3)
print(paste('Bayes decision =',d))
lines(x=c(d,d),y=c(0,dbeta(d,alphap,betap)),lty=3)
```

Example 4.2.2

Data, prior parameters, loss values

```
n=64
y=0
nprev=40
p=1/nprev
alpha=round(p*(nprev-1))
beta=round((1-p)*(nprev-1))
k0=2
k1=1
```

Posterior distribution and Bayes decision

```
alphap=alpha+y
betap=beta+n-y
plot(function(x) dbeta(x,alphap,betap),0,0.1,ylim=c(0,100),
xlab=expression(paste(theta)),ylab=expression(paste(pi)*
paste("(")*paste(theta)*paste('|') *paste(x) * paste(")")))
d=round(qbeta(k0/(k0+k1),alphap,betap),3)
print(paste('Bayes decision =',d))
lines(x=c(d,d),y=c(0,dbeta(d,alphap,betap)),lty=3)
```

Example 4.2.4

Data, prior parameters and loss values

```
n=100
y=25
lambda=2.5
mu=0.05
```

```
sigma2=0.0005
alpha=round(mu*((mu*(1-mu)/sigma2)-1),1)
beta=round((1-mu)*((mu*(1-mu)/sigma2)-1),1)
plot(function(x) dbeta(x,alpha,beta),0,0.3,
xlab=expression(paste(theta)),ylab=expression(paste(pi)*
paste("(") * paste(theta) * paste(")")),main='')
k0=2
k1=1
```

Approximated posterior distribution and Bayes decision

```
g=gamma(alpha+beta)/(gamma(alpha)*gamma(beta))
thetavalues=seq(0,1,0.001)
post=matrix(0,length(thetavalues))
for (i in 1:length(thetavalues)){
theta=thetavalues[i]
s=0
for (yb in 0:y){
s=s+choose(n,y-yb)*(theta^(y-yb))*((1-theta)^(n-y+yb))*
exp(-lambda)*(lambda^yb)/factorial(yb)
}
post[i]=g*theta^(alpha-1)*(1-theta)^(beta-1)*s
}
normpost=post/(sum(post*0.001))
plot(thetavalues,normpost,xlim=c(0,0.3),ylim=c(0,20),type='l',
xlab=expression(paste(theta)),ylab=expression(paste(pi)*
paste("(")*paste(theta) *paste('|') *paste(x) * paste(")")))

ord=k0/(k0+k1)
p=cumsum(normpost*0.001)
d=(thetavalues[sort(which(p<ord),decreasing=T)[1]]+
thetavalues[which(p>ord)[1]])/2
print(paste('Bayes decision =',d))
```

Example 4.2.5

Data, prior parameters

```
n=100
y=25
alpha=4.7
beta=89.3
a=3
b=2
```

Multiple-block M-H

```
n.iter=15000
burn.in=5000
acct=0
accl=0
tau2t=0.8
tau2l=1.8
```

```
thetavalues=matrix(0,nrow=n.iter,ncol=1)
lambdavalues=matrix(0,nrow=n.iter,ncol=1)
thetacurr=0.5
lambdacurr=10
thetavalues[1]=thetacurr
lambdavalues[1]=lambdacurr

for (i in 2:n.iter){
psicurr=log(thetacurr/(1-thetacurr))
phicurr=log(lambdacurr)

psiprop=rnorm(1,mean=psicurr,sd=tau2t)
thetaprop=exp(psiprop)/(1+exp(psiprop))

s=0
for (yb in 0:y){
s=s+choose(n,y-yb)*(thetaprop^(y-yb))*((1-thetaprop)^(n-y+yb))*
(lambdacurr^yb)/factorial(yb)
}
pipsiprop=(exp(psiprop)/(1+exp(psiprop))^2)*(thetaprop^(alpha-
1))*(1-thetaprop)^(beta-1)*(s)

s=0
for (yb in 0:y){
s=s+choose(n,y-yb)*(thetacurr^(y-yb))*((1-thetacurr)^(n-y+yb))*
(lambdacurr^yb)/factorial(yb)
}

pipsicurr=(exp(psicurr)/(1+exp(psicurr))^2)*(thetacurr^(alpha-
1))*(1-thetacurr)^(beta-1)*(s)


d=pipsiprop/pipsicurr
u=runif(1)
    if (u<d){
     thetacurr=thetaprop
     acct=acct+1
     }
    thetavalues[i]=thetacurr

phiprop=rnorm(1,mean=phicurr,sd=tau2l)
lambdaprop=exp(phiprop)

s=0
for (yb in 0:y){
s=s+choose(n,y-yb)*(thetacurr^(y-yb))*((1-thetacurr)^(n-y+yb))*
exp(-lambdaprop)*(lambdaprop^yb)/factorial(yb)
}
piphiprop=exp(phiprop)*lambdaprop^(a-1)*exp(-b*lambdaprop)*s

s=0
for (yb in 0:y){
s=s+choose(n,y-yb)*(thetacurr^(y-yb))*((1-thetacurr)^(n-y+yb))*
```

```
exp(-lambdacurr)*(lambdacurr^yb)/factorial(yb)
}

piphicurr=exp(phicurr)*lambdacurr^(a-1)*exp(-b*lambdacurr)*s

d=piphiprop/piphicurr
u=runif(1)
    if (u<d){
     lambdacurr=lambdaprop
     accl=accl+1
     }
    lambdavalues[i]=lambdacurr
}

print(paste('Acceptance rate theta - ',round(acct/n.iter,2)))
print(paste('Acceptance rate lambda - ',round(accl/n.iter,2)))

plot(thetavalues,type='l',xlab='Iterations',ylab='')
plot(lambdavalues,type='l',xlab='Iterations',ylab='')
acf(thetavalues[(burn.in+1):n.iter],type="correlation",
 main='',ci=0)
acf(lambdavalues[(burn.in+1):n.iter],type="correlation",
 main='',ci=0)

mean(thetavalues[(burn.in+1):n.iter])
mean(lambdavalues[(burn.in+1):n.iter])

l=n.iter-burn.in
ord=round(2*l/3,0)
d=round(sort(thetavalues[(burn.in+1):n.iter])[ord],4)
print(paste('Bayes decision = ',d))
```

Example 4.3.1

Data, prior parameters, loss values

```
n=50
y=2
alpha=1
beta=0
k0=1
k1=2
```

Sequential use of Bayes theorem and Bayes decision

```
alphap=alpha+y
betap=beta+n
alpha=alphap
beta=betap
n=35
y=1
alphap=alpha+y
betap=beta+n
```

```
d=round(qgamma(k0/(k0+k1),alphap,betap),3)
print(paste('Bayes decision =',d))
```

Example 4.3.3

Data, prior parameters, loss values

```
n=6
sumy=26
mu=6.5
s=3
alpha=(mu^2)/(s^2)
beta=mu/(s^2)
plot(function(x) dgamma(x,shape=alpha,scale=1/beta),0,20,
xlab=expression(paste(lambda[s])),ylab=expression(paste(pi)*
paste("(")*paste(lambda[s]) * paste(")")))
lambdab=0.05
k0=1
k1=2
```

Posterior distribution and Bayes decision

```
lambdavalues=seq(0,10,0.01)
post=matrix(0,nrow=length(lambdavalues))
for (i in 1:length(lambdavalues)){
lambdas=lambdavalues[i]
post[i]=exp(-n*(lambdas+lambdab))*((lambdas+lambdab)^sumy)*
lambdas^(alpha-1)*exp(-beta*lambdas)
}
normpost=post/(sum(post*0.01))

plot(lambdavalues,normpost,xlab=expression(paste(lambda[s])),
ylab=expression(paste(pi)*paste("(") * paste(lambda[s])
*paste('|')*paste(y) * paste(")")),type='l')

ord=k0/(k0+k1)
p=cumsum(normpost*0.01)
d=(lambdavalues[sort(which(p<ord),decreasing=T)[1]]+
lambdavalues[which(p>ord)[1]])/2
print(paste('Bayes decision =',d))
```

Example 4.4.1

Data and prior parameters

```
x=c(0.6066,0.5778)
n=length(x)
xbar=mean(x)
s2_hs=0.0229^2
mu=1
tau2=0.3^2
```

Posterior parameters

```
mux=(mu*s2_hs/n+tau2*xbar)/(s2_hs/n+tau2)
taux2=(tau2*s2_hs/n)/(tau2+s2_hs/n)
```

Sequential use of Bayes' theorem

```
x=c(0.4997,0.5197)
n=length(x)
xbar=mean(x)
s2_id=0.0463^2
mu=mux
tau2=taux2
mux=(mu*s2_id/n+tau2*xbar)/(s2_id/n+tau2)
taux2=(tau2*s2_id/n)/(tau2+s2_id/n)

plot(function(x) dnorm(x,mu,sqrt(tau2)), 0.2, 1,main = '',
xlab=expression(paste(theta)),ylab='Posterior distribution',
ylim=c(0,30))
plot(function(x) dnorm(x,mux,sqrt(taux2)), 0.2, 1,main = '',
xlab=expression(paste(theta)),ylab='Posterior distribution',
ylim=c(0,30),lty=2,add=TRUE)
```

Bayes decision

```
k0=1
k1=1
d=qnorm(k0/(k0+k1),mux,sqrt(taux2))
print(paste('Bayes decision =',round(d,4)))
```

5

Credible Intervals

5.1 INTRODUCTION

Interval estimation represents another common approach to statistical inference. Consider, for instance, a scientist who wishes to gain knowledge about a population parameter of interest θ (e.g. the proportion of individuals in a given population that have a given mt-DNA sequence, the proportion of red woollen fibres on a car seat or the mean of the weight of a consignment of seized ecstasy pills). All information that stems from prior beliefs and available data is contained in the posterior distribution and can be summarized through appropriate summary statistics (e.g. the posterior mean, median, quantile), as outlined in Chapter 4. Another way to summarize the information contained in the posterior distribution is to build a *confidence set* associated with which is a stated amount of probability, that is a subset C of the parameter space Θ in which parameter θ is located with that probability. For example, consider the weight of pills with illicit content (e.g. ecstasy) seized on an individual. The range of uncertainty around the point estimate can be assessed and visualized using confidence regions, commonly called confidence intervals in one dimensional problems.

The meaning of confidence sets is sometime a cause of trouble and confuses students (Albert 1992), lawyers (Kaye 1987a), and scientists (D'Agostini 2004a) as well as forensic scientists (Evett and Weir 1998). From a Bayesian viewpoint, once it is accepted that uncertainty about a parameter may be represented by a probability distribution, i.e. the unknown parameter is being treated as a random variable, then it is a straightforward matter to determine a subset such that the probability that θ belongs to that region is equal to a given amount, a so-called credible probability. For example, given a credible probability equal to 0.95, it is

Data Analysis in Forensic Science: A Bayesian Decision Perspective Franco Taroni, Silvia Bozza, Alex Biedermann, Paolo Garbolino and Colin Aitken

then possible to determine a subset C such that $P(\theta \in C) = 0.95$. The scientist would be able to say, on grounds of logic, that his degree of belief that the parameter is in fact in the realized subset is equal to 95%. Within the frequentist paradigm, such an assertion is not valid since the parameter is assumed fixed. There is no probability distribution associated to it. Formally, the frequentist confidence set is random, and the realized set is one of the possible realized values of the random set: the parameter being unknown but fixed, it will be inside the set with probability either 0 or 1. In contrast, the Bayesian paradigm allows one to say that θ is inside the specified set with some probability, not 0 or 1.

There is often confusion between *credible probability* and *coverage probability*. Credible probability reflects the information contained in the posterior distribution and allows the scientist to assert that the region contains the parameter with a given probability. So, when the scientist asserts a credible probability equal to 0.95, this means that he is 95% certain that the interval or region contains the true value of the parameter. Conversely, coverage probability reflects the uncertainty in the sampling procedure and is a frequentist procedure. A coverage of 95% means that, if it were possible to repeat the experiment in the same conditions, in a long sequence of identical trials, 95% of the realized intervals or regions would contain the parameter. A detailed comparison between frequentist and Bayesian methods is, however, beyond the scope of this book and is not pursued further.

5.2 CREDIBLE INTERVALS AND LOWER BOUNDS

Consider the posterior distribution $\pi(\theta \mid x)$ of the parameter θ. A subset C of Θ is said to be a $100(1-\alpha)\%$ *credible set* for θ if

$$1 - \alpha = P^{\pi(\theta|x)}(\theta \in C) = \begin{cases} \int_C \pi(\theta \mid x)\, d\theta & \text{continuous case} \\ \sum_{\theta \in C} \pi(\theta \mid x) & \text{discrete case.} \end{cases}$$

If the set is not disjoint it is called a *credible interval*. For disjoint subsets or for a vector parameter space the term *region* will be used. Thus C may be called a credible interval or region of probability $(1-\alpha)$. The construction of credible intervals is straightforward in principle, as will be shown in Example 5.2.1, but it should be noted that the credibility level α does not specify the extremes of the interval exactly; there are many credible intervals of probability $(1-\alpha)$. A simple way to select the credible interval at credibility level α consists in fixing the lower bound at the posterior quantile of order $\alpha/2$, denoted $\pi_{\alpha/2}$, and the upper bound at the posterior quantile of order $1-\alpha/2$, denoted $\pi_{1-\alpha/2}$, such that

$$P\left(\pi_{\alpha/2} \leq \theta \leq \pi_{1-\alpha/2}\right) = 1 - \alpha.$$

This method of selection uses as a criterion the equality of the tail-area probabilities. The resultant interval is not necessarily symmetric about the point estimate as the distribution may not be symmetric.

Suppose, for instance, it is of interest to obtain a $100(1-\alpha)\%$ credible interval for the mean θ of a Normally distributed random variable with known variance σ^2, $X \sim N(\theta, \sigma^2)$. A conjugate prior distribution for the unknown mean θ is the Normal distribution, $\theta \sim N(\mu, \tau^2)$, so that the posterior distribution $\pi(\theta \mid x)$ is Normal with parameters $\mu(x)$ and $\tau^2(x)$ as in (4.16) and (4.17). It follows that under the posterior distribution

$$\frac{\theta - \mu(x)}{\tau(x)} \sim N(0,1), \tag{5.1}$$

and that

$$P\left[z_{\alpha/2} \leq \frac{\theta - \mu(x)}{\tau(x)} \leq z_{1-\alpha/2}\right] = 1 - \alpha,$$

where $z_{\alpha/2}$ and $z_{1-\alpha/2}$ denote respectively the quantiles of order $\alpha/2$ and $1-\alpha/2$ of the standardized Normal distribution (i.e. a Normal distribution with zero mean and unit variance, as in Equation (5.1)). Note that, as the distribution is symmetric, $z_{1-\alpha/2} = -z_{\alpha/2}$ holds. So, a $100(1-\alpha)\%$ credible interval for θ is given by:

$$\left[\mu(x) - z_{1-\alpha/2}\tau(x), \mu(x) + z_{1-\alpha/2}\tau(x)\right]. \tag{5.2}$$

If both the mean θ and the variance σ^2 are unknown, and the noninformative prior

$$\pi(\theta, \sigma^2) = \frac{1}{\sigma}$$

is applied (a suggestion that very poor prior knowledge is available), then

$$\frac{\theta - \bar{x}}{s/\sqrt{n}} \sim \mathrm{St}(n-1),$$

where $s = \sqrt{\frac{1}{n-1}\sum_{i=1}^{n}(x_i - \bar{x})^2}$, and $\mathrm{St}(n-1)$ denotes the central *Student t* distribution (see Appendix B). The $100(1-\alpha)\%$ credible interval for μ is

$$\left[\bar{x} - t_{1-\alpha/2,n-1}s/\sqrt{n},\ \bar{x} + t_{1-\alpha/2,n-1}s/\sqrt{n}\right], \tag{5.3}$$

$t_{1-\alpha/2,n-1}$ denotes the quantile of order $1-\alpha/2$ of a central *Student t* distribution with $(n-1)$ degrees of freedom, and the interval corresponds to the same confidence interval as in the frequentist approach.

Example 5.2.1 *(Alcohol concentration in blood – continued). Consider again the scenario presented in Example 4.4.1. A sample of blood was analyzed using two independent analytical techniques, each of which produced two measurements. The alcohol concentration X (g/kg) in blood was assumed to be Normally distributed with variance known equal to* 0.0229^2*, for the procedure HS and to* 0.0463^2 *for procedure ID. The prior distribution of* θ *was assumed Normal with mean* $\mu = 1$ *and variance* $\tau^2 = 0.3^2$*. The posterior distribution* $\pi(\theta \mid x)$ *obtained was Normal with parameters* $\mu(x) = 0.577$ *and* $\tau^2(x) = 0.00021$*. A 95% equi-tailed credible interval for* θ *is obtained by substituting values in (5.2), where* $z_{1-\alpha/2} = 1.96$*, to give:*

$$\left[0.577 - 1.96\sqrt{0.00021}, 0.577 + 1.96\sqrt{0.00021}\right] = [0.548, 0.605].$$

Any interval, however, such that the probability that the parameter lies between the lower and the upper bound is equal to $1 - \alpha$ will result in a $100(1 - \alpha)\%$ credible interval.

The interval thus is not uniquely defined. Which interval should be chosen? Usually, the objective is to minimize the size (length) of the interval, that is to find the set C that satisfies both conditions:

$$1) \quad \int_C \pi(\theta \mid x)\, d\theta = 1 - \alpha$$

$$2) \quad \text{Size } C \leq \text{Size } C'$$

for any set C' satisfying $\int_{C'} \pi(\theta \mid x) d\theta \geq 1 - \alpha$. The interval should occupy the smallest possible volume in the parameter space. To find such an interval, one should include those points with the largest posterior density, that is the interval should have the property that the density for every point inside the interval is greater than that for every point outside the interval. The shortest credible interval for θ, for a given credibility level of α, is called a *highest posterior density* (HPD) interval (or region), and is the subset C_α of Θ of the form:

$$C_\alpha = \{\theta \in \Theta : \pi(\theta \mid x) \geq k(\alpha)\} \text{ where } \int_{C_\alpha} \pi(\theta \mid x)\, d\theta = 1 - \alpha.$$

The HPD interval consists of the values of the parameter for which the posterior density is highest.

If the posterior density is unimodal and symmetric, the HPD interval will also be symmetric: the HPD interval can simply be found by choosing an equal distance on either side of the mode. The credible interval for a Normal variable, see Example 5.2.1, is symmetric around the mean, and is, in fact, an HPD interval.

Credible and HPD intervals do not necessarily cover the same volume of the parameter space. The shape of the HPD interval is determined by the shape of the posterior distribution. So, if the posterior is asymmetric, the HPD region will not be symmetric about a Bayes estimator (e.g. the posterior mean). In this case, assuming unimodality for the posterior density, a $100(1-\alpha)\%$ HPD interval can be easily calculated numerically. One needs to write a routine that, in correspondence with several values of k:

1. Finds the solutions $a(k)$ and $b(k)$ to the equation $\pi(\theta \mid x) = k$ ($k <$ max $\pi(\theta \mid x)$);
2. Computes
$$\int_{a(k)}^{b(k)} \pi(\theta \mid x)\, \mathrm{d}\theta;$$
3. Finds the value of k, dependent on α but with the dependency suppressed in the notation for ease of reading, such that
$$\int_{a(k)}^{b(k)} \pi(\theta \mid x)\, \mathrm{d}\theta = 1 - \alpha.$$

There are several ways to write such a routine. A simple way is to start from the posterior mode θ^* and take k equal to $\{\pi(\theta^* \mid x) - \in\}$, for some small $\in$, as the initial value. An HPD region for a skewed distribution is depicted in Figure 5.1; note that it is not symmetric about the mode.

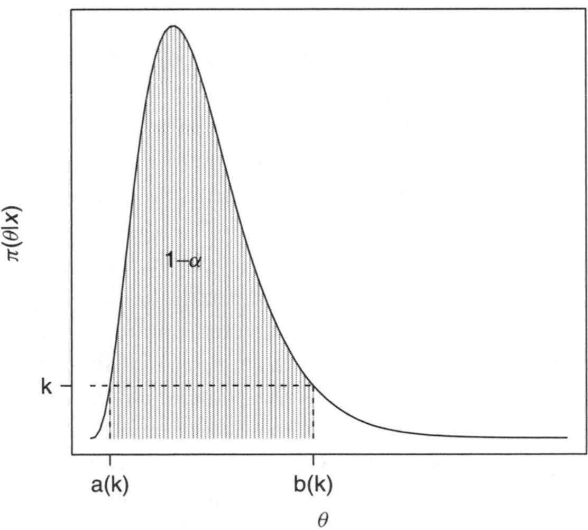

Figure 5.1 A $100(1-\alpha)\%$ HPD interval for θ where k is chosen such that $\int_{a(k)}^{b(k)} \pi(\theta \mid x)\, d\theta = 1 - \alpha$.

Example 5.2.2 *(Gunshot residue particles – continued). Consider again Example 4.3.1 involving the search for gunshot residue particles (GSR). In this case, the scientist intends to estimate probabilities for the number y of GSR particles (where* $y = 0, 1, 2, \ldots$*) on the hands of an individual unrelated to a shooting incident, based on data obtained from a new experiment. The GSR count y is assumed to follow a Poisson distribution with parameter* λ*. Consider a gamma prior distribution for* λ, $\text{Ga}(3, 50)$*, based on a scientist's knowledge (previous experiments). Prior beliefs about the parameter* λ *are updated by combining the prior with the results of a new experiment:* 1 *particle is found after analysing* $m = 35$ *individuals. The resulting gamma posterior distribution for* λ *is* $\text{Ga}(4, 85)$*. The 95% HPD interval for the parameter* θ *is obtained by implementing the routine outlined above. The mode of a gamma distribution* $\text{Ga}(\alpha, \beta)$ *is* $(\alpha - 1)/\beta$*, so the routine is initialized at* $\theta^* = (4 - 1)/85$ *and returns the interval* $[0.0083, 0.0939]$*. Note that when using Bayesian inference the experimental design does not affect the inference. It could have been that 35 individuals were studied and 1 particle was found. Alternatively, individuals could have been studied until 1 particle was found and this happened with the 35-th individual. In each situation, the inference is the same (see Example 3.3.4).*

It can happen that an HPD credible interval looks unusual. Consider for example a binomial random variable, $X \sim \text{Bin}(n, \theta)$, and its conjugate prior, $\theta \sim \text{Be}(\alpha, \beta)$. The posterior distribution $\pi(\theta \mid x)$ is the beta distribution $\text{Be}(\alpha + y, \beta + n - y)$. Since the beta distribution is not, in general, symmetric, the HPD region is one of four types, as shown in Figure 5.2, for a $\text{Be}(\alpha, \beta)$ distribution. The posterior HPD interval depends on the values of α, β, y, n.

While the HPD region for a beta distributed parameter θ might consist of disjoint intervals, such a situation might also occur in the presence of a multimodal posterior (e.g. a mixture of Normal distributions), in which case the computation of an HPD can be much more complicated. In such cases, the scientist might be tempted to abandon the HPD criterion and to look for more 'usual' connected intervals, that is common intervals with equal tails ($\alpha/2$). Berger (1988) discourages such a choice, mentioning that disconnected intervals often occur when there is discrepancy between the prior and the sampling distribution and that this phenomenon should question the choice of the prior or of the sampling distribution. In the presence of multimodal shapes where the computation of the HPD interval can be much more complicated, Press (2003) suggests that the highest mode be found first and that one then works down to lower modes. Methods for finding HPD regions for any given density are discussed in Hyndman (1996).

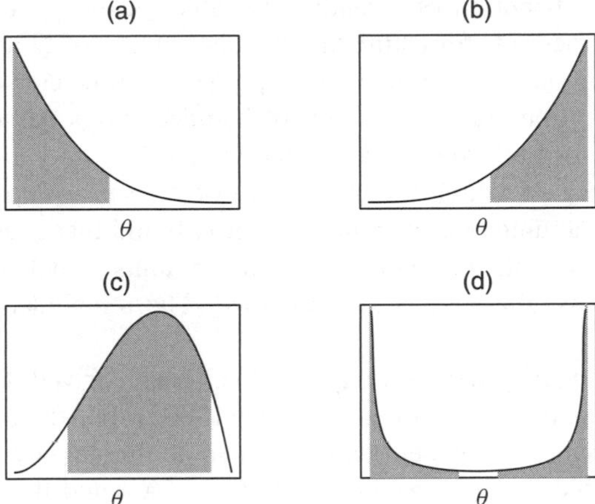

Figure 5.2 Beta HPD intervals, where (a) $\alpha \leq 1, \beta > 1$, (b) $\alpha > 1, \beta \leq 1$, (c) $\alpha > 1, \beta > 1$, and region (d) $\alpha \leq 1, \beta \leq 1$.

Example 5.2.3 *(Glass fragments – continued). Consider the same scenario illustrated earlier in Example 4.2.2, where $n = 64$ glass objects were surveyed with no positive outcomes ($y = 0$). The prior distribution was taken to be beta with parameters $\alpha = 1$ and $\beta = 38$, updated to $\alpha = 1$ and $\beta = 102$ for the posterior distribution. An equi-tailed 95% credible interval for θ is equal to $[0.00025, 0.03552]$. The HPD region differs slightly and can be obtained in a straightforward manner, since the posterior distribution is asymmetric and the HPD region will be of the type illustrated in Figure 5.2(a). It will be sufficient to compute the posterior quantile of order* 0.95, *the HPD region will therefore be $[0, 0.02894]$. This region is much more intuitively sensible than the two-tailed credible interval.*

Another context in which the estimate of a bound rather an interval is appropriate is that of the determination of a quantity of an illegal substance. This is of interest, for example, when the quantity is a factor in sentencing, e.g. sentencing guidelines of the USA in drug-related cases.

It is important when sentencing individuals engaged in illegal drug dealing to obtain valid estimates for the total quantity of drugs that have been handled, in particular for determining the overall length of the sentence to be imposed. The problem of estimation in such a context may be considered as a problem of estimating the distribution of a random variable Q, the total quantity of drugs handled. Different methods for estimating the distribution of Q are given in Aitken *et al.* (1997). In this section, the predictive approach will be considered. It involves a predictive distribution for the quantity of handled drugs based on independent prior distributions for both the mean and the variance.

Consider a setting in which a consignment of $N = m + n$ packages is seized. A number (m) of the units are examined and it is found that z units contain drugs and that $(m - z)$ do not. The contents of the z units which contain drugs are weighed and their weights $(x_1, \ldots, x_z)$ recorded. The remainder $(n = N - m)$ are not examined.

The probability density function $f(q)$ of Q has been derived, for small and large consignments respectively, in Aitken and Lucy (2002). For the purpose of illustration, consider a small consignment and let Y denote the unknown number of units not examined which contain drugs. Let $X = (X_1, \ldots, X_z)$ and $W = (W_1, \ldots, W_y)$ be the weights of the contents of the units examined and not examined, respectively, which contain drugs. It is assumed that these weights are normally distributed. The total weight Q is then given by

$$Q = z\bar{X} + Y\bar{W},$$

where $\bar{X} = \sum_{i=1}^{z} X_i/z$, and $\bar{W} = \sum_{j=1}^{y} W_j/y$.

Let Q be the total quantity of drugs in the consignment. The number of units examined equals m of which z $(\leq m)$ contain drugs. The mean and standard deviation of the quantity of drugs in the z units are denoted by $\bar{x}$ and s, respectively. The number of units not examined equals n, of which y (unknown) contain drugs and for which the mean quantity of drugs in these y units is $\bar{w}$. Thus, $Q = z\bar{x} + Y\bar{W}$ in which both Y and $\bar{W}$ are random variables.

A Bayesian approach to the lower bound for the quantity Q of drugs is derived as follows. First, condition on $Y = y$. Then

$$\begin{aligned} P(Q < q \mid y, z, \bar{x}, s, m, n) = \\ P(z\bar{x} + y\bar{w} < q \mid y, z, \bar{x}, s, m, n) \\ = P(\bar{W} < \frac{q - z\bar{x}}{y} \mid y, z, \bar{x}, s, m, n). \end{aligned}$$

Now, given $Y = y$,

$$\frac{\bar{W} - \bar{x}}{s\sqrt{\frac{1}{z} + \frac{1}{y}}} \sim \text{St}(z - 1).$$

For given values of $m, z, n, y, \bar{x}$ and s, lower bounds for $\bar{w}$ and hence q, can be determined from the formula

$$\bar{w} = \bar{x} + st_{\alpha,z-1}\sqrt{\frac{1}{z} + \frac{1}{y}}. \tag{5.4}$$

Let $T = (\bar{W} - \bar{x})/\{s\sqrt{(1/z) + (1/y)}\}$. Then,

$$P(\bar{W} < \frac{q - z\bar{x}}{y} \mid y, z, \bar{x}, s, m, n) =$$
$$P(T < \frac{q - (z + y)\bar{x}}{sy\sqrt{\frac{1}{z} + \frac{1}{y}}} \mid y, z, \bar{x}, s, m, n)$$

where $T \sim \mathrm{St}(z - 1)$. Let

$$t_{qy} = \frac{q - (z + y)\bar{x}}{sy\sqrt{\frac{1}{z} + \frac{1}{y}}}.$$

Now, combine this with the result for the conditional distribution for Q, given $Y = y$, and the marginal probability function for Y, to obtain

$$P(Q < q \mid z, \bar{x}, s, m, n) =$$
$$\sum_{y=0}^{n} Pr(T < \frac{q - (z + y)\bar{x}}{sy\sqrt{\frac{1}{z} + \frac{1}{y}}} \mid y, z, \bar{x}, s, m, n) Pr(Y = y)$$
$$= \sum_{y=0}^{n} Pr(T < t_{qy} \mid y, z, \bar{x}, s, m, n) Pr(Y = y). \quad (5.5)$$

The probability density function $f(q)$ of Q can be derived by differentiation of the distribution function. Let $f_{t,z-1}(.)$ denote the probability density function of the t distribution with $(z - 1)$ degrees of freedom. Then,

$$f(q) = \sum_{y=0}^{n} f_{t,z-1} \left\{ \frac{q - (z + y)\bar{x}}{sy\sqrt{\frac{1}{z} + \frac{1}{y}}} \right\} \left\{ sy\sqrt{\frac{1}{z} + \frac{1}{y}} \right\}^{-1} Pr(Y = y).$$

Example 5.2.4 *(Estimation of drug quantity). Consider an example from Tzidony and Ravreboy (1992) in which a seized drug exhibit contained* 26 (N) *street doses. A sample of six* $(m = 6)$ *units was taken and each was analyzed and weighed. Twenty* $(n = 20)$ *units were not examined. It was found that all six of the units examined contained drugs. The average net weight* $\bar{x}$ *of the powder in the six units was* 0.0425 g *with a standard deviation* s *of* 0.0073 g. *A 95% (frequentist) confidence interval for the total quantity* Q *in the* 26 *doses is* 1.105 ± 0.175 g *(Tzidony and Ravreboy 1992). This interval incorporates a finite population correction factor* $\sqrt{(N - m)/N}$ *to allow for the relatively* *large sample size ($m = 6$) compared with the consignment size ($N = 26$). Results from a Bayesian analysis using (5.4) are given in Table 5.1.*

Table 5.1 Estimates of quantities q g of drugs from (5.4), in a consignment of $m+n$ units, according to various possible burdens of proof, expressed as percentages $P = 100 \times P(Q > q \mid m, z, n, \bar{x}, s)$ in 26 packages when 6 packages are examined ($m = 6$, $n = 20$) and $z = 6$, 5, or 4 are found to contain drugs. The mean ($\bar{x}$) and standard deviation (s) of the quantities found in the packages examined which contain drugs are 0.0425 g and 0.0073 g. The parameters for the beta prior are $\alpha = \beta = 1$. Numbers in brackets are the corresponding frequentist lower bounds using the finite population correction factor, $n/(m+n)$. (Aitken CGG and Lucy D 2002 Estimation of the quantity of a drug consignment from measurements on a sample. Journal of the Forensic Sciences 47, 968–975.)

Percentage P	Number of units examined which contain drugs			Possible burden of proof (Illustrative)
	6	5	4	
99	0.617 (0.876)	0.435 (0.683)	0.290 (0.519)	
97.5	0.689 (0.930)	0.501 (0.744)	0.345 (0.575)	
95	0.750 (0.968)	0.559 (0.785)	0.397 (0.613)	Beyond reasonable doubt
90	0.818 (1.005)	0.628 (0.823)	0.461 (0.647)	
70	0.944 (1.067)	0.770 (0.885)	0.603 (0.704)	Clear and convincing
60	0.982 (1.087)	0.819 (0.904)	0.655 (0.721)	
50	1.015 (1.105)	0.862 (0.921)	0.704 (0.737)	Balance of probabilities

5.3 DECISION-THEORETIC EVALUATION OF CREDIBLE INTERVALS

The use of decision theory in interval estimation presents some drawbacks that makes this approach less appealing than it appears, for example, in point estimation, though interval estimators are used extensively in decisions. One reason explaining the widespread preference for Bayesian credible intervals (without a decision-theoretic approach), is the difficulty in the evaluation of an appropriate loss function in these contexts.

What characteristics should a loss function have so as to be considered appropriate for the definition of a credible interval? In the previous section it has been outlined that an optimal interval should meet two conditions: it should have a high credible probability and be of the shortest size. If both of these requirements are met by a loss function, then the optimal interval estimator can be found using a decision approach.

Initially, start by specifying the elements of the decision process in interval estimation. The *decision space* $\mathcal{D}$ will consist of subsets of the parameter space Θ. Possible decisions will be called C (in conformity with the notation introduced earlier in this chapter). For any given loss function $L(C,\theta)$ and prior distribution $\pi(\theta)$, the *Bayes estimator* C is a solution of the minimization problem

$$\min \bar{L}(C, \pi(\theta \mid x)) = \min \int_{\Theta} L(C,\theta)\pi(\theta \mid x)\, d\theta.$$

The loss function $L(C,\theta)$ must necessarily take account of two elements: a measure of whether the realized subset includes the value of θ (correctness of coverage), and a measure of its size. As a measure of the correctness, an *indicator function* $I_C(\theta)$ is considered, which takes value 1 if the subset contains the parameter, and 0 otherwise:

$$I_C(\theta) = \begin{cases} 1 & \theta \in C \\ 0 & \theta \notin C. \end{cases}$$

If the subset is an interval, the length can be taken as a measure of its size: $Length(C) =$ length of C[1]. The loss function should reflect the fact that an optimal credible interval should include the parameter θ and should have a short size, that is it should have $E\,[I_C(\theta)]$ high and length $Length(C)$ small. A loss function satisfying these requirements is the *linear loss function*:

$$L(C,\theta) = bLength(C) - I_C(\theta), \tag{5.6}$$

where b is a positive constant that reflects the relative weight that is given to the shortness of the interval. The smaller b is, the less the size of the interval matters (the more the correctness matters). In fact, if $b = 0$ and only correctness matters, the best decision will be the real line, $C = (-\infty, +\infty)$, which has credible probability equal to 1. Vice versa, the greater b is, the more the size of the interval matters. In the next example, it will be shown that for values of b greater than a given threshold, the Bayes decision corresponds to a point estimate.

Example 5.3.1 *(Normal interval estimator). Consider a random sample from a Normal distribution, $X \sim N(\theta,\sigma^2)$, and assume σ^2 known. Let $\pi(\theta) = N(\mu,\tau^2)$, then*

$$\pi(\theta \mid x) = N(\mu(x), \tau^2(x)).$$

[1]In general, the credible set is not necessarily an interval, so a measure of the size of any set C is the volume, denoted $Vol(C)$.

The posterior distribution being symmetric around the posterior mean $\mu(x)$*, define an interval estimator* C *symmetric around the mean,* $C = [\mu(x) - c\tau(x), \mu(x) + c\tau(x)], c > 0$*. This interval has length* $2c\tau(x)$*. The linear loss function will be:*

$$L(C, \theta) = b2c\tau(x) - I_C(\theta).$$

The posterior expected loss will be:

$$\begin{aligned}\bar{L}(C, \pi(\theta \mid x)) &= \int_\Theta (b2c\tau(x) - I_C(\theta))\, \pi(\theta \mid x)\, d\theta \\ &= b2c\tau(x) \int_\Theta \pi(\theta \mid x)\, d\theta - \int_\Theta I_C(\theta) \pi(\theta \mid x)\, d\theta \\ &= b2c\tau(x) - P^{\pi(\theta|x)}\, [\theta \in C].\end{aligned}$$

The last term $P^{\pi(\theta|x)}\, [\theta \in C]$ *can be expressed as*

$$\begin{aligned}P\, [\mu(x) - c\tau(x) \le \theta \le \mu(x) + c\tau(x)] &= P\left[-c \le \frac{\theta - \mu(x)}{\tau(x)} \le c\right] \\ &= 1 - 2P\left[\frac{\theta - \mu(x)}{\tau(x)} \le c\right].\end{aligned}$$

The best interval has a length that minimizes the posterior expected loss. To find the minimun, the first derivative of the posterior expected loss is calculated and then set equal to zero:

$$\frac{\partial}{\partial c}\bar{L}(C, \pi(\theta \mid x)) = 2b\tau(x) - \frac{2}{\sqrt{2\pi}} \exp\left(-\frac{c^2}{2}\right) = 0.$$

The solution depends on the value of b*. In particular,*

- *if* $b\tau(x) > \frac{1}{\sqrt{2\pi}}$*, the first derivative is positive for any* $c \ge 0$*, so the expected posterior loss is minimized with an interval of length* 0 $(c = 0)$*. The best interval is the point estimator* $C = [\mu(x), \mu(x)]$*.*
- *if* $b\tau(x) < \frac{1}{\sqrt{2\pi}}$*, the expected posterior loss is minimized at* $b\tau(x) = \frac{1}{\sqrt{2\pi}} \exp\left(-\frac{c^2}{2}\right)$*, that is for* $c = \sqrt{-2 \log\left(b\tau(x)\sqrt{2\pi}\right)}$*. If* b *is taken equal to* $\frac{1}{\exp\left(\frac{1}{2}z^2_{\alpha/2}\right)\tau(x)\sqrt{2\pi}}$*, then* $c = z_{\alpha/2}$ *and the interval that minimizes the expected loss is the usual* $100(1 - \alpha)\%$ *credible interval.*

A drawback of the linear loss function, Equation (5.6), is the difficulty in choosing b. It has been observed in the previous example that a choice that might seem reasonable could lead to results that are not intuitive. In fact, for $b\tau(x) > 1/\sqrt{2\pi}$, the procedure proposes a single point estimate (or an interval of length zero), indicating certainty. An increasing value of the variance $\tau^2(x)$, indicating more uncertainty, should lead to increased uncertainty in the set estimator. However, to the contrary, the optimal interval collapses to a single point. *Student t* intervals, as in (5.3), lead to another example of disconcerting results if a linear loss function is adopted. Casella *et al.* (1993b) showed a peculiar behaviour of the resulting Bayes set in the sense that its size decreases as uncertainty increases.

This problem is connected to the asymmetry of the linear loss function. The two criteria, size and coverage, are unequally penalized since the indicator function varies between 0 and 1, while the volume can increase to infinity, and this favours small credible sets. It is possible to derive Bayes sets that are trivial since they may be either empty (the penalty for large sizes increases too rapidly), or equal to the entire parameter space (the penalty increases too slowly). For this reason, this function does not provide a coherent basis for decision-based set estimation. A class of loss functions that avoids this problem and that allows the experimenter to balance correctly size and coverage has been proposed by Casella *et al.* (1993b) and has the form

$$L(C,\theta) = S\,[Vol(C)] - I_C(\theta),$$

where $S(\cdot)$ is a size function. There are several classes of size functions. The class

$$S\,[Vol(C)] = \frac{Vol(C)}{Vol(C)+b}, \qquad b > 0, \tag{5.7}$$

leads to the *rational loss*

$$L(C,\theta) = \frac{Vol(C)}{Vol(C)+b} - I_C(\theta). \tag{5.8}$$

Note that both terms are bounded by one: volume and coverage are weighted equally, and this is necessary to avoid counterintuitive Bayes sets.

Conditions for the existence of nontrivial Bayes sets are given in Casella *et al.* (1993a). For example, it is shown that for a Normal characteristic, and the size function (5.7), there exists a smallest non-empty Bayes set for any $b > 0$, while there exists no largest bounded set (the maximum value is infinity). It is observed, however, that the behaviour of the Bayes sets varies with the density function and the size function, and that there is not a loss that provides a nontrivial Bayes set for any kind of problem. The behaviour of the loss function changes depending on whether or not the parameter space is bounded. For a bounded parameter space, the loss function

$$L(C,\theta) = \frac{Vol(C)}{Vol(\Theta)} - I_C(\theta)$$

provides, for any distribution $f(x \mid \theta)$ and for any prior distribution $\pi(\theta)$, a non-trivial Bayes set. Conversely, in the general case of unbounded parameter spaces, some loss functions provide results that are not coherent.

Size and coverage are combined in a single-valued loss function, instead of being treated separately, and this represents an advantage of the decision-theoretic approach that appears to be a powerful tool for set estimation allowing complementary criteria to be balanced. Even allowing for this, the difficulties in the choice of the loss function make the decision-theoretic approach less appealing as an approach for the construction of a credible interval. It must be emphasized that the decision framework is also not without difficulty in point estimation. Different loss functions provide different Bayes decisions from the same posterior distribution. An inconvenience in set estimation is that the choice of the loss function is less intuitive and needs more careful thinking: size and coverage must be appropriately balanced, not least to avoid trivial solutions. For these reasons, a decision-theoretic approach to interval estimation is not considered in further detail here.

5.4 R CODE

A symbol '$*$', '$+$', ',' and so on at the end of a line indicates that the command continuous to the following line. The absence of such a symbol indicates the end of a command.

Example 5.2.2

Data, prior and posterior parameters

```
n=35
y=1
alpha=3
beta=50
alphap=alpha+y
betap=beta+n
```

HPD interval

```
mode=round((alphap-1)/betap,3)
m=round(qgamma(0.99,alphap,betap),3)
x=seq(mode,m,0.00001)
fx=dgamma(x,alphap,betap)
incr=0.0001
a=mode-incr
q=dgamma(a,alphap,betap)
ind=which(fx<=q)[1]
```

```
b=x[ind]
p=pgamma(b,alphap,betap)-pgamma(a,alphap,betap)

while (p<=0.95){
a=a-incr
q=dgamma(a,alphap,betap)
ind=which(fx<q)[1]
b=x[ind]
p=pgamma(b,alphap,betap)-pgamma(a,alphap,betap)
}
print(paste('HPD interval =[ ',round(a,4),paste(','),round(b,4),
paste(']')))
```

Example 5.2.3

Data, prior and posterior parameters

```
n=64
y=0
alpha=1
beta=38
alphap=alpha+y
betap=beta+n-y
```

Equi-tailed credible interval and HPD interval

```
z=0.05
l=round(qbeta(z/2,alphap,betap),5)
u=round(qbeta(1-z/2,alphap,betap),5)
print(paste('Equi-tailed credible interval =[ ',l,paste(','),
u,paste(']')))

u=round(qbeta(1-z,alphap,betap),5)
print(paste('HPD interval =[ ',0,paste(','),u,paste(']')))
```

6

Hypothesis Testing

6.1 INTRODUCTION

The concept of testing hypotheses is fundamental in statistics. It was initially developed by two statisticians, J. Neyman (1894–1981) and E.S. Pearson (1895–1980). A *null hypothesis* or working hypothesis is established and data are collected to test this hypothesis with the use of a statistic, known as a *test statistic*. Extreme values of the test statistic, with respect to the null hypothesis, lead the experimenter to reject the null hypothesis in favour of an alternative hypothesis. A value of the test statistic is deemed extreme if the probability of obtaining the observed value of the statistic or a value further removed from the value expected if the null hypothesis were true is small. Conventionally, a small value for this probability is taken to be 0.05 or 0.01. There are two errors associated with this approach, known as type 1 and type 2 errors. A type 1 error is to reject the null hypothesis when the null hypothesis is true. The probability of a type 1 error is known as the *significance level* of a test. A type 2 error is to fail to reject the null hypothesis when the alternative hypothesis is true (and the null hypothesis is false). A related probability is the *power* of a test. This is the probability that the null hypothesis is rejected correctly, i.e. when the alternative is true. Much of the theory associated with hypothesis testing within the Neyman–Pearson paradigm is concerned with developing tests of high power and low significance level; i.e. tests with a high probability of correctly rejecting a null hypothesis and a low probability of incorrectly rejecting a null hypothesis.

These ideas sit uncomfortably with the Bayesian paradigm with which this text is concerned. There is an asymmetry associated with the hypotheses. Data have (or evidence has) to be collected against the null hypothesis before it is rejected.

Data Analysis in Forensic Science: A Bayesian Decision Perspective Franco Taroni, Silvia Bozza, Alex Biedermann, Paolo Garbolino and Colin Aitken

Only when there are sufficient data is the null hypothesis rejected in favour of the alternative. This rejection may happen even if the data are more likely under the null hypothesis than the alternative hypothesis. Hypothesis testing assumes the hypotheses are fixed and it is the data which vary. Thus, one looks at the probability of the data if the null (or alternative) hypothesis is true. For inferential purposes it is of more interest to consider the probability of the null or alternative hypotheses given the data which have been observed. Finally, no consideration is given in the Neyman–Pearson paradigm to the consequences of a decision to reject or not a null hypothesis.

The following parts in this chapter explain the inferential procedure of the Bayesian paradigm and associated consequences (or *losses*) when testing hypotheses.

Data are often collected in order to answer questions of forensic interest such as

(a) 'Is the mutation rate for the chromosome X STR marker DXS7132 greater than 0.0038?'
(b) 'Is the proportion of illicit pills in a large seizure of pills greater than a fixed threshold, say 0.6?'

Answers to such questions can be obtained through a methodology that takes the implicit null value as the *status quo* or *null hypothesis* and tests it against the alternative implied by the question with data collected for the purpose. For the above questions, the null hypotheses would be as follows.

(a) The mutation rate for the chromosome X STR marker DXS7132 is 0.0038.
(b) The proportion of illicit pills in a large seizure of pills is 0.6.

The alternative hypotheses would then be as follows.

(a) The mutation rate for the chromosome X STR marker DXS7132 is greater than 0.0038.
(b) The proportion of illicit pills in a large seizure of pills is greater than 0.6.

The statistical *testing of a hypothesis* involves a decision about the plausibility of a hypothesis based on some data.

In forensic science as well as in other disciplines such as law, medicine or physics, many scientists are attracted by, and reason according to, the *falsificationist* scheme[1]. Therein, the aim is to reason with the use of data about the acceptability of a theory or a hypothesis, either by confirming it or by disconfirming it. This scheme of reasoning implies that a hypothesis should yield verifiable predictions, which can be checked to be true or false. If the empirical consequence of a hypothesis is shown in an experiment to be false, then the hypothesis is refuted[2]. Thus it is very

[1]The origin of falsificationism is commonly ascribed to Cournot (1843) and Popper (1959).

[2]As noted by D'Agostini (2004b), falsificationism is nothing but an extension of the proof by contradiction to the experimental method. He noted that '[t]he proof by contradiction of standard dialectics and mathematics consists in assuming true a hypothesis and in looking for (at least) one of its logical consequences that is manifestly false. If a false consequence exists, then the hypothesis

tempting for a scientist to test a theory by setting up a hypothesis that claims the opposite of what he actually believes, that is a hypothesis he believes a priori to be false, and then state – through appropriate experimental data – that this hypothesis, known as the *null hypothesis*, is false.

The statistical formalization of this procedure is due to a statistician, R.A. Fisher (1890–1962), who proposed a *test of significance*, the implementation of which allows for the rejection of a null hypothesis if an appropriate test statistic exceeds some particular value, based on a pre-assigned significance level. Neyman and Pearson (1928a,b) subsequently extended this approach with the introduction of the notions of alternative hypotheses and of type I and type II errors that could be made in testing statistical hypotheses. The underlying ideas stipulate that, after the determination of an appropriate test statistic, a practitioner computes the so-called *p-value* (or *observed significance*) which represents the probability, assuming that the null hypothesis is true, that one would obtain a value of the test statistic that is as extreme as, or more extreme than, that obtained from the data. Following this *classical* or *frequentist* line of reasoning (known as 'classical' because of its widespread use over many years of the twentieth century and not because of any longevity greater than that, known as 'frequentist' because the inferences rely on the relative frequencies of events), the null hypothesis can thus be rejected if the p-value is less than some pre-specified *significance level* (type I error rate). Otherwise, whenever the p-value is greater than the significance level, no conclusion can be drawn until additional evidence becomes available. Note that the significance level is arbitrary. In practice, a 5% level is commonly used as the threshold for the assessment of the significance of departures from a null hypothesis. This p-value is sometimes interpreted (wrongly) as the probability that the null hypothesis is true.

The discussion of p-values is not pursued in further detail here because, from the principles advocated throughout this book, there are several difficulties associated with their use. A noteworthy difficulty is the well-known problem of the fallacy of the 'transposed conditional' (Lindley 2006), which arises when the use of a p-value introduces confusion about (a) a probability about some aspect of the data, assuming the null hypothesis to be true, and (b) the probability the hypothesis is true, assuming some aspect of the data. The point is that probability statements about hypotheses lie within the province of the Bayesian approach. In the frequentist concept, hypotheses do not have probabilities associated with them[3] (Lindley 2000b).

---

under test is considered false and its opposite true' (D'Agostini 2004b, p. 5). This means that, as we should expect, once a hypothesis is refuted, no further evidence can ever confirm it, unless the refuting evidence or some portion of the background assumptions (knowledge) is revoked (Howson and Urbach 1996, p. 119). Application in practice of the *falsificationist* reasoning scheme faces several problems that conflict with a probabilistic view of the world. Falsificationism can be seen as a particular case of Bayes' theorem (see Section 2.3.1).

[3]The correct interpretation of the p-value is much more tortuous. Probabilities in the frequentist approach must be based on repetition under identical conditions. So, if one were to repeat an analysis many times, using data each time, and if the null hypothesis were actually true, then on only 5% of those occasions one would (falsely) reject the null hypothesis. This definition refers to repetition of the experiment. As mentioned by O'Hagan (2004, p. 42), 'to interpret a p-value as the probability that the

The main emphasis will be placed here on the Bayesian approach to hypothesis testing as developed by Jeffreys (1961). This approach is soundly based and intuitively more satisfactory than the Fisherian and Neyman–Pearson methods. It concentrates on the application of Bayes' theorem to answer the relevant scientific question through the computation of the posterior probability of a hypothesis of interest. It is often argued that there should be no testing since the posterior distribution contains all the knowledge of interest and is thus sufficient for making conclusions about any question under investigation. Notice however that such a line of reasoning refers to a *partial* Bayesian approach. A *full* Bayesian point of view deals with decision making for which purpose loss (or utility) functions are introduced. In the forthcoming sections, both of these approaches will be developed and illustrated through examples.

One of the appealing features of Bayesian methods is that they allow one to overcome the difficulties that arise with classical (frequentist) hypothesis testing. A difficulty already noted is that users may tend to view the p-value as the probability of the null hypothesis. That is, when $p = 0.05$, it is tempting to state that there is only a 5% probability that the null hypothesis is true. The p-value cannot, however, measure the probability of the truth of the null hypothesis because its calculation assumes the null hypothesis is true. As mentioned by Goodman (2005, p. 284) one 'can't have a measure assuming something to be true while simultaneously measuring how likely the same thing is to be false'[4]. This is an important difference with respect to Bayesian analysis, as distinct from frequentist analysis, in that Bayesian analysis allows the definition of a (prior) probability (i.e., a probability that is evaluated, usually subjectively, prior to observation of the data) for each hypothesis (null and alternative), that is an expression of the personal degree of uncertainty about a hypothesis' truthfulness, on the basis of which a posterior distribution (i.e., a probability that is evaluated posterior to observation of the data) for the hypotheses can be inferred. Also, Bayesian testing procedures do not require a pre-assigned significance level. However, it is emphasized that.

> Some Bayesians actually think that there should be no testing, or, at least, that there should be no point null hypothesis testing [...] But pragmatic considerations are such that the Bayesian toolbox must also include testing devices, if only because users of Statistics have been accustomed to testing as a formulation of their problems [...] (Robert 2001, pp. 223–224).

A further concept considered in the next section is the notion of *'evidence'* in the context of statistical evidence. According to Goodman and Royall (1988, pp. 1568–1574), *evidence* may be defined as a property of data that makes one alter

---

null hypothesis is true is not only wrong but also dangerously wrong. The danger arises because this interpretation ignores how plausible the hypothesis might have been in the first place'. Examples are given in O'Hagan (2004).

[4]For a survey of the pitfalls of classical hypothesis testing in forensic science and in litigation cases, and for some recommendations for improvements, see Kaye (1986a,b).

one's beliefs about how the world is working. Another way to say this is that evidence is the basis upon which inferences are derived. The value of evidence is measured by a ratio of likelihoods, that is the likelihood of the evidence (data) if the null hypothesis is true and the likelihood of the evidence (data) if the alternative hypothesis is true. The value of experimental evidence can therefore be measured by how more probable that evidence makes the null hypothesis relative to the alternative hypothesis than it was before the evidence was considered, conditioned on prior information (Good 1950). Within such a setting, one hypothesis will be preferred to another if the value of the evidence favours it.

In what follows, different approaches to testing hypotheses will be addressed, including, notably, estimation of the posterior distribution of the hypothesis of interest and the decision-theoretic approach in which utilities associated with the hypotheses are considered. A more general discussion of complications with frequentist methods is not within the main scope of this book, but the interested reader is referred to Press (2003), Carlin and Louis (1998) and Leonard and Hsu (1999) for technical overviews, to Goodman and Royall (1988), Goodman (1999), Marden (2000) and Winkler (2001) for general (not technical) treatments and to Howson and Urbach (1996) for a philosophical and historical discourse.

6.2 BAYESIAN HYPOTHESIS TESTING

6.2.1 Posterior odds and Bayes factors

Consider an unknown quantity X, such as the proportion of ecstasy pills in a seizure, and suppose $f(x \mid \theta)$ is a suitable probability model for X, where the unknown parameter θ belongs to the parameter set Θ. Suppose also that the parameter set is partitioned into two non-overlapping sets Θ_0 and Θ_1 such that $\Theta = \Theta_0 \cup \Theta_1$. A question that may be of interest is whether the true but unknown value of the parameter θ belongs to Θ_0, or to Θ_1, that is to test the hypothesis

$$H_0 : \theta \in \Theta_0,$$

usually called the *null hypothesis*, against the hypothesis

$$H_1 : \theta \in \Theta_1,$$

usually called the *alternative hypothesis*. A hypothesis is called *simple* if there is only one possible value for the unknown parameter, say $\Theta_0 = \{\theta_0\}$; if a hypothesis is not simple it is called *composite*. Let $\pi_0 = P(\theta \in \Theta_0)$ and $\pi_1 = P(\theta \in \Theta_1)$ denote one's prior probabilities for the truth of the null hypothesis and the alternative hypothesis, respectively. Suppose a random sample $x = (x_1, \ldots, x_n)$ is available. Observational data will rarely provide conclusive evidence about the questions of interest, but they do allow prior beliefs about the null and the alternative hypothesis to be updated and enhanced. Stated otherwise, the acceptance of a null hypothesis by a scientist does not mean it is true, but that, given the available information, it is more probable than the alternative.

First, consider a basic case of the test of a simple null hypothesis $H_0 : \theta = \theta_0$ against a simple alternative hypothesis $H_1 : \theta = \theta_1$. The parameter sets in this case are $\Theta_0 = \{\theta_0\}$ and $\Theta_1 = \{\theta_1\}$. Denote the prior probabilities by $\pi_0 = P(\theta = \theta_0)$ and $\pi_1 = P(\theta = \theta_1)$. If $\pi_0 + \pi_1 = 1$, the ratio π_0/π_1 of the prior probabilities of the null and alternative hypotheses is called the *prior odds*[5] of H_0 to H_1. The odds indicate whether the null hypothesis is more or less likely than the alternative (prior odds being larger or smaller than 1), or whether the hypotheses are almost equally likely (prior odds close to 1). The posterior probability of the null hypothesis in the light of the data and prior probabilities is denoted α_0 and can be easily computed with an application of Bayes' theorem:

$$\alpha_0 = P(\theta = \theta_0 \mid x) = \frac{f(x \mid \theta_0)\pi_0}{f(x \mid \theta_0)\pi_0 + f(x \mid \theta_1)\pi_1}. \tag{6.1}$$

The posterior probability α_1 of the alternative hypothesis is computed analogously by

$$\alpha_1 = P(\theta = \theta_1 \mid x) = \frac{f(x \mid \theta_1)\pi_1}{f(x \mid \theta_0)\pi_0 + f(x \mid \theta_1)\pi_1}. \tag{6.2}$$

The ratio of the posterior probabilities α_0/α_1 is called the *posterior odds* of H_0 to H_1, and is equal to the product of the *likelihood ratio* and the prior odds in favour of H_0, that is

$$\frac{\alpha_0}{\alpha_1} = \frac{f(x \mid \theta_0)}{f(x \mid \theta_1)}\frac{\pi_0}{\pi_1}. \tag{6.3}$$

These elements allow one to test a null hypothesis, without fixing an arbitrary level of significance. Following Jeffreys' hypothesis testing criterion, the null hypothesis is accepted (rejected) if the posterior odds are greater (lower) than unity. In other words, the null hypothesis are rejected or accepted on the basis of its posterior probability being greater or smaller than that of the alternative hypothesis. Notice that the acceptance (or rejection) of the null hypothesis is not meant as an assertion of its truth (or falsity), only that it is more (or less) probable than the alternative hypothesis (Press 2003).

Testing simple versus simple hypotheses is only a particular setting amongst many others. Practitioners may face, for instance, the more general situation of testing a *composite hypothesis*. When a parameter θ is continuous, one or both

[5] As mentioned in Section 2.3.2, the ratio of the probabilities of two mutually exclusive and exhaustive events (which of necessity add to 1) is called odds in favour of the event whose probability is in the numerator of the ratio. The word 'odds' is sometimes used loosely in reference to the ratio of the probabilities of two mutually exclusive events whose probabilities add up to something less than 1.

of the two hypotheses may be composite. As an example, consider the testing of $H_0 : \theta \in \Theta_0$ versus $H_1 : \theta \in \Theta_1$, and let $\pi(\theta)$ denote the prior probability density. Accordingly, in contrast to the discrete probabilities above of $\pi_0 = P(\theta = \theta_0)$ and $\pi_1 = P(\theta = \theta_1)$, the prior probabilities π_0 and π_1 are now:

$$\pi_0 = P(\theta \in \Theta_0) = \int_{\Theta_0} \pi(\theta)d\theta \quad ; \quad \pi_1 = P(\theta \in \Theta_1) = \int_{\Theta_1} \pi(\theta)d\theta. \quad (6.4)$$

The posterior probability of the null hypothesis can be easily computed as

$$\begin{aligned}\alpha_0 = P(\theta \in \Theta_0 \mid x) &= \int_{\Theta_0} \pi(\theta \mid x)d\theta \\ &= \int_{\Theta_0} f(x \mid \theta)\pi(\theta)d\theta/m(x),\end{aligned}$$

where $m(x)$ is the normalizing constant

$$m(x) = \int_{\Theta} f(x \mid \theta)\pi(\theta)d\theta.$$

Similarly, the posterior probability of the alternative hypothesis is of the form

$$\alpha_1 = \int_{\Theta_1} f(x \mid \theta)\pi(\theta)d\theta/m(x).$$

Hence the posterior odds are

$$\frac{\alpha_0}{\alpha_1} = \frac{\int_{\Theta_0} f(x \mid \theta)\pi(\theta)d\theta}{\int_{\Theta_1} f(x \mid \theta)\pi(\theta)d\theta}. \quad (6.5)$$

The ratio of the posterior odds to the prior odds, that is

$$BF = \frac{\alpha_0/\alpha_1}{\pi_0/\pi_1}, \quad (6.6)$$

is called the *Bayes factor* in favour of H_0. The Bayes factor measures the change produced by the evidence in the odds when going from the prior to the posterior distribution in favour of one scientific theory as opposed to another (Kass and Raftery 1995; Lavine and Schervish 1999). If $\pi_0 = \pi_1 = 1/2$, then the prior odds are equal to 1 and the Bayes factor is equal to the posterior odds.

In the case of testing a simple null hypothesis versus a simple alternative hypothesis, it can be easily observed from (6.3) that the Bayes factor is just the likelihood ratio of H_0 to H_1,

$$BF = \frac{f(x \mid \theta_0)\pi_0}{f(x \mid \theta_1)\pi_1} \times \frac{\pi_1}{\pi_0} = \frac{f(x \mid \theta_0)}{f(x \mid \theta_1)}. \quad (6.7)$$

A likelihood ratio, of say k, corresponds to evidence strong enough to cause a k-fold increase in the prior odds, regardless of whether the prior odds are actually available in a specific problem or not (Royall 1997, p. 13). In this case, the Bayes factor depends only upon the sample data and reflects the extent to which the data favour one hypothesis over another, without prior information. The hypothesis better supported by the data is the hypothesis which better models the data.

In the more general case of testing composite hypotheses, the Bayes factor becomes more complicated than a simple likelihood ratio and will depend on the prior input. It is useful to rewrite the prior density $\pi(\theta)$ in the following form. Let $\pi_{H_0}(\theta)$ denote the restriction of the prior density on Θ_0, and $\pi_{H_1}(\theta)$ denote the restriction of the prior density on Θ_1, that is

$$\pi_{H_0}(\theta) = \frac{\pi(\theta)}{\pi_0} \quad \text{for } \theta \in \Theta_0; \tag{6.8}$$

$$\pi_{H_1}(\theta) = \frac{\pi(\theta)}{\pi_1} \quad \text{for } \theta \in \Theta_1. \tag{6.9}$$

Densities $\pi_{H_0}(\theta)$ and $\pi_{H_1}(\theta)$ are proper and describe how the prior probability is spread over the two hypotheses. In other words they are the conditional densities of θ given H_0 and H_1, respectively. Therefore, the prior density, $\pi(\theta)$, can be written as

$$\pi(\theta) = \begin{cases} \pi_0\pi_{H_0}(\theta) & \text{if} \quad \theta \in \Theta_0 \\ \pi_1\pi_{H_1}(\theta) & \text{if} \quad \theta \in \Theta_1 \end{cases}. \tag{6.10}$$

The posterior probabilities are easily rewritten as

$$\alpha_0 = \pi_0 \int_{\Theta_0} f(x \mid \theta)\pi_{H_0}(\theta)d\theta / m(x)$$

and

$$\alpha_1 = \pi_1 \int_{\Theta_1} f(x \mid \theta)\pi_{H_1}(\theta)d\theta / m(x).$$

The Bayes factor is then of the form

$$BF = \frac{\int_{\Theta_0} f(x \mid \theta)\pi_{H_0}(\theta)d\theta}{\int_{\Theta_1} f(x \mid \theta)\pi_{H_1}(\theta)d\theta}. \tag{6.11}$$

The Bayes factor is now the ratio of weighted likelihoods under the postulated hypotheses, and it appears that it no longer depends only upon the sample data. The reason for this is that the prior enters via the weights $\pi_{H_0}(\theta)$ and $\pi_{H_1}(\theta)$. Graphical displays of the Bayes factor as a function of the prior parameters can be used to present scientific results to different users with different prior opinions (Dickey 1973). An example can be found later in Section 6.4.3, Figure 6.12, following consideration of further complications in which other parameters are fixed.

Example 6.2.1 *(Prior density restriction for a beta uniform variable). Suppose $\theta \sim Be(1,1)$, and it is desired to test $H_0 : \theta \leq \theta_0$ against $H_1 : \theta > \theta_0$. The restriction $\pi_{H_0}(\theta)$ of the prior density on $\Theta_0 = [0, \theta_0]$ is given by*

$$\pi_{H_0}(\theta) = \frac{\pi(\theta)}{\pi_0} = \frac{1}{\int_0^{\theta_0} 1 d\theta} = \frac{1}{\theta_0}, \qquad for\, \theta \leq \theta_0,$$

and similarly the restriction $\pi_{H_1}(\theta)$ of the prior density on $\Theta_1 = (\theta_0, 1]$

$$\pi_{H_1}(\theta) = \frac{\pi(\theta)}{\pi_1} = \frac{1}{\int_{\theta_0}^{1} 1 d\theta} = \frac{1}{1-\theta_0}, \qquad for\, \theta > \theta_0.$$

It can be easily verified that these are density functions.

A guide for interpreting Bayes factors offered by Jeffreys (1961) is shown in Table 6.1.

Table 6.1 Verbal scale for interpreting the support of the null hypothesis over the alternative hypothesis (Jeffreys 1961).

BF	Evidence in favour of H_0
1 to 3.2	Not worth more than a bare mention
3.2 to 10	Substantial
10 to 100	Strong
> 100	Decisive

It goes without saying that the range of the scale and its interpretation may depend on the specific context. Multiple scales have been proposed and their principal role is to offer a rough guide to support interpretation.

Both the use of scales, however, and the discussion of likelihood ratios applied to real-case scenarios are sometimes viewed cautiously. Analogies to other measuring settings, such as temperature, may thus be helpful. As mentioned by Goodman and Royall (1988, pp. 1571–1572), for instance, few would argue that a thermometer does not provide a measure of thermal energy that may, in at least some sense, be called objective. Notwithstanding, a thermometer represents only a general index of the subjective experience of 'heat'. In particular, how hot one actually feels on a 40°C day depends on factors such as humidity, wind, clouds, one's ability to sweat as well as one's acclimatization. In the same way as temperature is a measure of thermal energy and a guide to the sensation of heat, the likelihood ratio is a measure of evidence and a guide to belief. Neither the feeling of a given quantity of degrees nor the meaning of a likelihood ratio of a given value can be described exactly in words. Both scales acquire their meaning through use and experience.

In the context of forensic science applications, the quantitative value has also been thought to be given a qualitative interpretation (Evett 1987, 1990; Evett *et al.* 2000). Table 6.2 summarizes a scale proposed by Evett *et al.* (2000) with H_p denoting the proposition put forward by the prosecution.

Table 6.2 Verbal scale for expressing the evidential value *BF* in support of the prosecutor's proposition H_p over the alternative (defence) proposition (Evett *et al.* 2000).

BF	Evidence in favour of H_p
1 to 10	Limited evidence to support
10 to 100	Moderate evidence to support
100 to 1000	Moderately strong evidence to support
1000 to 10000	Strong evidence to support
$>$ 10000	Very strong evidence to support

Note that this scale also works with values for Bayes factors smaller than unity, that is in support of the alternative hypothesis, which is normally the one favoured by the defence. When dealing with DNA evidence, however, where very large Bayes factors may be obtained, the above-mentioned verbal scale may be inadequate. This may also occur in cases where the forensic scientist seeks to combine multiple items of evidence. For this reason, forensic scientists may resort to the logarithm of the Bayes factor which has been conceptualized as the *weight of the evidence* (Good 1950) (see Section 2.4.1). In this way, as subsequent experiments become available, Bayes factors will multiply whereas the weights of evidence add as one naturally thinks about evaluation in association with scales of justice. A useful discussion on this topic is given in Kaye (1986b), Schum (1994) and Aitken and Taroni (2004). Notice also that – as mentioned by Singpurwalla (2006, p. 31) – with the use of verbal scales, some Bayesians have declined to specify prior odds. In effect they have chosen to use Bayes factors as an alternative to frequentist significance probabilities. While this strategy may be appropriate in the case of simple hypotheses, it is inappropriate in the case of composite hypotheses. With composite hypotheses the Bayes factor also depends on how the prior mass is spread over the two hypotheses. In particular, it cannot be interpreted as a summary of the evidence provided by the data alone, because it requires knowledge of prior probabilities, as illustrated in Equation (6.11). However, some authors observe that, for given priors, the Bayes factor is reasonably stable and that it can effectively be interpreted as the measure of the support given only by the data. The advantage of this interpretation is that, by starting from it, a practitioner can determine his 'personal' posterior odds by multiplication of the Bayes factor with his 'personal' prior odds (Berger 1988).

6.2.2 Decision-theoretic testing

Hypothesis testing is a special case of decision making (Edwards *et al.* 1963; Lindley 1961). Accordingly, hypotheses should be tested on the basis of the consequences of making wrong decisions. When testing hypotheses $H_0 : \theta \in \Theta_0$ versus $H_1 : \theta \in \Theta_1$, for instance, the decision space includes decisions d_0 and d_1, where d_i denotes acceptance of H_i. The cost of possible erroneous decisions should be taken into account by specifying a loss function for each alternative.

For the ease of argument, suppose that a correct decision incurs no loss and no gain, that is it produces a zero loss. The loss function for a two-action problem can be described by a two-way table that has entries of zeros on the diagonal. An example of such a function is given by the *'0–1' loss function* introduced earlier in Sections 2.3.2 and 3.4.2, Equation (3.21), summarized here in Table 6.3.

Table 6.3 The '0–1' loss function.

	$\theta \in \Theta_0$	$\theta \in \Theta_1$
d_0	0	1
d_1	1	0

Following the discussion presented in Section 3.4, a decision maker should compute the expected loss for each decision and choose the one which minimizes the loss. The Bayes decision is to choose d_1, which is equivalent to rejection of the null hypothesis when

$$P(\theta \in \Theta_0 \mid x) < P(\theta \in \Theta_1 \mid x), \tag{6.12}$$

or to choose d_0 otherwise which is to accept H_0, otherwise. The Bayesian procedure chooses the hypothesis with the largest posterior probability, and, thus, with a symmetric loss function the decision which minimizes the expected loss.

A generalization of the above loss function is to penalize errors differently when the null hypothesis is true from when it is false. This is managed with the *'0–k_i' loss function* which was introduced in Equation (3.22) of Section 3.4.2, a summary of which is given in Table 6.4.

Table 6.4 The '0–k_i' loss function.

	$\theta \in \Theta_0$	$\theta \in \Theta_1$
d_0	0	k_0
d_1	k_1	0

Under this loss setting the null hypothesis should be rejected when

$$\frac{P(\theta \in \Theta_0 \mid x)}{P(\theta \in \Theta_1 \mid x)} < \frac{k_0}{k_1}. \tag{6.13}$$

The larger k_0/k_1 is, that is the more a wrong answer under the alternative hypothesis ($d_0(\theta \in \Theta_0)$ when $\theta \in \Theta_1$) is penalized, the larger the posterior probability of H_0 needs to be in order for H_0 to be accepted.

A threshold for the interpretation of the Bayes factor can be obtained by multiplying both sides of Equation (6.13) by the prior odds π_1/π_0, that is

$$\frac{\pi_1}{\pi_0} \times \frac{P(\theta \in \Theta_0 \mid x)}{P(\theta \in \Theta_1 \mid x)} < \frac{k_0}{k_1} \times \frac{\pi_1}{\pi_0}.$$

Therefore, when applying a '0–k_i' loss, the optimal decision is d_1 or, equivalently, rejecting H_0, whenever

$$BF < \frac{k_0}{k_1}\frac{\pi_1}{\pi_0}, \tag{6.14}$$

and to accept $H_0(d_0)$ otherwise. Similarly, for a '0–1' loss, the optimal decision is d_1 whenever

$$BF < \frac{\pi_1}{\pi_0},$$

and to accept $H_0(d_0)$ otherwise.

Loss functions, as described above, with their simplicity, do not take into account the possible severity of a wrong decision. Consider Θ_0 to be the set $\theta \leq \theta_0$. Situations may be encountered where the effect of making an incorrect decision, such as d_0, and so accepting $H_0 : \theta \leq \theta_0$, depends on whether the true value of θ is close to θ_0, or not. For a given positive quantity ϵ, the true value might in fact be $(\theta_0 + \epsilon)$ or $(\theta_0 + 100\epsilon)$ so that the consequences of a wrong decision would, accordingly, be less or more serious (Bernardo and Smith 2000). In such a setting it would be preferable to build a loss function that incorporates a measure of the distance between the decision and the true value of the parameter of interest. Such a function has been introduced in Section 3.4.2, Equation (3.23), and is summarized here in Table 6.5. The function $f_i(\theta)$ is a positive function defined on Θ_i. It can be linear, quadratic, or something else depending on the context. A linear loss function is developed in the next section.

Table 6.5 The '0–$f_i(\theta)$' loss function.

	$\theta \in \Theta_0$	$\theta \in \Theta_1$
d_0	0	$f_0(\theta)$
d_1	$f_1(\theta)$	0

6.3 ONE-SIDED TESTING

6.3.1 Background

One-sided hypothesis testing applies when the parameter set Θ is a subset of the real line, and when Θ_0 is entirely to one side of Θ_1, that is either

$$\theta_0 < \theta_1 \ \text{ for } \ \theta_0 \in \Theta_0, \ \theta_1 \in \Theta_1$$

or

$$\theta_0 > \theta_1 \ \text{ for } \ \theta_0 \in \Theta_0, \ \theta_1 \in \Theta_1.$$

Consider, for instance, testing $H_0 : \theta \leq \theta_0$ against $H_1 : \theta > \theta_0$ (*i.e.* $\Theta_0 = \{\theta : \theta \leq \theta_0\}, \Theta_1 = \{\theta : \theta > \theta_0\}$). Recalling that $\pi(\theta \mid x)$ is the posterior distribution of the parameter of interest θ, the posterior probability of the null hypothesis is

$$\alpha_0 = P\left(\theta \leq \theta_0 \mid x\right) = \int_{-\infty}^{\theta_0} \pi(\theta \mid x) d\theta, \tag{6.15}$$

illustrated by the shaded area in Figure 6.1.

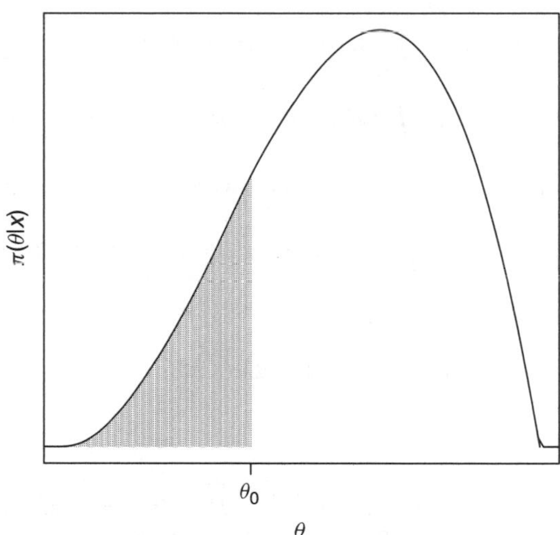

Figure 6.1 Posterior probability of the null hypothesis ($\theta \leq \theta_0$) given data x.

An example for such a setting could be one involving a consignment of pills where one is interested in testing if the proportion of pills containing an illegal substance is lower or greater than a certain threshold. In such a case the posterior probability of the null hypothesis becomes $\alpha_0 = \int_0^{\theta_0} \pi(\theta \mid x) d\theta$ since it is not possible to have a negative proportion and $0 \leq \theta_0 \leq 1$.

A loss function as defined by Equation (3.23) (with $f_i(\theta) = k_i \mid \theta - \theta_0 \mid$), that is one that takes into account the distance between the decisions (d_0, d_1) and the true value of θ, may be considered:

$$L(d_i,\theta) = \begin{cases} 0 & \text{if } \theta \in \Theta_i, \\ k_i(\theta-\theta_0) & \text{if } \theta \notin \Theta_i \text{ and } \theta > \theta_0, \\ k_i(\theta_0-\theta) & \text{if } \theta \notin \Theta_i \text{ and } \theta < \theta_0, \end{cases} , \tag{6.16}$$

for $\Theta_0 = [0,\theta_0]$ and $\Theta_1 = (\theta_0, 1]$ and $k_i > 0$. This is an asymmetric linear loss that preferentially penalizes a wrong decision as the difference between the true value θ and θ_0 increases. The loss is equal to zero when the decision is correct.

Imagine, for instance, a scenario where a person is stopped because of a suspicion of driving under the influence of alcohol. A blood sample is taken to estimate the concentration of alcohol. One is interested in testing whether or not the alcohol concentration exceeds a given legal limit (e.g. 0.5 g/kg). More formally, the aim is to test $H_0 : \theta \leq 0.5$ g/kg versus $H_1 : \theta > 0.5$ g/kg. A false acceptance of the hypothesis H_0 (letting the person drive) would be less or more serious depending on the real – but unknown – concentration of alcohol in blood, and should therefore be penalized accordingly.

Given the stated loss function, the Bayesian posterior expected loss for decision d_0, that is accepting $H_0 : \theta \leq \theta_0$, is:

$$\begin{aligned} \bar{L}(d_0, \pi(\theta \mid x)) &= \int_{\Theta_1} k_0(\theta-\theta_0)\pi(\theta \mid x)d\theta \\ &= \int_{\Theta_1} k_0\theta\pi(\theta \mid x)d\theta - \int_{\Theta_1} k_0\theta_0\pi(\theta \mid x)d\theta. \end{aligned} \tag{6.17}$$

Similarly, the Bayesian posterior expected loss for decision d_1 is:

$$\bar{L}(d_1, \pi(\theta \mid x)) = \int_{\Theta_0} k_1\theta_0\pi(\theta \mid x)d\theta - \int_{\Theta_0} k_1\theta\pi(\theta \mid x)d\theta. \tag{6.18}$$

The null hypothesis should be rejected if

$$\bar{L}(d_0, \pi(\theta \mid x)) > \bar{L}(d_1, \pi(\theta \mid x)).$$

6.3.2 Proportion

Suppose $X \sim \text{Bin}(n,\theta)$ and the aim is to test $H_0 : \theta \leq \theta_0$ against $H_1 : \theta > \theta_0$. The parameter sets thus are $\Theta_0 = [0,\theta_0]$ and $\Theta_1 = (\theta_0, 1]$ and a conjugate beta prior density for the unknown parameter θ may be used, that is $\theta \sim \text{Be}(\alpha,\beta)$. A random sample $(x_1,\ldots,x_n)$ from this distribution is observed, with $y = \sum_{i=1}^n x_i$. Following the earlier discussion of Example 3.3.2, the posterior distribution $\pi(\theta \mid x)$ is of the form $\text{Be}(\alpha^* = \alpha + y, \beta^* = \beta + n - y)$. The posterior probability of H_0 is

$$\alpha_0 = P(\theta \leq \theta_0 \mid x) = \int_0^{\theta_0} \pi(\theta \mid x)d\theta,$$

which can be easily computed since $\pi(\theta \mid x)$ is in a closed form.

Next, consider a linear loss function as in Equation (6.16), and compute the Bayesian posterior expected loss for each decision. The first term of Equation (6.17) can be rewritten as

$$\begin{aligned}\int_{\Theta_1} k_0\theta\pi(\theta \mid x)d\theta &= k_0\int_{\theta_0}^{1}\frac{\theta\cdot\theta^{\alpha+y-1}(1-\theta)^{\beta+n-y-1}}{B(\alpha+y,\beta+n-y)}d\theta\\ &= \frac{k_0B(\alpha+y+1,\beta+n-y)}{B(\alpha+y,\beta+n-y)}\int_{\theta_0}^{1}\frac{\theta^{\alpha+y}(1-\theta)^{\beta+n-y-1}}{B(\alpha+y+1,\beta+n-y)}d\theta.\end{aligned}$$

The integral can be readily calculated since the density of a beta variable with parameters $\alpha^* = \alpha + y + 1$ and $\beta^* = \beta + n - y$ can be recognized. Simple algebra shows that $\frac{B(\alpha+y+1,\beta+n-y)}{B(\alpha+y,\beta+n-y)} = \frac{\alpha+y}{\alpha+\beta+n}$. Therefore the Bayesian posterior expected loss from (6.17) is equal to

$$\begin{aligned}\bar{L}(d_0,\pi(\theta \mid x)) = k_0\frac{\alpha+y}{\alpha+\beta+n}&P^{\pi}\left(\theta > \theta_0 \mid \alpha^* = \alpha+y+1, \beta^* = \beta+n-y\right)\\ &- k_0\theta_0P^{\pi}\left(\theta > \theta_0 \mid \alpha^* = \alpha+y, \beta^* = \beta+n-y\right)\end{aligned} \tag{6.19}$$

Similarly, the Bayesian expected loss from (6.18) for decision d_1 is:

$$\begin{aligned}\bar{L}(d_1,\pi(\theta \mid x)) &= k_1\theta_0P^{\pi}\left(\theta < \theta_0 \mid \alpha^* = \alpha+y, \beta^* = \beta+n-y\right)\\ &- k_1\frac{\alpha+y}{\alpha+\beta+n}P^{\pi}\left(\theta < \theta_0 \mid \alpha^* = \alpha+y+1, \beta^* = \beta+n-y\right).\end{aligned} \tag{6.20}$$

Example 6.3.1 *(Ion mobility spectrometer). Consider a detection apparatus, such as an ion mobility spectrometer (IMS). Such devices have been developed and marketed for the detection of substances such as explosives and illicit drugs. Target surfaces of interest such as banknotes, tablets, luggage, clothing and so on may be vacuumed and particles thus collected analyzed. When a detectable quantity of a target substance is present, then the IMS would indicate that event by a sonar signal. Suppose that a given device fails to detect the presence of a target substance (* e.g. *cocaine) in, for example, 10% of the screened surfaces which do bear the substance, but may present particular retention properties (matrix effects). A question that may be of interest is how the performance of such an apparatus compares to that of a given other device. In order to investigate this issue, a random sample of* $n = 50$ *surfaces known to be contaminated are screened with the alternative apparatus. Assume, for illustration, that among these samples,* $y = 4$ *yielded a negative result. Let* $X_i = 1$ *denote the event of a negative result,* $X_i = 0$ *the event of a positive result and* θ *the probability of interest, that is the probability of not detecting the target substance with the alternative* *apparatus. If $\theta \leq 0.1$, the performance of the alternative device is better than the IMS, therefore the competitive hypotheses of interest are $H_0 : \theta \leq 0.1$ versus $H_1 : \theta > 0.1$. Given a prior density of type Be$(1, 1)$, the posterior density becomes a Be$(5, 47)$. Approximate posterior probabilities of the null and the alternative hypotheses are found to be $\alpha_0 \simeq 0.59$ and $\alpha_1 \simeq 0.41$ (see Figure 6.2 (left))*[6]*. The posterior odds are $\alpha_0/\alpha_1 = 1.42$. Based on this calculation, one may slightly increase one's belief in the proposition according to which the alternative device gives a better performance. Given that the prior probabilities are $\pi_0 = 0.1$ $(P(\theta \leq 0.1))$ and $\pi_1 = 0.9$, the prior odds are $\pi_0/\pi_1 = 0.11$. The Bayes factor, which measures the change in the prior odds to the posterior odds equals* 12.9 (1.42/0.11). *This value suggests moderate support in favour of H_0. From a decision-theoretic point of view, by taking a linear loss as in (6.16) with $k_0 = k_1 = 1$, the Bayesian posterior expected losses may be computed from (6.19) and (6.20) and are equal to*

$$\bar{L}(d_0, \pi(\theta \mid x)) = 0.014$$
$$\bar{L}(d_1, \pi(\theta \mid x)) = 0.018.$$

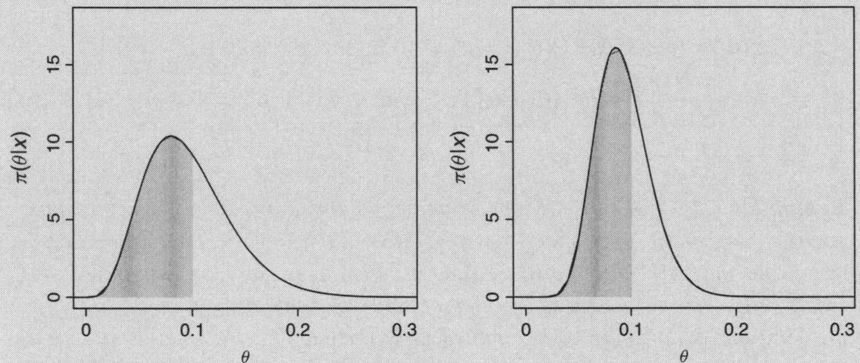

Figure 6.2 The posterior probabilities of the null hypothesis tested in Example 6.3.1. The probability of $H_0 : \theta \leq 0.1$ is represented by the grey shaded area. The figure on the left shows the Be$(5, 47)$ posterior density and $P(H_0 : \theta \leq 0.1) = 0.59$. The plot on the right depicts the Be$(12, 118)$ posterior density and $P(H_0 : \theta \leq 0.1) = 0.65$. Note that the right-hand end of the displayed horizontal scale is 0.3. The theoretical maximum value for this scale is 1 but the probability density function is very close to zero for $\theta > 0.3$ and so is not displayed.

[6]Note that α_0 and α_1 represent the posterior probabilities of the null and the alternative hypothesis, and must not be confused with parameter α of the Be(α, β) distribution.

The optimal decision is d_0 since it minimizes the loss. The null hypothesis is not rejected.

Alternatively, suppose that the prior knowledge could be modelled through a $Be(8, 72)$ prior distribution. The posterior probability of the null hypothesis would then become $\alpha_0 = 0.65$ (as shown in Figure 6.2 (right)), whereas the Bayes factor would equal 1.56, which means limited support in favour of H_0. The effect of the change of the value of the prior odds of 0.11 arising from a $Be(1, 1)$ prior distribution to a prior odds of 1.17 arising from a $Be(8, 72)$ distribution is to change the posterior odds from 1.42 to 1.83. The posterior density, $Be(12, 118)$, is in fact more peaked. The range of non-negligible parameter values for $Be(12, 118)$ is less than the corresponding range for $Be(5, 47)$. Thus $Be(12, 118)$ represents more information. It can be verified that the optimal decision with this more informative prior is still d_0.

Comparison of two proportions

Imagine that items are drawn from two populations of sizes n_1 and n_2, respectively, and that each item can be classified either as a success or a failure. The success rate in the first population is denoted θ_1. Similarly the success rate in the second population is denoted θ_2. This situation can be described in terms of a 2×2 table as shown in Table 6.6.

Table 6.6 Occurrences of successes (y) and failures ($n - y$) in two populations (population 1 and population 2) of size n_1 and n_2, respectively.

	Population 1	Population 2
Successes	y_1	y_2
Failures	$n_1 - y_1$	$n_2 - y_2$
Total	n_1	n_2

A typical request would be to determine the extent to which data support the hypothesis that the success probability in the first population is lower than that in the second population. Formally, the aim is to test $H_0 : \theta_1 \leq \theta_2$ against $H_1 : \theta_1 > \theta_2$, or, equivalently, $H_0 : \theta_1 - \theta_2 \leq 0$ against $H_1 : \theta_1 - \theta_2 > 0$. Consider, for example, the proportion of a given allele in two distinct populations. It is almost certain that there is a difference between the two populations, but it may not be known in which direction. A comparable problem may be encountered in the presence of two treatment groups where one seeks to know which of two treatments has a higher success rate. The Bayesian approach to such settings requires one, initially, to choose a prior density over the unit space for θ_1 and θ_2 that will be weighted by the likelihood function and then normalized to obtain a posterior distribution. There are several approaches to modelling prior beliefs about θ_1 and θ_2. They mainly differ depending on whether θ_1 and θ_2 are believed to be dependent or not. If they are judged to be independent, inference problems concerning θ_1 are independent of inference problems concerning θ_2.

For the proportions θ_1 and θ_2, assume two informative and independent beta priors, $\theta_i \sim \mathrm{Be}(\alpha_i, \beta_i)$, $i = 1, 2$. Further, suppose two independent samples of size n_1 and n_2 are available for θ_1 and θ_2, written $x_1 = (x_{11}, \ldots, x_{1n_1})$ and $x_2 = (x_{21}, \ldots, x_{2n_2})$. The posterior distribution of θ_i is still beta with parameters

$$\alpha_i^* = \alpha_i + y_i \quad \text{and} \quad \beta_i^* = \beta_i + n_i - y_i \quad \text{for} \quad y_i = \sum_{j=1}^{n_i} x_{ij},\ i = 1, 2.$$

It is often convenient to work with log-odds, defined by

$$\Lambda = \log\left(\frac{\theta_i}{1 - \theta_i}\right) \qquad\qquad i = 1, 2,$$

where logarithms are taken to base e. This quantity is close to a Fisher's z distribution (see Appendix B) with mean

$$E(\Lambda) \approx \log\left[(\alpha_i^* - 1/2)/(\beta_i^* - 1/2)\right], \qquad\qquad i = 1, 2,$$

and variance

$$Var(\Lambda) \approx \alpha_i^{*^{-1}} + \beta_i^{*^{-1}},\ i = 1, 2,$$

which is approximately Normal (Lee 2004). The comparison between the two proportions can be based on the quantity given by the difference of the two log-odds

$$\log\left(\frac{\theta_1}{1 - \theta_1}\right) - \log\left(\frac{\theta_2}{1 - \theta_2}\right) = \log\left(\frac{\theta_1(1 - \theta_2)}{\theta_2(1 - \theta_1)}\right), \qquad (6.21)$$

to give a log-odds ratio which is approximately Normal with mean

$$\mu_{12} = \log\left(\frac{(\alpha_1^* - 1/2)(\beta_2^* - 1/2)}{(\beta_1^* - 1/2)(\alpha_2^* - 1/2)}\right),$$

and variance

$$\tau_{12}^2 = \alpha_1^{*^{-1}} + \beta_1^{*^{-1}} + \alpha_2^{*^{-1}} + \beta_2^{*^{-1}}.$$

The approximation can be considered acceptable only if all the entries in the associated 2×2 table are at least 5. The log-odds ratio is a sensible measure of the degree to which the two populations differ: the null hypothesis is rejected if and only if the log-odds ratio is positive. Nevertheless nothing can be said about the probability of the null hypothesis being true since knowledge of the posterior distribution of the log-odds ratio does not imply knowledge of the posterior distribution of the difference $\theta = \theta_1 - \theta_2$. The exact distribution of the difference of two beta variables was established by Pham-Gia and Turkkan (1993) (a reproduction is available in Johnson *et al.* 1995 and is denoted the beta difference distribution $\text{BDI}(\alpha_1, \beta_1; \alpha_2, \beta_2)$). This distribution is closed under sampling: if independent samples of size n_1 and n_2 are available for θ_1 and θ_2, the posterior distribution still has the same form. The exact probability of the null hypothesis $H_0 : \theta_1 < \theta_2$ is

$$P(\theta_1 < \theta_2) = \sum_{x=\max(\alpha_2-\beta_1,0)}^{\alpha_2-1} \frac{\binom{\alpha_2+\beta_2-1}{x}\binom{\alpha_1+\beta_1-1}{\alpha_1+\alpha_2-1-x}}{\binom{\alpha_1+\beta_1+\alpha_2+\beta_2-2}{\alpha_1+\alpha_2-1}}$$

(see Altham 1969), and an expression for the more general hypothesis $P(\theta_1 \leq c\theta_2)$ $(0 < c \leq 1)$ is given in Weisberg (1972). However, for sufficiently large n_1 and n_2 – Pham-Gia and Turkkan (2003) suggest a number of observations larger than 30 – and θ_1 and θ_2 not too close to 0 or 1, the beta posterior distribution of each proportion can be approximated by a Normal distribution with the same mean and the same variance, that is $\pi(\theta_i \mid x_i) \approx N(\mu_i, \tau_i^2)$, $i = 1, 2$, where

$$\mu_i = \frac{\alpha_i^*}{\alpha_i^* + \beta_i^*}, \tag{6.22}$$

and

$$\tau_i^2 = \frac{\alpha_i^* \beta_i^*}{(\alpha_i^* + \beta_i^*)^2(\alpha_i^* + \beta_i^* + 1)}. \tag{6.23}$$

The priors and the samples being independent, the posteriors are also independent and the distribution of the difference $\theta = \theta_1 - \theta_2$ is approximately Normal with mean $\mu = \mu_1 - \mu_2$ and variance $\tau^2 = \tau_1^2 + \tau_2^2$. The posterior probability of the null hypothesis ($\theta < \theta_0 = 0$; i.e. $\theta_1 < \theta_2$) can be easily computed from (6.15) with a Normal distribution for π with $\theta = \theta_1 - \theta_2$ and θ_0 taken equal to 0.

Whenever non-informative independent priors seem to be more appropriate, a development from Howard (1998) is available for obtaining the posterior probability of the null hypothesis and a Normal approximation for several classes of vague priors. However, such independence can in some cases be rather unlikely and practitioners would more reasonably expect positive correlation between θ_1 and θ_2. That is to say, knowledge of a proportion may influence prior beliefs about the value of the second proportion.

For the purpose of illustration, consider a scenario where seized items (e.g. individual white tablets in a consignment of white tablets) are tested in different laboratories with each item able to be tested in only one place. Possible results are successes or failures and, as usual, they are considered to be a realization of a Bernoulli process with success probabilities θ_1 and θ_2. In such a situation, it is reasonable to assume that knowledge about θ_1 will lead to a revision of the prior expectation about θ_2. Howard (1998) proposes a prior of the form:

$$\pi(\theta_1, \theta_2) \propto e^{-(1/2)u^2} \theta_1^{\alpha_1 - 1}(1-\theta_1)^{\beta_1 - 1}\theta_2^{\alpha_2 - 1}(1-\theta_2)^{\beta_2 - 1}, \tag{6.24}$$

where

$$u = \frac{1}{\rho}\log\left(\frac{\theta_1(1-\theta_2)}{\theta_2(1-\theta_1)}\right).$$

The parameters $\alpha_1, \beta_1, \alpha_2, \beta_2$ reflect prior beliefs about the values of θ_1 and θ_2, and the parameter ρ indicates prior beliefs about the dependence between the two proportions. Note that Howard (1998) uses the notation σ to denote prior beliefs in the dependence between the two proportions. However, σ is used in this book to denote standard deviation so the notation ρ has been used here to denote dependence. Parameters should be understood as a measure of personal beliefs. Vague prior beliefs are modelled by $\alpha_1 = \beta_1 = \alpha_2 = \beta_2 = 1$. The contour plots of Howard's dependent prior, Equation (6.24), are shown in Figure 6.3 for different

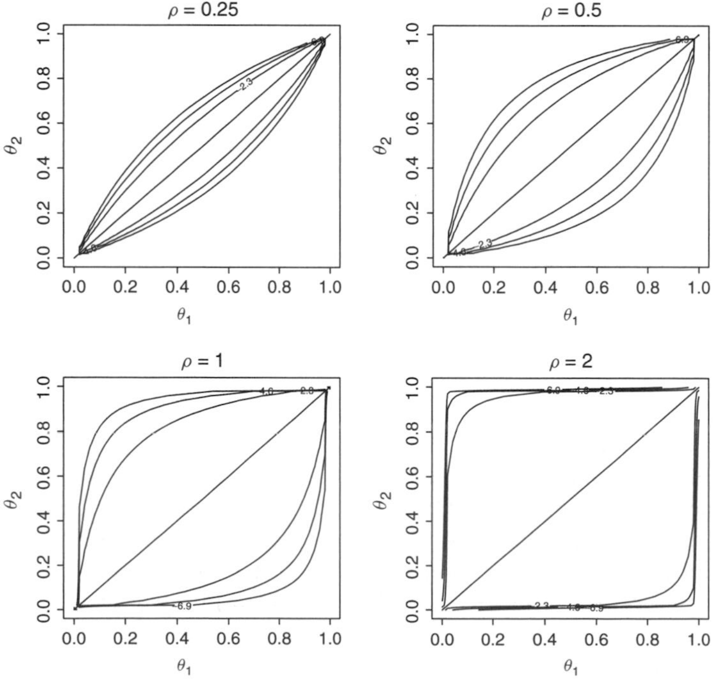

Figure 6.3 Contour plots of Howard's dependent prior distribution for $\alpha_1 = \beta_1 = \alpha_2 = \beta_2 = 1$ and different values of the dependence parameter $\rho = 0.25$, 0.5, 1, 2. Note that the line $\theta_1 = \theta_2$ is for illustration only and is not a contour of the distributions.

values of the dependence parameter ρ[7]. Notice that, the smaller the value of ρ, the more informative is the prior in the sense that the joint distribution is more peaked. Also, the bulk of the distribution is shifted towards the line $\theta_1 = \theta_2$ as ρ decreases. For independence, $\rho = \infty$. As an example of less information, consider, for instance, a dependent prior with $\rho = 1$. Such a choice means that if θ_1 takes, for example, the value 0.8, it is almost certain that θ_2 is in the range $(0.1, 0.99)$. If it is felt that this range of values is too wide (or even too narrow), than ρ can be adjusted to reflect prior beliefs more accurately.

The posterior density has the same functional form as Equation (6.24) with updated parameters

$$(\alpha_1 + y_1, \beta_1 + n_1 - y_1, \alpha_2 + y_2, \beta_2 + n_2 - y_2, \rho) .$$

Note that ρ is unchanged. A further complication might apply whenever, given the scenario discussed so far about items tested in different laboratories, there are also items tested in both laboratories (complete observations), as an addition to items tested in one and only one of the laboratories (marginal observations). In such a case and in general where the practitioner must deal with both marginals and complete observations, prior dependence can be modelled by using a Dirichlet-beta prior as suggested by Antelman (1972).

Example 6.3.2 *(Pitfalls of intuition). There are different practices among forensic scientists for expressing the value of scientific evidence. Some may sound more correct than others, yet some are fallacious. A particularly well known and common pitfall of intuition is, for instance, the so-called 'Prosecutor's fallacy' (Thompson and Schumann 1987). Imagine a researcher who is interested in comparing two populations of students and in detecting whether there is a difference in the proportion of correct answers to questions relating to the presentation of DNA evidence in a court of justice (Taroni and Aitken, 1998). In particular, the students were presented with different explanations of the value of DNA evidence (Taroni and Aitken 1999). The target students came from two European schools of forensic science; the first group is one which teaches probabilistic foundations of evidence interpretation and the second does not. With respect to a specific case in which a DNA expert's testimony constitutes a 'Probability (another match) error' (see Aitken and Taroni (2004) for further details), the two groups of students reacted in the following way: 10 out of 13 students from the school teaching evidence interpretation, correctly detected the fallacy that was committed by the DNA expert. In the second group of students, 6 among 32 tested individuals correctly*

[7]Contour plots are obtained using the routines in the Learn-Bayes R package (Albert 2007).

answered the question. Consider then testing the following pair of hypotheses: $H_0 : \theta_1 \leq \theta_2$ and $H_1 : \theta_1 > \theta_2$, where θ_1 and θ_2 represent the proportions of correct answers in the first and second population of students, respectively. The sample estimates are 10/13 = 0.77 and 6/32 = 0.19, respectively. Assume for the first population a Be(3.5, 2) density for describing the prior beliefs on θ_1, and for the second population a Be(3.5, 3.5) density for describing prior beliefs on θ_2. The two prior densities reflect a priori confidence in a higher probability of success among students having a stronger background on evidence interpretation (Figure 6.4).

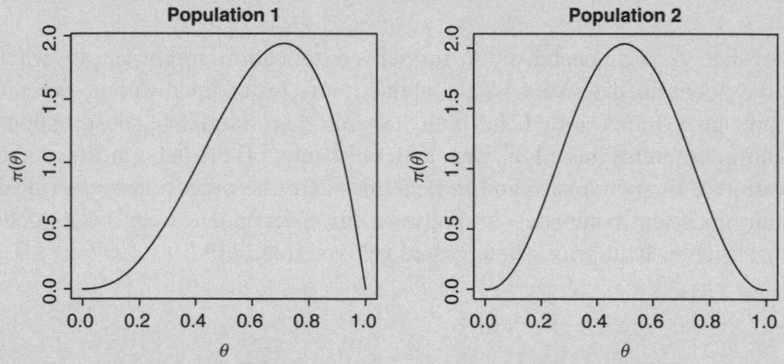

Figure 6.4 Prior distributions of the success rate θ in Example 6.3.2 for population 1 (Be(3.5, 2) on the left) and population 2 (Be(3.5, 3.5) on the right).

The distribution of the log-odds ratio (lOR), Equation (6.21), is approximately $N(2.23, 0.41)$ and $P(lOR > 0) = 0.9997$. The null hypothesis is rejected in favour of the alternative hypothesis. Thus, it is accepted that the first group gives more correct answers than the second group. A plot of the posterior distribution of the log-odds ratio is given in Figure 6.5. Alternatively, the beta posterior can be approximated by a Normal distribution. Equations (6.22) and (6.23) give $\mu = \mu_1 - \mu_2 = 0.49$ and $\tau^2 = \tau_1^2 + \tau_2^2 = 0.015$, and

$$P(\theta \leq 0 \mid \theta \sim N(0.49, 0.015)) \approx 0.$$

Therefore, the null hypothesis $H_0 : \theta_1 - \theta_2 \leq 0$ is rejected.

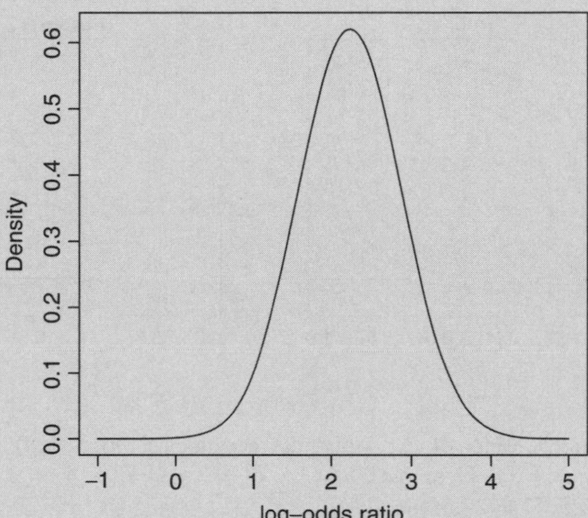

Figure 6.5 Posterior Normal distribution of the log-odds ratio (Equation (6.21)) in Example 6.3.2. for the difference in responses for the two groups of students. A value of 0 for the log-odds ratio corresponds to no difference in the responses.

The same conclusions can be obtained when assuming the dependent prior density given by Equation (6.24), with parameters $\alpha_1 = 3.5$, $\beta_1 = 2$, $\alpha_2 = 3.5$, $\beta_2 = 3.5$ *fixed as above and parameter* $\rho = 1$ *(knowledge of answers from one group of students will influence belief about the answers from the other group of students). The contour plot of the prior, depicted in Figure 6.6 (left), reflects a weak dependence a priori between the two proportions. The posterior probability of the null hypothesis* $H_0 : \theta_1 \leq \theta_2$ *can be computed as in Albert (2007, p. 68). The posterior probability of the null hypothesis is* 0.0012, *as confirmed by the contour plot of the posterior distribution that is shifted almost entirely in the lower triangle (Figure 6.6 (right)). The posterior probability is necessarily sensitive to the prior choice of* ρ. *Table 6.7 shows this for different values of the dependence parameter. However, in all cases there is very strong evidence that there is a higher probability of success amongst the students that were taught evidence interpretation. Similar arguments can be applied in the contexts of drug sampling, pirated CDs and computer pornographic images.*

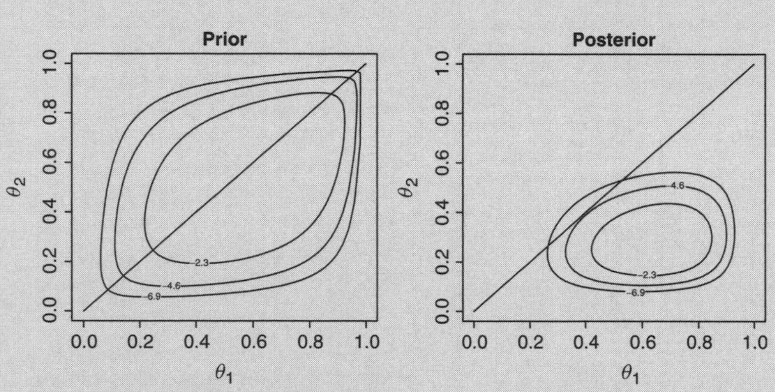

Figure 6.6 Contour plots of the Howard's dependence prior (left) and posterior (right) in Example 6.3.2. Note that $\alpha_1 = 3.5$, $\beta_1 = 2$, $\alpha_2 = 3.5$, $\beta_2 = 3.5$; $\alpha_1^* = 13.5$, $\beta_1^* = 5, \alpha_2^* = 9.5, \beta_2^* = 29.5$; $\rho = 1$.

Table 6.7 Posterior probability of the null hypothesis, H_0 (the proportion of correct answers in the first group is less than the proportion of correct answers in the second group), for different values of ρ, the prior belief in the dependence between the two proportions. The accuracy of the calculations of the probability of H_0 depends on the number of grid points used for the construction of the contour plot. The R code for this example used 10000 grid points. Repetition of the algorithm from the same starting point will give small differences in the output.

Parameter ρ	Probability of H_0
0.5	0.0104
1	0.0012
2	0.0003

6.3.3 A note on multinomial cases (k categories)

Suppose one has a multinomial population with k categories and a natural ordering on the categories. For example, medical patients being treated by a particular method may be rated in terms of their response as being much improved, somewhat improved, the same, and so on (Weisberg 1972). Let $\theta_1, \ldots, \theta_k$ be the probabilities associated with category j, $j = 1, \ldots, k$, for the multinomial population. If $y = (y_1, \ldots, y_k)$ is the vector of the counts of the number of observations that fall in each category, then

$$f(y_1, y_2, \ldots, y_k \mid \theta_1, \ldots, \theta_k) \propto \prod_{j=1}^{k} \theta_j^{y_j}, \qquad \sum_{j=1}^{k} \theta_j = 1.$$

As it was outlined in Section 4.2.5, the conjugate prior distribution for a multinomial population with k categories is the Dirichlet distribution, $(\theta_1, \ldots, \theta_k) \sim D_k\,(\alpha_1, \ldots, \alpha_k)$, that is

$$f\,(\theta_1, \ldots, \theta_k) \propto \prod_{j=1}^{k} \theta_j^{\alpha_j - 1}.$$

The posterior distribution is $D_k\,(\alpha_1 + y_1, \ldots, \alpha_k + y_k)$.

It is of interest to evaluate the difference between the proportion of observations falling in different categories, say category 1 and 2, and therefore the quantity $\theta_1 - \theta_2$. For example, it could be of interest knowing the difference between people who were rated as much improved and those who were rated as somewhat improved. Values from the posterior distribution of $\theta_1 - \theta_2$ can be simulated drawing several values $(\theta_1, \ldots, \theta_k)$ from the posterior Dirichlet distribution and computing $\theta_1 - \theta_2$ for each draw.

Another point of interest is the determination as to whether there is a difference between two populations of patients, for example patients treated by two different methods. Let θ_{1j} and θ_{2j} be the probabilities associated with category j, $j = 1, \ldots, k$, for the two populations, and $y_i = (y_{i1}, y_{i2}, \ldots, y_{ik})$ be the vector of the counts of the number of observations that fall in each category in population i, $i = 1, 2$. For each population, it will be assumed $(\theta_{i1}, \ldots, \theta_{ik}) \sim D_k\,(\alpha_{i1}, \ldots, \alpha_{ik})$. The posterior distribution is $D_k\,(\alpha_{i1} + y_{i1}, \ldots, \alpha_{ik} + y_{ik})$. Let proposition $H_0 : P_1 < P_2$ 'Population 1 has a lower response than population 2'. The derivation of the posterior probability of $(P_1 < P_2)$ is not straightforward, and can be performed through ad hoc algorithms (see for example the proposal of Weisberg (1972)). However, proposition $P_1 < P_2$ is equivalent to

$$\sum_{j=1}^{m} \theta_{1j} \geq \sum_{j=1}^{m} \theta_{2j}, \qquad m = 1, \ldots, k - 1. \tag{6.25}$$

Several draws can be taken from the posterior Dirichlet distributions and the proportion of time on which (6.25) is true is computed.

***Example 6.3.3** (Detection of nandrolone metabolites – continued from Example 3.3.6). Nandrolone is an efficient drug to increase muscle mass. This anabolic steroid is administered via intramuscular injections and its major metabolites, 19-norandrosterone (NA) and 19-noretiocholoanolone (NE) can be detected in urine. The level of nandrolone (NA-metabolite), grouped in five ordered categories, in urine samples coming from samples of 358 amateur football players and 126 students are shown in Table 6.8. The prior parameters* $\alpha_{i1}, \ldots, \alpha_{ik}$*;* $i = 1, 2$*;* $k = 5$*, are taken to be all equal to 1 for purposes of illustration. Given prior knowledge, values for the prior parameters can be determined from subjective beliefs about the prior expectations and variances. The stronger the belief, the higher the value of* α_{i0}*, where* $\alpha_{i0} = \sum_{j=1}^{k} \alpha_{ij}, i = 1, 2$*. From the expression for the variance of a Dirichlet distribution given in Appendix B it can be seen that the higher the value of* α_{i0}*, the smaller the value of the variance and hence the greater the precision attached to the estimates. The individual* α_{ij} *values can then be chosen to reflect subjective beliefs about the relative values of the proportions* $\theta_{i1}, \ldots, \theta_{ik}$*, subject to* $\sum_{i=1}^{k} \alpha_{ij} = \alpha_{i0}$*.*

Table 6.8 Results, in five levels $k = 1, \ldots, 5$, of the measurements of NA (ng/ml) in a sample of amateur football players ($i = 1$) and a sample of students ($i = 2$).

Population	Level				
	0.0 – 0.2 ($k = 1$)	0.2 – 0.5 ($k = 2$)	0.5 – 1.0 ($k = 3$)	1.0 – 2.0 ($k = 4$)	2.0 – 3.0 ($k = 5$)
Football players ($i = 1$)	336	9	5	5	3
Students ($i = 2$)	126	0	0	0	0

Let P_1 *denote the response from the football players and* P_2 *the response from the students. Given values for the prior parameters and the data, simulations from the resultant posterior Dirichlet distribution enable probabilities for the proposition* $P_1 < P_2$ *to be computed. For* $m = 1, \ldots, 4$*, these probabilities are* $(0.97, 0.89, 0.83, 0.70)$*; see also Figure 6.7. Since all the students fall into the first category, that of the lowest level of nandrolone, the probability is very high that the football players' response (i.e. the football players' levels of nandrolone) is higher than that of the students (since no student has a level of nandrolone in a higher category than that of a football player).*

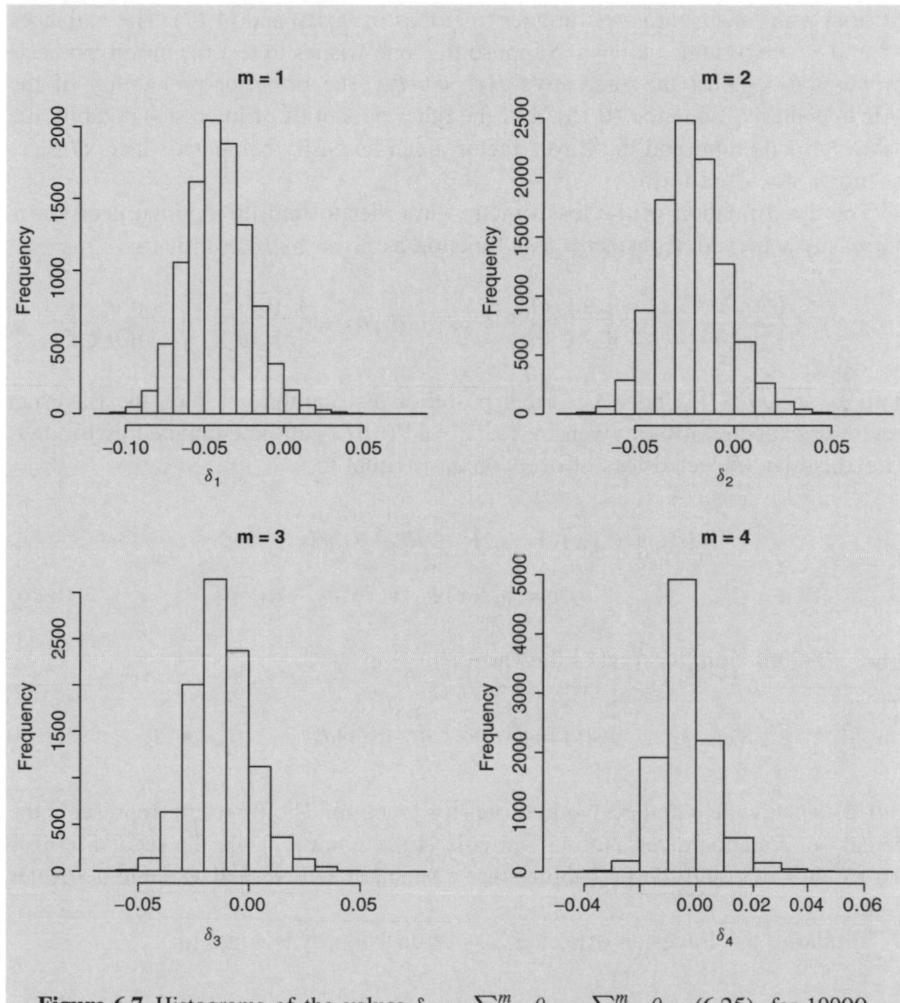

Figure 6.7 Histograms of the values $\delta_m = \sum_{j=1}^{m} \theta_{1j} - \sum_{j=1}^{m} \theta_{2j}$, (6.25), for 10000 simulations from the posterior Dirichlet distribution for the detection of nandrolone metabolite. Example, with $\alpha_{i1}, \ldots, \alpha_{ik}$ $(i = 1, 2;\ k = 5)$ all equal to 1.

6.3.4 Mean

Consider a continuous measurement X with Normal density and known variance, $X \sim N(\theta, \sigma^2)$, and assume for the unknown mean μ a Normal density, $\theta \sim N(\mu, \tau^2)$. Distributions are conjugate, and the posterior distribution $\pi(\theta \mid x)$ is

Normal with mean $\mu(x)$ and variance $\tau^2(x)$, as in (4.16) and (4.17). The variances σ^2 and τ^2 are assumed known. Suppose that one wishes to test the null hypothesis $H_0 : \theta \leq \theta_0$ against the alternative $H_1 : \theta > \theta_0$. The posterior probability of the null hypothesis, Equation (6.15), and the other quantities of interest – notably, the posterior odds ratio and the Bayes factor – can be easily computed since $\pi(\theta \mid x)$ is known in closed form.

The determination of the loss function in order to find the optimal decision is not easily achieved. Consider a loss function as given by (6.16), that is

$$L(d_0, \theta) = \begin{cases} 0 & \text{if } \theta \in \Theta_0 \\ k_0(\theta - \theta_0) & \text{if } \theta \in \Theta_1 \end{cases} \quad ; \quad L(d_1, \theta) = \begin{cases} 0 & \text{if } \theta \in \Theta_1 \\ k_1(\theta_0 - \theta) & \text{if } \theta \in \Theta_0 \end{cases},$$

with k_0 and $k_1 > 0$. Under a Normal posterior distribution $\pi(\theta \mid x)$, the Bayesian posterior expected losses given by (6.17) and (6.18) can be computed as follows. The Bayesian expected loss of decision d_0 is equal to

$$\begin{aligned} \bar{L}(d_0, \pi(\theta \mid x)) &= k_0 \int_{\theta > \theta_0} (\theta - \theta_0)\pi(\theta \mid x) d\theta \\ &= k_0 \tau(x) \Psi_0 \left[\tau(x)\left(\theta_0 - \mu(x)\right)\right], \end{aligned} \tag{6.26}$$

where the function $\Psi_0(\cdot)$ takes the form

$$\Psi_0(t) = \phi(t) - t \int_t^{\infty} \phi(s) ds,$$

and ϕ denotes the standard Normal density function. The first term represents the density of a standardized Normal variable at the point t, while the second term is the product of t and the probability that a standardized Normal variable is greater than t.

Similarly, the Bayesian expected loss of decision d_1 is equal to:

$$\begin{aligned} \bar{L}(d_1, \pi(\theta \mid x)) &= k_1 \int_{\theta \leq \theta_0} (\theta_0 - \theta)\pi(\theta \mid x) d\theta \\ &= k_1 \tau(x) \Psi_1 \left[\tau(x)\left(\theta_0 - \mu(x)\right)\right], \end{aligned} \tag{6.27}$$

where the function $\Psi_1(\cdot)$ takes the form

$$\Psi_1(t) = \phi(t) + t \int_0^{t} \phi(s) ds.$$

The decision d_0 is therefore optimal when

$$k_0 \Psi_0 \left[\tau(x)\left(\theta_0 - \mu(x)\right)\right] < k_1 \Psi_1 \left[\tau(x)\left(\theta_0 - \mu(x)\right)\right]. \tag{6.28}$$

Notice that if $k_0 = k_1$, *i.e.*, the loss is symmetric, it will be sufficient to observe whether θ_0 is greater or lower than the posterior mean $\mu(x)$. In particular, it turns out that decision d_0 is optimal if $\theta_0 > \mu(x)$. The loss function is symmetric so Equation (6.28) reduces to

$$-t \int_t^\infty \phi(s)ds < t \int_0^t \phi(s)ds,$$

which is true if and only if $t > 0$, that is $\theta_0 > \mu(x)$. Further details are available from Bernardo and Smith (2000, p. 396).

Example 6.3.4 *(Alcohol concentration in blood – continued). Consider again the scenario described earlier in Example 4.4.1. A person suspected of driving under the influence of alcohol is stopped. A sample taken from that individual is submitted to a forensic laboratory. The sample means provided by the toxicology laboratory are as follows:* 0.5191 *(using the HS method) and* 0.5093 *(using the ID method). Recalling that the prior density was* $N(1, 0.3^2)$*, the sequential use of Bayes' theorem gives a posterior density* $N(0.5183, 0.0145^2)$*. One is interested in knowing whether the estimated quantity of alcohol is greater than the legal threshold of* $0.5\,g/kg$*. Thus, one determines the probability of this event. To do so, the null hypothesis* $H_0 : \theta \leq 0.5$ *is tested against the alternative hypothesis* $H_1 : \theta > 0.5$.

The prior probabilities of the null and the alternative hypotheses are taken to be $\pi_0 = 0.05$ *and* $\pi_1 = 0.95$*, so the prior odds are* $\pi_0/\pi_1 \simeq 0.05$*. This indicates that the null hypothesis is, a priori, unlikely. The posterior probabilities of the null and the alternative hypothesis are* $\alpha_0 = 0.1034$ *and* $\alpha_1 = 0.8966$ *(Figure 6.8). The corresponding posterior odds are* $\alpha_0/\alpha_1 = 0.1153$ *and the Bayes factor* $BF = 2.3$*. This value suggests limited evidence in favour of* H_0 *that the alcohol level is* $< 0.5\,g/kg$*. Notice that the probability of the null hypothesis has increased from* 0.05 *to* 0.1034.

Consider further an asymmetric linear loss function as in Equation (6.16), with $k_0 = 2$ *and* $k_1 = 1$*. Following the discussion presented in Section 3.4.2, the weight* k_0 *is chosen so as to be greater than* k_1 *in order to express the understanding that an error of second type (accepting* H_0 *that* $\theta < 0.5$ *when it is false) is more serious. Bayesian expected losses are computed as in (6.24) and (6.25) and give*

$$\bar{L}(d_0, \pi(\theta \mid x)) = 0.011,$$
$$\bar{L}(d_1, \pi(\theta \mid x)) = 0.005.$$

Accordingly, the optimal decision is d_1*, since its expected loss is the smaller, and the null hypothesis is rejected. The decision is made that the true quantity*

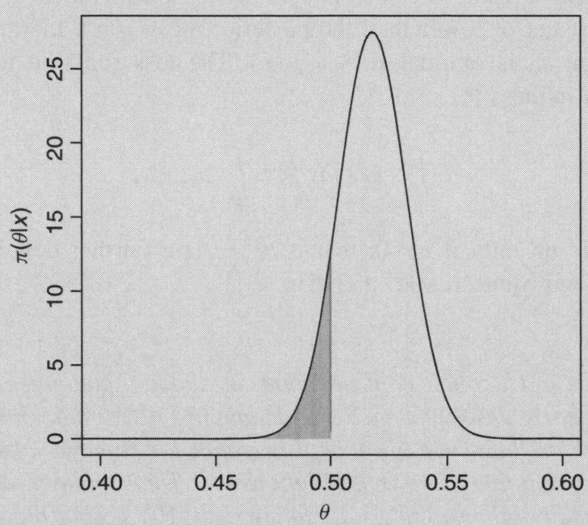

Figure 6.8 Posterior distribution of the alcohol concentration in Example 6.3.4. The posterior probability of the null hypothesis ($\theta < 0.5$) is highlighted by the grey shaded area.

of alcohol is greater than the legal threshold of 0.5 g/kg. *This result has been reached despite the evidence giving support to* H_0. *This illustrates the importance of the choice of the prior distribution and of the loss function.*

Compare this result with that in Example 4.4.1. In both examples the posterior probability for $\theta > 0.5$ *is greater than 0.5. However, in this example, where testing of a hypothesis is the motivation, the evidence supports* $\theta < 0.5$. *This paradoxical result arises because the prior density has mean 1 whereas the sample means are 0.5191 (HS) and 0.5093 (ID). Evidential support for a particular hypothesis may still mean that the alternative hypothesis has a higher probability of being true.*

Comparison of two means

Consider two groups of observations, such as the continuous elliptical Fourier descriptors of mandibular outline in the lateral view for females and males (Schmittbuhl *et al.* 2002, 2001). The probability distribution in each group is assumed Normal with known variance, $X_i \sim N(\theta_i, \sigma_i^2)$, $i = 1, 2$. Two samples, one for females and one for males, are taken independently of each other, $x_1 = (x_{11}, \ldots, x_{1n_1})$ and $x_2 = (x_{21}, \ldots, x_{2n_2})$, and it is of interest to test whether the first group has a lower mean than the second group, that is to test the null hypothesis $H_0 : \theta_1 - \theta_2 \leq 0$ against the alternative hypothesis $H_1 : \theta_1 - \theta_2 > 0$. Independent Normal prior densities are assumed for both means, $\theta_i \sim N(\mu_i, \tau_i^2)$, $i = 1, 2$. The posterior distributions $\pi(\theta_1 \mid x_1)$ and $\pi(\theta_2 \mid x_2)$ of the two group means are Normal, with means and variances obtained using the updating formulae, Equations (4.16) and (4.17), that is $\pi(\theta_i \mid x_i) = \mathrm{N}\left(\mu_i(x_i), \tau_i^2(x_i)\right)$, $i = 1, 2$. The priors and the samples are independent, so the posteriors are also independent and the difference between the two means $\theta = \theta_1 - \theta_2$ is Normally distributed with mean equal to the difference of the means, and the variance equal to the sum of the variances, that is $\pi\ (\theta_1 - \theta_2 \mid x_1, x_2) = \mathrm{N}\left(\mu_1(x_1) - \mu_2(x_2), \tau_1^2(x_1) + \tau_2^2(x_2)\right)$.

This is the most favourable scenario since in most experimental situations variances are unknown, and for inference they may be equal or unequal. Consider the situation where the variances σ_1^2 and σ_2^2, although unknown, can be assumed to be equal, say $\sigma_1^2 = \sigma_2^2 = \sigma^2$. Suppose further that independent local uniform priors are assumed for θ_1, θ_2 and $\log \sigma$. The difference $\theta = \theta_1 - \theta_2$ has, a posteriori, a non-central *Student t* distribution

$$\mathrm{St}\left(n_1 + n_2 - 2, \bar{x}_1 - \bar{x}_2, s^2\left(\frac{1}{n_1} + \frac{1}{n_2}\right)\right),$$

where $\bar{x}_i = \sum_{j=1}^{n_i} x_{ij}/n_i$ denotes the sample mean and

$$s^2 = \frac{\sum_{i=1}^{2}\sum_{j=1}^{n_i}\left(x_{ij} - \bar{x}_i\right)^2}{n_1 + n_2 - 2},$$

the pooled variance estimate (Box and Tiao 1973). Equivalently,

$$\frac{(\theta_1 - \theta_2) - (\bar{x}_1 - \bar{x}_2)}{s\sqrt{\frac{1}{n_1} + \frac{1}{n_2}}} \sim \mathrm{St}\,(n_1 + n_2 - 2)\,, \tag{6.29}$$

the central *Student t* distribution with $n_1 + n_2 - 2$ degrees of freedom.

Example 6.3.5 *(Identity card analysis). Sophisticated measuring techniques are available to estimate the width of specific characters (letters or numbers) printed on an identity card. A forensic examiner would like to compare width measurements made on a particular character, say the vertical line of a number* 1 *(e.g. one of the characters of the serial number of an identity card), printed on an authentic identity card and on an identity card of questioned authenticity, respectively. The mean, θ_1, of the measurements made on an original card may be lower than the mean, θ_2, of a card of questioned authenticity. It thus may be of interest to test hypothesis $H_0 : \theta_1 \leq \theta_2$ versus $H_1 : \theta_1 > \theta_2$. The prior probabilities for H_0 and H_1 are taken to be equal, $\pi_0 = \pi_1 = 0.5$.*

Suppose that $n_1 = 10$ *observations are obtained from the authentic identity card and* $n_2 = 10$ *observations from the questioned identity card. The sample means are equal to* $\bar{x}_1 = 201.854$ *and* $\bar{x}_2 = 202.182$ microns *for the first and the second group of observations respectively. The sample variance is* $s^2 = 31.47$*. The posterior probability of the null hypothesis is computed according to Equation (6.29) with local uniform priors and is equal to* $\alpha_0 \simeq 0.55$ *and hence* $\alpha_1 \simeq 0.45$*. The resulting posterior odds ratio is* $\alpha_0/\alpha_1 = 1.23\,(0.5513/0.4487)$*. This value is equal to the Bayes factor, BF, since uniform priors have been chosen. The data thus suggest slight support of the null hypothesis.*

Assuming a symmetric '0–k_i*' loss, the optimal decision is* d_0*, decide that the card is authentic. In fact,*

$$BF = 1.23 > \frac{k_0}{k_1}\frac{\pi_1}{\pi_0} = 1.$$

6.4 TWO-SIDED TESTING

6.4.1 Background

Practitioners often face the problem of testing a simple null hypothesis $H_0 : \theta = \theta_0$ versus a composite alternative hypothesis $H_1 : \theta \neq \theta_0$. Such a practice is known as a test of a *point null hypothesis* (against a *two-sided alternative hypothesis*) and is characterized by the fact that even though the parameter θ may be continuous, Θ_0 is simple, that is $\Theta_0 = \{\theta_0\}$ and Θ_1 is its complement ($\theta \neq \theta_0$).

For the moment consider the determination of an interval, known as a *credible interval*, for the parameter θ. Assume a non-informative prior density for θ, $\pi(\theta) \propto$ constant, so that there is no particular reason to believe that θ takes value θ_0 rather than any value θ_1 in the neighbourhood of θ_0. Under this assumption, Lindley (1965) proposed the development of a credible interval for the parameter θ at some given level of credibility α as a test of a point null hypothesis. In this way, if θ_0 is not included in the $(100 - \alpha)\%$ credible interval, there is evidence against the null hypothesis and it is rejected. Examination of the interval will provide a good indication of the magnitude of the distance between the true value of θ and θ_0. Conversely, if θ_0 is included in the interval, hypothesis H_0 is not to be rejected. This procedure is only partially Bayesian since it still envisages the choice of a level of credibility. In addition, it does not assign a probability value for the alternative hypothesis. An argument against the use of confidence regions is given by Berger and Delampady (1987). They write that a point can be outside the interval and not be so strongly contraindicated by the data because the likelihood of a point that is outside the interval is often not much smaller than the average likelihood of the parameter θ in the credible interval. Moreover, they argue that only a measure such as the Bayes factor can indicate the strength of evidence against a particular point θ_0, while the credible interval indicates the magnitude of the possible discrepancy.

Hereafter, a full Bayesian procedure is developed. A test of the null hypothesis $\theta = \theta_0$ is to be developed. This is a different approach from consideration of a credible interval. In particular, a continuous prior density on the parameter space Θ cannot be assumed because that would result in a probability of zero for the hypothesis $H_0 : \theta = \theta_0$ and hence a posterior probability of zero. Thus, there would be no possibility of learning from experience. To avoid this problem, the prior density is built as a mixture of a discrete component that assigns to the null hypothesis a prior mass of π_0, and a continuous component that spreads the remaining mass $\pi_1 = 1 - \pi_0$ over Θ_1 according to the probability density $\pi_{H_1}(\theta)$. The posterior probability α_0 of the null hypothesis is

$$\alpha_0 = \frac{\pi_0 f(x \mid \theta_0)}{\pi_0 f(x \mid \theta_0) + \pi_1 \int_{\Theta_1} f(x \mid \theta)\pi_{H_1}(\theta)d\theta},$$

and, similarly, the posterior probability α_1 of the alternative hypothesis is

$$\alpha_1 = \frac{\pi_1 \int_{\Theta_1} f(x \mid \theta)\pi_{H_1}(\theta)d\theta}{\pi_0 f(x \mid \theta_0) + \pi_1 \int_{\Theta_1} f(x \mid \theta)\pi_{H_1}(\theta)d\theta}.$$

It follows that the posterior odds are:

$$\frac{\alpha_0}{\alpha_1} = \frac{\pi_0 f(x \mid \theta_0)}{\pi_1 \int_{\Theta_1} f(x \mid \theta)\pi_{H_1}(\theta)d\theta}, \tag{6.30}$$

and the Bayes factor, which allows a judgement as to how well the data support a precise hypothesis H_0, is

$$BF = \frac{f(x \mid \theta_0)}{l_1(x)}, \tag{6.31}$$

where

$$l_1(x) = \int_{\Theta_1} f(x \mid \theta)\pi_{H_1}(\theta)d\theta$$

denotes the weighted likelihood of θ under H_1. The posterior probability of the null hypothesis can also be written as:

$$\alpha_0 = \left[1 + \frac{\pi_1 \int_{\Theta_1} f(x \mid \theta)\pi_{H_1}(\theta)d\theta}{\pi_0 f(x \mid \theta_0)}\right]^{-1} = \left[1 + \frac{\pi_1}{\pi_0}\frac{1}{BF}\right]^{-1}. \tag{6.32}$$

The computation of the Bayes factor requires the specification of the prior density for θ conditional on H_1, $\pi_{H_1}(\theta)$, while the specification of a prior density function for θ, conditional on H_0 is not necessary since H_0 specifies θ as $\theta = \theta_0$ precisely. If $\pi_{H_1}(\theta)$ is given, the scientist can then compute the posterior probability α_0 of H_0. If $\pi_0 = \pi_1 = 1/2$,

$$\alpha_0 = \left(1 + \frac{1}{BF}\right)^{-1} = \left(\frac{BF}{1+BF}\right) \simeq BF$$

for small BF (see Example 6.4.2).

Some authors argue that there should be no point-null hypothesis testing. Though it is very commonly suggested in practice, it will rarely be the case that one entertains the hypothesis that θ takes exactly value θ_0. Consider the following example (Lee 2004, p. 124). A chemical is analyzed with respect to some aspect, described by a parameter θ, of its reaction with a known chemical. It is of interest to test whether or not an unknown chemical is a specific compound, with a reaction strength θ_0 known to an absolute accuracy of ϵ. In such circumstances, the precise null hypothesis is better represented as the interval null hypothesis $H_0 : \theta \in \Theta_0 = (\theta_0 - \epsilon, \theta_0 + \epsilon)$, where $\epsilon > 0$ is a constant such that any value in Θ_0 can be considered indistinguishable from θ_0. So, given a continuous prior density $\pi(\theta)$ for parameter θ, the prior probabilities π_0 and π_1, and the conditional densities $\pi_{H_0}(\theta)$ and $\pi_{H_1}(\theta)$ of θ, are defined as in Section 6.2.1, Equations (6.4), (6.8) and (6.9). Typically, $\pi_{H_0}(\theta)$ will be a sharply peaked density near θ_0, while $\pi_{H_1}(\theta)$ will be rather diffuse. One can work with the interval as the null hypothesis, or approximate the interval hypothesis with the point null hypothesis. The latter option may be preferable, because of the difficulties in specifying ϵ and $\pi_{H_0}(\theta)$. As a second example, consider testing the null hypothesis H_0 : *females and males of a species are the same in terms of a given characteristic, say* A. This is not meant as a precise hypothesis. There may be a difference, although negligible. The point null hypothesis is a good approximation to many realistic scenarios. Ideally, for a good approximation, the posterior probabilities obtained with and without approximation should be nearly equal, a case that is guaranteed when the likelihood function is nearly constant on Θ_0. An example for a Normal population with known variance is discussed in Berger (1988) and Lee (2004), where a bound for the variation of the likelihood over the range of indistinguishable values in Θ_0 is computed.

At this point, a cautionary note may be useful on the point that Bayesian and frequentist methods may lead to radically different answers. Examples may be found, for instance, where frequentist methods would reject a null hypothesis at a significance level of $\alpha = 0.05$ whereas from a Bayesian point of view, the posterior probability of the null hypothesis is substantial, that is to say that the null hypothesis should not be rejected. It might be argued that the discrepancy between p-values and posterior probabilities in testing precise hypotheses is sensitive to the chosen prior density, but it can be shown that it is a general phenomenon. Berger and Delampady (1987), for example, present settings in which lower bounds on the posterior probability are computed for a Normal population and a binomial population over wide classes of prior densities and find that even these lower bounds are much larger than the p-value. In Lee (2004), bounds are computed for a Normal population that does not depend on the prior distribution. From a decision-theoretic point of view, however, diverging results may be regarded as irrelevant, as noted by Berger and Delampady (1987, p. 330):

> A frequent attempt to dismiss the conflict between p-values and Bayes factors is to argue that neither is relevant: one should instead quantify losses in incorrectly accepting or rejecting H_0 and perform a decision analysis.

Consider the '0–k_i' loss function, summarized in Table 6.4. According to Equation (6.14), the optimal decision is to reject the null hypothesis when

$$BF = \frac{f(x \mid \theta_0)}{\int_{\Theta_1} f(x \mid \theta)\pi_{H_1}(\theta)d\theta} < \frac{k_0}{k_1}\frac{\pi_1}{\pi_0}. \tag{6.33}$$

6.4.2 Proportion

The problem considered in this subsection is that of testing a binomial parameter. Suppose, as usual, that y represents the number of successes in n trials, with θ being the probability of success. A binomial variable $Y \sim \text{Bin}(n, \theta)$ can thus be defined. Suppose further that it is of interest to test $H_0 : \theta = \theta_0$ against $H_1 : \theta \neq \theta_0$, where $0 < \theta_0 < 1$. Assuming a uniform prior density for the proportion θ, $\pi(\theta) = 1$, the weighted likelihood $l_1(y)$ under H_1 is equal to:

$$l_1(y) = \int_{\Theta_1} \binom{n}{y} \theta^y (1-\theta)^{n-y} d\theta = \binom{n}{y} \mathrm{B}\,(y+1, n-y+1),$$

where $\Theta = (0, 1)$. Therefore, the Bayes factor turns out to be equal to

$$BF = \frac{\theta_0^y (1-\theta_0)^{n-y}}{\mathrm{B}(y+1, n-y+1)}. \tag{6.34}$$

If the prior probabilities π_0 and π_1 of the null and alternative hypotheses are available, then α_0, the posterior probability of the null hypothesis, is given by

$$\begin{aligned} \alpha_0 &= \left[1 + \frac{\pi_1}{\pi_0}\frac{1}{BF}\right]^{-1} \\ &= \left[1 + \frac{\pi_1}{\pi_0}\frac{\mathrm{B}(y+1, n-y+1)}{\theta_0^y (1-\theta_0)^{n-y}}\right]^{-1}. \end{aligned} \tag{6.35}$$

For cases where prior knowledge is available and allows the specification of a more informative Be(α, β) distribution, the weighted likelihood $l_1(y)$ with $\Theta_1 = (0, 1)$ is

$$l_1(y) = \int_{\Theta_1} \binom{n}{y} \frac{\theta^{\alpha+y-1}(1-\theta)^{\beta+n-y-1}}{B(\alpha, \beta)} d\theta = \binom{n}{y} \frac{\mathrm{B}\,(\alpha+y, \beta+n-y)}{\mathrm{B}(\alpha, \beta)},$$

and the Bayes factor is

$$BF = \theta_0^y (1-\theta_0)^{n-y} \frac{\mathrm{B}(\alpha, \beta)}{\mathrm{B}(\alpha + y, \beta + n - y)}. \tag{6.36}$$

Example 6.4.1 *(Allele mutation). Consider a DNA allele-counting experiment to identify the proportion θ of a certain type of phenomenon (e.g. the mutation rate) within a specified population. A well-defined null hypothesis is entertained, that is $\theta = 0.1$, so that the aim is to test $H_0 : \theta = 0.1$. As there is no specific alternative hypothesis, $H_1 : \theta \neq 0.1$ is chosen. For a count of $n = 500$ maternally transmitted alleles, $y = 53$ are found to be of the specified mutated type. What is the strength of this evidence against H_1?*

Assuming a uniform prior, the Bayes factor can be computed by applying Equation (6.34) and is found to be $BF = 26.28$. This result suggests a moderate support of hypothesis H_0. In addition, assuming $\pi_0 = \pi_1 = 1/2$, the posterior probability of the null hypothesis is equal to $\alpha_0 = 0.96$. The same conclusions are obtained from a decision-theoretic approach by taking a '0–k_i' loss function. In order to reject H_0, it follows from Equation (6.33) that an asymmetric loss is necessary with k_0 approximately 27 times larger than k_1, which is a rather unrealistic loss function.

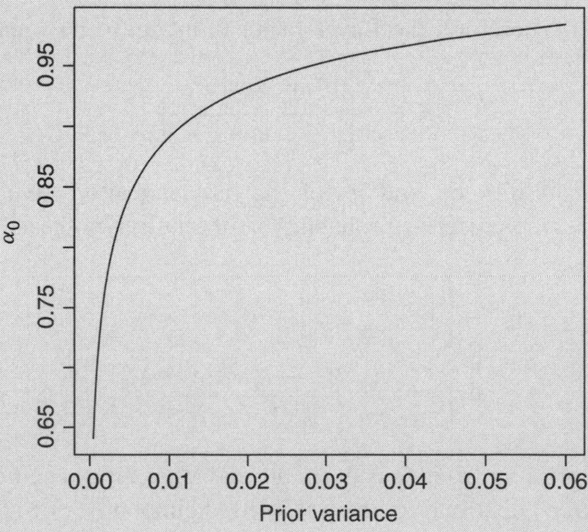

Figure 6.9 Plot of the posterior probability α_0 of the null hypothesis as the prior variance increases from 0.0004 to 0.06.

For the purpose of illustration, imagine that available knowledge suggests a prior mean equal to 0.1. *The beta hyperparameters thus need to be chosen such that* $\alpha/(\alpha+\beta)=0.1$. *Assuming uncertainty about the prior variance, a sensitivity analysis may be conducted in order to analyze the impact of different values of prior variability on the posterior probability of the null hypothesis. This is illustrated in Figure 6.9 which shows that a small variance for the prior density under the alternative hypothesis does not influence the optimal decision, which remains* d_0.

Comparison of two proportions

Consider two populations. It is of interest to test whether the proportions of elements sharing a given property in the two populations are equal or not. The quantities of interest are denoted with θ_1 and θ_2 for the first and the second population, respectively. The competing hypotheses are $H_0 : \theta_1 = \theta_2$ against $H_1 : \theta_1 \neq \theta_2$, more conveniently rewritten as $H_0 : \theta_1 - \theta_2 = 0$ and $H_1 : \theta_1 - \theta_2 \neq 0$. Beta prior distributions, $\pi(\theta_i) = \text{Be}(\alpha_i, \beta_i)$, $i = 1, 2$, are assumed for the parameters θ_1 and θ_2.

Let there be two independent random samples from the two populations,

$$x_1 = (x_{11}, \ldots, x_{1n_1}) \text{ and } x_2 = (x_{21}, \ldots, x_{2n_2}),$$

where $x_{ij} = 1$ if the j-th member of the sample from the i-th population is a success, and $= 0$ otherwise. The sum of the observed successes $y_i = \sum_{j=1}^{n_i} x_{ij}$, $i = 1, 2$, can be considered as a realization of a binomial variable with parameters n_i and θ_i, $Y_i \sim \text{Bin}(n_i, \theta_i)$. The posterior densities of the parameters θ_1 and θ_2 are $\text{Be}(\alpha_i^*, \beta_i^*)$, $i = 1, 2$, with parameters α_i^* and β_i^*, obtained following the updating rules given in Example 3.3.2.

A natural choice for the probability density under the alternative hypothesis would then be the distribution of the difference of two beta densities. This distribution is not straightforward to handle but a Normal approximation may be used (see Section 6.3.2).

The approximate posterior probability of the difference $\theta_1 - \theta_2$ can be written as

$$\pi(\theta_1 - \theta_2 \mid x_1, x_2) \approx N\left(\mu_1 - \mu_2, \tau_1^2 + \tau_2^2\right)$$

where

$$\mu_i = \frac{\alpha_i^*}{\alpha_i^* + \beta_i^*} \quad ; \quad \tau_i^2 = \frac{\alpha_i^* \beta_i^*}{(\alpha_i^* + \beta_i^*)^2(\alpha_i^* + \beta_i^* + 1)}, \qquad i = 1, 2.$$

One possibility for a test of the null hypothesis is to see whether the parameter value specified by the null hypothesis (that is, 0) lies inside the $100(1-\alpha)\%$ credible interval (see Section 5.2, and Bolstad 2004):

$$\left[\mu_1 - \mu_2 - z_{1-\alpha/2}\sqrt{\tau_1^2 + \tau_2^2}; \mu_1 - \mu_2 + z_{1-\alpha/2}\sqrt{\tau_1^2 + \tau_2^2}\right]. \tag{6.37}$$

Another possibility for the test of the hypothesis of interest is that of simulations. Draw m pairs of samples at random from the beta posteriors $\pi(\theta_i \mid x_i) = \text{Be}\left(\alpha_i^*, \beta_i^*\right)$, $i = 1, 2$, and compute the difference in estimates of the parameters θ_1 and θ_2 for each draw. The sample quantiles of order $\alpha/2$ and $1 - \alpha/2$ provide a (simulated) $100(1 - \alpha)\%$ credible interval for $\theta_1 - \theta_2$. The null hypothesis will be rejected if the value $(\theta_1 - \theta_2 = 0)$ under the null hypothesis lies outside the interval.

However, as mentioned at the beginning of the section, a full Bayesian solution would require probabilities to be assigned to the competing hypotheses. Moreover, a fixed credibility level would not be needed. Consider, for example, the testing of the hypotheses $\theta_1 = \theta_2$ versus $\theta_1 \neq \theta_2$ by computing the posterior odds and the Bayes factor as given by Equations (6.30) and (6.31). Since the posterior distribution has been approximated by a Normal distribution, it is convenient to consider the Normal approximation of the binomial distribution of the data, that is

$$Y_i \approx N(n_i\theta_i, n_i\bar{x}_i(1 - \bar{x}_i)) \qquad i = 1, 2,$$

where $\bar{x}_i$ is the proportion of successes in the i-th sample and is an estimate of $\theta_i, i = 1, 2$. The variance $n_i\theta_i(1 - \theta_i)$ has been estimated using the sample mean $\bar{x}_i$ which is a consistent estimator of θ_i[8].

Consider the variable $Z_i = Y_i/n_i$ which is Normally distributed with mean θ_i and variance $\bar{x}_i(1 - \bar{x}_i)/n_i$. The difference $Z = Z_1 - Z_2$ is still, approximately, Normally distributed with mean $\theta_1 - \theta_2$:

$$Z \approx \text{N}\left(\theta_1 - \theta_2, s^2\right),$$

with

$$s^2 = \frac{\bar{x}_1(1 - \bar{x}_1)}{n_1} + \frac{\bar{x}_2(1 - \bar{x}_2)}{n_2}.$$

The Bayes factor can be written as

$$BF = \frac{f(z \mid (\theta_1 - \theta_2 = 0))}{\int_{\Theta_1} f(z \mid \theta_1 - \theta_2)\pi_{H_1}(\theta_1 - \theta_2)d(\theta_1 - \theta_2)} = \frac{f(z \mid 0)}{l_1(z)}.$$

It can be shown that the weighted likelihood ratio under H_1, $l_1(z)$, is $\text{N}(\mu_1 - \mu_2, s^2 + \tau_1^2 + \tau_2^2)$ (Aitken and Taroni 2004, p. 324). At this point, then, all the elements are available to compute the Bayes factor which can be found to be

$$BF = \frac{\left(s^2\right)^{-1/2} \exp\left[-\frac{1}{2s^2}z^2\right]}{\left(s^2 + \tau_1^2 + \tau_2^2\right)^{-1/2} \exp\left[-\frac{1}{2\left(s^2+\tau_1^2+\tau_2^2\right)}\left(z - (\mu_1 - \mu_2)\right)^2\right]}. \qquad (6.38)$$

[8]The property of consistency for an estimator requires that the estimator converges to the parameter of interest as the sample size becomes infinite. Informally, as the sample size increases, a consistent estimator will become arbitrarily close to the parameter of interest with high probability. For a formal definition, see the concept of convergence in probability (Casella and Berger 2002, for example).

The optimal decision can be identified from Equation (6.33).

Example 6.4.2 *(Detection of nandrolone metabolites – continued). As in Example 6.3.3, consider a setting in which urine samples are taken from football players before and after a football game to find out whether there is a significant difference before and after physical effort, and therefore, whether traces of nandrolone detected after effort may have an endogenous origin (natural generation), as claimed by some players. For this purpose, a total of 137 urine samples belonging to amateur football players were collected before and after effort (Robinson et al. 2001) and it is known that they have not been taking the anabolic steroid. Analytical results are shown in Table 6.9 in terms of presence or absence of the metabolite in the urine samples.*

Table 6.9 Results of the quantification of nandrolone in amateur football players before and after physical effort.

	Before	After
Presence	0	8
Absence	137	129

The two groups (before and after) are described by two independent binomial variables, with probability of success (say, the absence of the NA traces) θ_1 and θ_2. Based on previous experience, a $Be(20, 2)$ prior distribution is introduced for both proportions. The aim may then be to test $H_0 : \theta_1 - \theta_2 = 0$ versus $H_1 : \theta_1 - \theta_2 \neq 0$. The posterior distributions are Be(157,2) and Be(149, 10), respectively.

First, consider the approach based on determination of a credible interval. The posterior densities of the parameters are approximately Normal with means

$$\mu_1 = \frac{157}{159} = 0.987 \quad ; \quad \mu_2 = \frac{149}{159} = 0.937,$$

and variances

$$\tau_1^2 = 0.000077 \quad ; \quad \tau_2^2 = 0.00037.$$

The posterior probability of the difference can be approximated with a Normal distribution of the form $\pi(\theta_1 - \theta_2 \mid x_1, x_2) \approx N(0.05, 0.021^2)$. The 95% credible interval, Equation (6.36), is

$$[0.05 - 1.96 \cdot 0.021\,;\, 0.05 + 1.96 \cdot 0.021] = [0.009\,;\, 0.09]\,.$$

At a credibility level of $\alpha = 0.05$*, the null hypothesis is rejected because the interval does not cover the point value* 0*. The same conclusion may be obtained by drawing* $m = 1000$ *independent samples from the posterior distributions* $Be(157, 2)$ *and* $Be(149, 10)$*. The sampling distribution of the difference between draws is illustrated in Figure 6.10. The sample quantiles of order 0.025 and 0.975 are equal to* 0.01 *and* 0.09*, respectively, giving a 95% quantile interval which does not include zero and hence results in a rejection of the null hypothesis. Thus, there is evidence of a difference in the probability of finding nandrolone in the urine before a game and the probability of finding it after a game. This is the outcome of a test of a two-sided alternative hypothesis. Further inspection of the data shows that the level of nandrolone before the game is lower than the level after the game; i.e., there is evidence of an endogenous origin for nandrolone. This result has not arisen solely because of the choice of the prior. A uniform prior also provides a posterior credible interval which supports this conclusion (with a Be(1, 1) prior the credible interval is* [0.014 ; 0.1]*).*

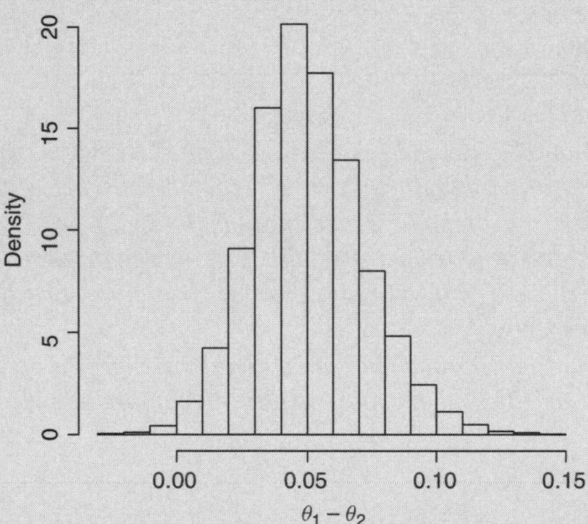

Figure 6.10 Sampling posterior distribution of $\theta_1 - \theta_2$ obtained by taking m draws from a Be(157, 2) and a Be(149, 10).

In a full Bayesian perspective, the Bayes factor (6.38) can be found to be $BF = 0.02 = 1/50$*. This result provides moderate supports for the hypothesis* H_1*, that there is a difference in nandrolone level between the two groups. When assuming the competing hypotheses to be equally likely, a priori,*

$\pi_0 = \pi_1 = 1/2$, then the posterior probability of the null hypothesis is

$$\alpha_0 = \left(1 + \frac{1}{0.02}\right)^{-1} \simeq 0.02.$$

The null hypothesis that there has not been a change in the level of nandrolone can thus be rejected. Given any symmetric '0–k_i' loss function, the optimal decision is d_1, there has been a change in the level of nandrolone. Figure 6.11 shows that only a strong prior belief in the hypothesis H_0 being true ($\pi_0 > 0.98$) would produce a posterior probability α_0 greater than 0.5 in favour of no change in the level of nandrolone, and thus a different decision.

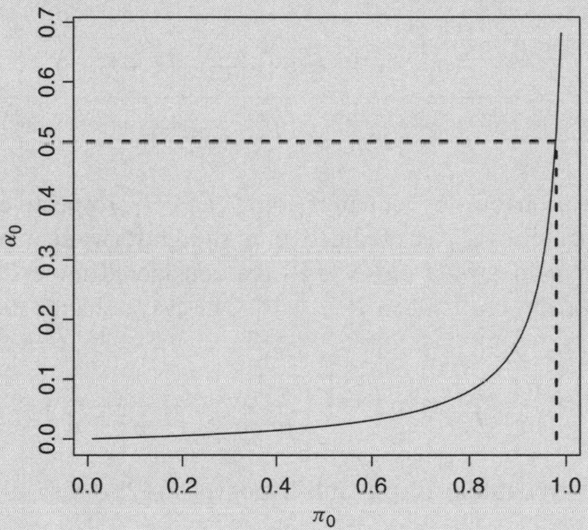

Figure 6.11 Posterior probability α_0 of the null hypothesis H_0 of no difference in nandrolone levels before and after a game of football given varying prior beliefs π_0 in H_0 being true. The dashed horizontal line at $\alpha_0 = 0.5$ indicates the optimal decision threshold when $\pi_0 = 0.98$ (given a symmetric loss function).

6.4.3 Mean

Consider a continuous measurement, such as the weight X of ecstasy pills, that is Normally distributed with known variance σ^2, $X \sim N(\theta, \sigma^2)$. Suppose that it is of interest to test the null hypothesis $H_0 : \theta = \theta_0$ against the alternative hypothesis $H_1 : \theta \neq \theta_0$. The probability density of θ under the alternative hypothesis $\pi_{H_1}(\theta)$ is taken to be Normal, $\pi_{H_1}(\theta) = N(\mu, \tau^2)$. As has been observed in Section 4.4.1, the distribution of the sample mean of n Normally distributed random variables, $X \sim$

$N(\theta, \sigma^2)$, is still Normal with the same mean θ and variance σ^2/n, $\bar{X} \sim N(\theta, \sigma^2/n)$. Suppose then that a random sample $(x_1, \ldots, x_n)$ is observed. Considering that the sample mean is a sufficient statistic for θ and recalling from Section 3.3.2, Equation (3.12), that the posterior distribution depends on the data only through the sufficient statistic, the probability density function $f(x_1, \ldots, x_n \mid \theta)$ reduces to

$$f(x_1, \ldots, x_n \mid \theta) = \frac{1}{\sqrt{2\pi\sigma^2/n}} \exp\left(-\frac{1}{2\sigma^2/n}(\bar{x} - \theta)^2\right).$$

It can be shown (see for example Aitken and Taroni (2004, p. 324)) that the weighted likelihood of the mean $\bar{X}$ of n measurements under hypothesis H_1, $l_1(\bar{x})$, is $N\left(\mu, \tau^2 + \sigma^2/n\right)$. Therefore, the Bayes factor in Equation (6.31) turns out to be equal to:

$$BF = \frac{(\sigma^2/n)^{-1/2} \exp\left(-\frac{1}{2\sigma^2/n}(\bar{x} - \theta_0)^2\right)}{(\tau^2 + \sigma^2/n)^{-1/2} \exp\left(-\frac{1}{2(\tau^2 + \sigma^2/n)}(\bar{x} - \mu)^2\right)}.$$

Then, if the prior probabilities π_0 and π_1 of H_0 and H_1, respectively, are available, the posterior probability α_0 is obtained in a straightforward way as in Equation (6.32). Whenever values of θ close to θ_0 are considered more likely, then a reasonable choice for the prior mean is $\mu = \theta_0$. The Bayes factor then reduces to

$$BF = \left(1 + \frac{n\tau^2}{\sigma^2}\right)^{1/2} \exp\left(-\frac{1}{2}\left(\frac{1}{\sigma^2/n} - \frac{1}{\tau^2 + \sigma^2/n}\right)(\bar{x} - \theta_0)^2\right), \quad (6.39)$$

and the posterior probability of the null hypothesis is

$$\alpha_0 = \left[1 + \frac{\pi_1}{\pi_0}\left(1 + \frac{n\tau^2}{\sigma^2}\right)^{-1/2} \exp\left(-\frac{1}{2}\left(\frac{1}{\tau^2 + \sigma^2/n} - \frac{1}{\sigma^2/n}\right)(\bar{x} - \theta_0)^2\right)\right]^{-1}.$$

As far as the choice of the prior under H_1 is concerned, the standard deviation τ is supposed to be considerably greater than the width of the interval of values of the parameter considered indistinguishable from θ_0 (Lee 2004).

Example 6.4.3 *(Scales accuracy test). Suppose that it is of interest to test the accuracy of a laboratory scale. Measurements X are assumed to be Normally distributed with known variance, $X \sim N(\theta, \sigma^2 = 0.01^2)$.*

For this purpose, a weight standard of 1000 mg is used. It is, thus, of interest to test whether the standard weight will be found to be equal to

1000 mg or whether it will be found either smaller or larger than 1000 mg. In other words, one is interested in testing the null hypothesis $H_0 : \theta = 1000$ against the alternative hypothesis $H_1 : \theta \neq 1000$. Assume the null hypothesis is believed, a priori, to be as likely as the alternative, that is $\pi_0 = \pi_1 = 1/2$, and that the prior density under the alternative hypothesis is $N(1000, 0.1^2)$. Thirty measurements are taken with a mean $\bar{x} = 999.996$ mg. The Bayes factor then is:

$$BF = \left(1 + \frac{30 \cdot 0.1^2}{0.01^2}\right)^{1/2}$$
$$\exp\left(-\frac{1}{2}\left(\frac{1}{0.01^2/30} - \frac{1}{0.01^2/30 + 0.1^2}\right)(-0.004)^2\right) = 4.97$$

and the posterior probability α_0 in favour of $H_0 = \left(1 + \frac{1}{4.97}\right)^{-1} = 0.83$. These values suggest substantial evidence in favour of H_0 that the mean is 1000 mg. Figure 6.12 presents the Bayes factor obtained with different prior opinions about the value of τ^2. Of course in practice the prior belief π_0 that the weight of a standard is as specified should be very close to 1, and π_1 correspondingly very close to 0. The Bayes factor is independent of the prior odds and is the factor by which the prior odds changes as a result of the experiment. The R code at the end of the chapter allows for the user to choose their own values of π_0.

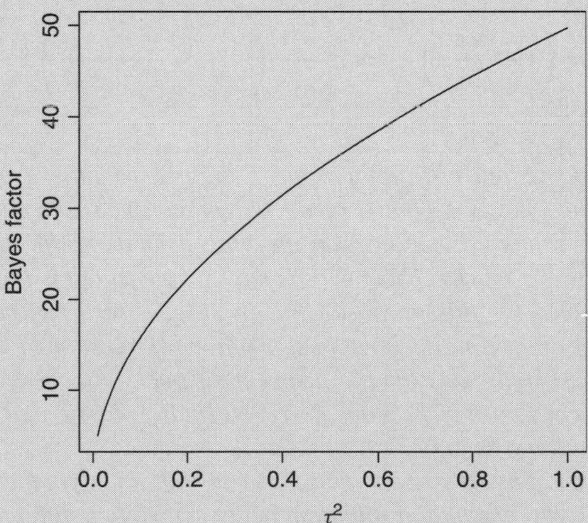

Figure 6.12 The relationship of the Bayes factor in (6.39) to the increase of the value of the prior variance τ^2 from 0 to 1 ($n = 30, \sigma^2 = 0.01^2, \bar{x} = 999.996, \theta_0 = 1000, \pi_0 = \pi_1 = 1/2$).

Assuming a symmetric '0–k_i' loss function, the term $\frac{k_0}{k_1}\frac{\pi_1}{\pi_0}$ in Equation (6.33) becomes 1. *Therefore, the optimal decision is d_0, accepting the null hypothesis that the mean weight is* 1000 *mg, since the Bayes factor is greater than* 1.

Comparison of two means

Consider two groups of observations from two Normal populations with known and equal variance, $X_1 \sim N(\theta_1, \sigma^2)$ and $X_2 \sim N(\theta_2, \sigma^2)$ and suppose that it is of interest to test the equality of the means θ_1 and θ_2. Stated otherwise, the two competing hypotheses are $H_0 : \theta_1 - \theta_2 = 0$ against $H_1 : \theta_1 - \theta_2 \neq 0$. Assume for the parameters of interest a conjugate Normal prior, $\pi(\theta_i) = N(\mu, \tau^2/2)$, $i = 1, 2$, so that the prior probability density under the alternative hypothesis can be taken to be $\pi_{H_1}(\theta_1 - \theta_2) = N(0, \tau^2)$. Two random samples then are observed, one for each population, $x_1 = (x_{11}, \ldots, x_{1n_1})$ and $x_2 = (x_{21}, \ldots, x_{2n_2})$. Thus, $\bar{X} = \bar{X}_1 - \bar{X}_2 \sim N(\theta_1 - \theta_2, \sigma^2/n_1 + \sigma^2/n_2)$, and the Bayes factor is

$$BF = \left(1 + \frac{n_1 n_2 \tau^2}{\sigma^2(n_1 + n_2)}\right)^{1/2} \exp\left[-\frac{1}{2}\left(\frac{1}{\frac{\sigma^2}{n_1} + \frac{\sigma^2}{n_2}} - \frac{1}{\frac{\sigma^2}{n_1} + \frac{\sigma^2}{n_2} + \tau^2}\right)\bar{x}^2\right].$$

Note that if $n_1 = n_2 = n$, it reduces to

$$BF = \left(1 + \frac{n\tau^2}{2\sigma^2}\right)^{1/2} \exp\left[-\frac{1}{2}\left(\frac{1}{\frac{2\sigma^2}{n}} - \frac{1}{\frac{2\sigma^2}{n} + \tau^2}\right)\bar{x}^2\right]. \qquad (6.40)$$

Example 6.4.4 *(Firearms examination). Forensic scientists commonly seek to determine the source of a bullet recovered on a crime scene. Among the initial steps in the examination process is the analysis of the widths in millimetres of land and groove marks. Their widths may be measured using microscope devices conceived for this purpose. Imagine that a series of measurements of the width of groove marks have been made on a questioned bullet and on a known bullet (fired, for example, using a suspect's firearm), respectively. These measurements may be considered Normally distributed with known standard deviation $\sigma = 0.02$.*

It may then be of interest to compare the means θ_1 and θ_2 of the two series of measurements, with particular attention being drawn to a difference in width. More specifically, a scientist may intend to test the null hypothesis $H_0 : \theta_1 - \theta_2 = 0$ versus the alternative hypothesis $H_1 : \theta_1 - \theta_2 \neq 0$. The prior density under the alternative hypothesis is taken to be Normal with zero mean and *standard deviation τ fixed on the basis of prior knowledge from an available database, as suggested in Example 4.4.3. In particular, differences between measurements greater, in absolute value, than* 0.15 *millimetres are believed extremely unlikely, so that a value of $\tau = 0.05$ is chosen so that 0.15 is three standard deviations from the mean.*

Suppose then that a total of $n = 36$ measurements are taken from a questioned and a known bullet. The sample mean width of groove marks $\bar{x}_1$ of the questioned bullet is equal to $\bar{x}_1 = 1.9314$ millimetres, while the sample mean $\bar{x}_2$ of the known bullet is $\bar{x}_2 = 1.9508$ millimetres. The Bayes factor, Equation (6.40), is equal to $BF = 0.002 = 1/500$ and provides moderately strong evidence for the alternative hypothesis, H_1. With prior probabilities $\pi_0 = \pi_1 = 1/2$, the posterior probability for H_0 is approximately equal to BF, which is 0.002. *There is consequently a very large posterior probability* (0.998) *that the mean widths of groove marks for the known and questioned bullets are different and hence that the bullets come from different firearms.*

6.5 R CODE

A symbol '$*$', '+', ',' and so on at the end of a line indicates that the command continuous to the following line. The absence of such a symbol indicates the end of a command.

Example 6.3.1

Prior odds

```
alpha=1
beta=1
theta0=0.1
pi0=pbeta(theta0,alpha,beta)
pi1=pbeta(theta0,alpha,beta,lower.tail=FALSE)
podds=pi0/pi1
print(paste('Prior odds =',round(podds,2)))
```

Posterior odds and Bayes factor

```
n=50
y=4
alphap=alpha+y
betap=beta+n-y
alpha0=pbeta(theta0,alphap,betap)
alpha1=pbeta(theta0,alphap,betap,lower.tail=FALSE)
postodds=alpha0/alpha1
print(paste('Posterior odds =',round(postodds,2)))
print(paste('Bayes factor =',round(postodds/podds,2)))
```

Decision-theoretic approach

```
k0=1
k1=1
ld0=k0*((alpha+y)/(alpha+beta+n))*pbeta(theta0,alpha+y+1,
betap,lower.tail=FALSE)-k0*theta0*pbeta(theta0,alphap,betap,
lower.tail=FALSE)
ld1=k1*theta0*pbeta(theta0,alphap,betap)-k1*((alpha+y)/
(alpha+beta+n))*pbeta(theta0,alpha+y+1,betap)
l=c(ld0,ld1)
print(paste('Optimal decision: decision',which(l==min(l))-1))
```

Example 6.3.2

Data, prior and posterior information for population 1 and population 2

```
alpha1=3.5
beta1=2

alpha2=3.5
beta2=3.5

par(mfrow=c(1,2))
plot(function(x) dbeta(x,alpha1,beta1),0,1,main='Population 1',
xlab=expression(paste(theta)),ylab='Prior density')
plot(function(x) dbeta(x,alpha2,beta2),0,1,main='Population 2',
xlab=expression(paste(theta)),ylab='Prior density')

n1=13
y1=10
alpha1p=alpha1+y1
beta1p=beta1+n1-y1

n2=32
y2=6
alpha2p=alpha2+y2
beta2p=beta2+n2-y2
```

Comparison between the two proportions

```
print('Log-odds ratio')
mu=log((alpha1p-1/2)*(beta2p-1/2)/((beta1p-1/2)*(alpha2p-1/2)))
tau=alpha1p^(-1)+beta1p^(-1)+alpha2p^(-1)+beta2p^(-1)
p=pnorm(-mu/sqrt(tau),0,1,lower.tail=FALSE)
print(paste('Probability log-odds ratio greater than zero =',
round(p,4)))
plot(function(x) dnorm(x,mu,sqrt(tau)),-1,5,ylab='Density',
xlab='log-odds ratio')

print('Approximated distribution')
mu1=(alpha1+y1)/(alpha1+beta1+n1)
mu2=(alpha2+y2)/(alpha2+beta2+n2)
mu=mu1-mu2
tau1=(alpha1p*beta1p/((alpha1+beta1+n1)^2*
```

```
(alpha1+beta1+n1+1)))
tau2=(alpha2p*beta2p/((alpha2+beta2+n2)^2*
(alpha2+beta2+n2+1)))
tau=tau1+tau2
alpha0=pnorm(0,mu,sqrt(tau))-pnorm(-1,mu,sqrt(tau))
print(paste('Posterior probability of the null hypothesis =',
alpha0))

print(paste('Howard prior'))
library(LearnBayes)
plo=0.0001
phi=0.9999
n.grid.points=10000
par=c(alpha1,beta1,alpha2,beta2)
rho=2
parp=c(alpha1p,beta1p,alpha2p,beta2p)
s=simcontour(howardprior,c(plo,phi,plo,phi),c(parp,rho),
 n.grid.points)
alpha0=sum(s$x<s$y)/n.grid.points
print(paste('Posterior probability of the null hypothesis =',
alpha0))

par(mfrow=c(1,2))
mycontour(howardprior,c(plo,phi,plo,phi),c(par,rho))
title(main='Prior',xlab=expression(paste(theta[1])),
ylab=expression(paste(theta[2])))
lines(c(0,1),c(0,1))
mycontour(howardprior,c(plo,phi,plo,phi),c(parp,rho))
title(main='Posterior',xlab=expression(paste(theta[1])),
ylab=expression(paste(theta[2])))
lines(c(0,1),c(0,1))
```

Example 6.3.3

Data, prior and posterior information for population 1 and population 2

```
f=c(336,9,5,5,3)
s=c(126,0,0,0,0)
af=c(1,1,1,1,1)
as=c(1,1,1,1,1)
afpost=af+f
aspost=as+s
```

Simulations from the posterior Dirichlet distribution

```
library(MCMCpack)
n=10000
rf=rdirichlet(n,afpost)
rs=rdirichlet(n,aspost)

d=matrix(0,nrow=n,ncol=1)
par(mfrow=c(2,2))
for (i in 1:4){
```

```
for (j in 1:n){
d[j]=sum(rf[j,1:i])-sum(rs[j,1:i])
}
hist(d,xlab=expression(paste(theta[1])*
paste(' - ')*paste(theta[2])),main='')
print(length(which(d<0))/n)
}
```

Example 6.3.4

Prior odds

```
s=c(0.0229^2,0.0463^2)
mu=1
tau=.3
theta0=.5
pi0=pnorm(theta0,mu,tau)
pi1=pnorm(theta0,mu,tau,lower.tail=FALSE)
podds=pi0/pi1
print(paste('Prior odds =',round(podds,3)))
```

Posterior odds and Bayes factor

```
m=c(0.5191,0.5093)
n=length(m)
for (i in 1:2){
mu=(mu*s[i]/n+tau^2*m[i])/(s[i]/n+tau^2)
tau=sqrt((tau^2*s[i]/n)/(tau^2+s[i]/n))
}
print(paste('Posterior mean =',round(mu,4)))
print(paste('Posterior standard deviation =',round(tau,4)))

alpha0=pnorm(theta0,mu,tau)
alpha1=pnorm(theta0,mu,tau,lower.tail=FALSE)
postodds=alpha0/alpha1
B=postodds/podds
print(paste('Posterior odds =',round(postodds,4)))
print(paste('Bayes factor =',round(B,2)))
```

Decision-theoretic approach

```
k0=2
k1=1
t=(theta0-mu)*tau
ld0=(k0*tau*(dnorm(t)-t*pnorm(t,lower.tail=FALSE)))
ld1=(k1*tau*(dnorm(t)+t*(0.5-pnorm(t))))
l=c(ld0,ld1)
print(paste('Optimal decision: decision',which(l==min(l))-1))
```

Example 6.3.5

Data

```
auth=c(203.33,204.06,202.62,200.65,199.77,203.25,
```

```
202.28,201.22,200.3,201.06)
quest=c(206.17,195.78,197.44,200.61,196.19,213.18,
217.23,197.35,203.29,194.58)
n1=length(auth)
n2=length(quest)
xbar1=mean(auth)
xbar2=mean(quest)
s=(sum((quest-xbar2)^2)+sum((auth-xbar1)^2))/(n1+n2-2)
```

Posterior odds

```
tq=-(xbar1-xbar2)/(sqrt(s*(1/n1+1/n2)))
alpha0=pt(tq,n1+n2-2)
alpha1=pt(tq,n1+n2-2,lower.tail=FALSE)
postodds=alpha0/alpha1
print(paste('Posterior odds =',round(postodds,2)))
```

Example 6.4.1

Input values

```
theta0=0.1
n=500
y=53
```

Bayes factor and posterior odds

```
B=(theta0^y)*(1-theta0)^(n-y)/beta(y+1,n-y+1)
print(paste('Bayes factor =',round(B,2)))
pi0=.5
pi1=1-pi0
alpha0=(1+(pi1/pi0)*(1/B))^(-1)
print(paste('Posterior probability of the null hyp =',
round(alpha0,2)))
```

Example 6.4.2

Data, prior and posterior parameters

```
alpha=c(20,20)
beta=c(2,2)
n=c(137,137)
y=c(137,129)
alphap=alpha+y
betap=beta+n-y
```

Credible interval

```
mup=alphap/(alphap+betap)
taup=(alphap*betap)/((alphap+betap)^2*(alphap+betap+1))
cl=.05
z=round(qnorm(cl/2,lower.tail=FALSE),2)
u=(mup[1]-mup[2])+z*sqrt(taup[1]+taup[2])
l=(mup[1]-mup[2])-z*sqrt(taup[1]+taup[2])
```

```
print(paste((1-cl)*100,'% credibility interval: [',round(l,3),
';',round(u,3),']'))
```

Simulations

```
m=10000
theta1=rbeta(m,alphap[1],betap[1])
theta2=rbeta(m,alphap[2],betap[2])
d=theta1-theta2
hist(d,freq=FALSE,xlab=expression(paste(theta[1])*paste(' - ')*
paste(theta[2])),main='')
cl=.05
q=c(quantile(d,cl/2),quantile(d,1-cl/2))
print(q)
```

Bayes factor and posterior probability of the null hypothesis

```
z=y[1]/n[1]-y[2]/n[2]
s2=((y[1]/n[1])*(1-y[1]/n[1])/n[1])+((y[2]/n[2])*
(1-y[2]/n[2])/n[2])
Bfnum=(s2)^(-1/2)*exp(-(1/(2*s2))*z^2)
v=s2+taup[1]+taup[2]
Bfden=v^(-1/2)*exp(-(1/(2*v))*(z-mup[1]+mup[2])^2)
BF=Bfnum/Bfden
print(paste('Bayes factor =',round(BF,2)))
alpha0=(1+1/BF)^(-1)
print(paste('Posterior probability of the null hyp =',
round(alpha0,2)))

pi0=seq(0.01,0.99,0.01)
pi1=1-pi0
alpha0=matrix(0,nrow=length(pi0),ncol=1)
for (i in 1:length(pi0)){
alpha0[i]=(1+(pi1[i]/pi0[i])*(1/BF))^(-1)
}
p=pi0[which(alpha0>=.5)][1]
print(paste('Lower limit of prior probability necessary to
 change decision:',p))
```

Example 6.4.3

Input values

```
n=30
barx=999.996
sigma2=0.01^2
theta0=1000
mu=theta0
tau2=0.1^2
pi0=0.5
pi1=1-pi0
priorodds=pi0/pi1
print(paste('Prior odds =',priorodds))
```

Bayes factor and posterior probability of the null hypothesis

```
B=sqrt(1+n*tau2/sigma2)*exp(-(1/2)*
(1/(sigma2/n)-1/(sigma2/n+tau2))*(barx-theta0)^2)
alpha0=(1+1/B)^(-1)
print(paste('Bayes factor =',round(B)))
print(paste('Posterior probability of the null hypothesis =',
round(alpha0,2)))
```

alternatively, it can be checked

```
alpha0=(1+(pi1/pi0)*(1+(n*tau2/sigma2))^(-1/2)*exp(-(1/2)*
(1/(sigma2/n+tau2)-1/(sigma2/n))*(barx-theta0)^2))^(-1)
alpha1=1-alpha0
postodds=alpha0/alpha1
B=postodds/priorodds
```

Example 6.4.4

Input values

```
n=36
sigma2=0.02^2
barx1=1.9314
barx2=1.9508
barx=barx1-barx2
tau2=0.05^2
pi0=.5
pi1=1-pi0
```

Bayes factor and posterior probability of the null hypothesis

```
B=sqrt(1+n*tau2/(2*sigma2))*
exp(-(1/2)*(1/(2*sigma2/n)-1/(2*(sigma2/n)+tau2))*barx^2)
alpha0=(1+(pi1/pi0)*(1/B))^(-1)
print(paste('Bayes factor =',B))
print(paste('Posterior probability of the alternative
 hypothesis =',1-alpha0))
```

7 Sampling

7.1 INTRODUCTION

Operational laboratories providing forensic science services commonly encounter consignments of discrete units whose characteristics may be of interest within a criminal investigation. Typical examples include consignments that consist of individual items, such as tablets, bags or electronic data storage devices where each item of such a consignment may, or may not, contain something illegal. That illegal content could be drugs or pornographic images, for instance. Forensic science laboratories may be called on to inspect individual items of a consignment in order to gather information from which may be drawn an inference about the proportion, θ, of items that contain something illegal. Only rarely, however, are circumstances such that laboratories are in the advantageous situation that the entirety of a consignment may be inspected – a setting in which certainty about the proportion could obviously be obtained (assuming error-free analyses).

Most often, laboratories face a high workload together with constraints of time and finance. In addition, consignments may be very large or consist of items that may contain hazardous substances whose analysis may require sophisticated, expensive and time-consuming analytical procedures. For practical reasons, both scientists and their clients thus seek to agree on ways to confine analyses to a subset or sample of items. This results in incomplete information and subsequent inferences about the composition of the whole consignment are affected by uncertainty. However, this uncertainty may be quantified probabilistically. Some of the commonly used approaches for the so-called 'sampling problems' are compiled in a booklet issued by the ENFSI Drug Working Group (ENFSI 2004). All of these approaches consider sampling essentially as a problem of the derivation of probabilistic statements about the composition of a consignment in order to answer the

Data Analysis in Forensic Science: A Bayesian Decision Perspective Franco Taroni, Silvia Bozza, Alex Biedermann, Paolo Garbolino and Colin Aitken

question of interest: 'how big a sample should be taken?' Although the handling of uncertainty through probability is an essential aspect of sampling scenarios, the situation actually faced by a customer of forensic expertise is one that contains elements that allow the outcome to be considered as a *problem of decision making*. Subsequent sections of this chapter will examine this point in futher detail.

Experimental design can be thought of in various different ways. According to the so-called 'classical (or frequentist) approach', the sample information is considered solely in order to draw an inference about the proportion of interest, θ. Nevertheless, other relevant aspects of the problem, arising from sources other than the statistical investigation, may be available.

A first source of non-sample information is the prior information, which reflects a reasoner's personal degree of belief about the composition of a consignment (e.g. the proportion of units containing illegal substances) and which is measured by a probability distribution for θ. A second source of non-sample information is given by the possible consequences of decisions. These are quantified on the basis of personal preferences by determining the loss that would be incurred for each possible decision and for each possible value of the quantity of interest, θ.

These ingredients are combined with the aim of making an optimal decision about the problem at hand, that is the size of sampling that should be done in order to gather knowledge of the desired accuracy about the value of θ. In this chapter, particular consideration will be given to a sequential way of proceeding, in which the size of a sample is not predetermined. The proposed procedure involves – after each observation of a sampled item – a decision about whether it is advisable to draw a conclusion about the value of θ (e.g. the composition of a consignment) or whether more observations should be gathered.

7.2 SAMPLING INSPECTION

7.2.1 Background

Consider a population which consists of discrete items. Each item may be categorized in one and only one of two ways, e.g. effective or defective, legal or illegal. In all other respects the items are identical. For example, for tablets this would imply identity in size, colour, texture and logo. A statistical investigation is conducted for the purpose of obtaining information about θ. The outcome is denoted X. The set of possible outcomes of X is the sample space and is denoted $\mathcal{X}$. A random sample of items $(X_1, \ldots, X_n)$ is drawn for inspection where the X_i are considered as independent observations from a common density $f(x \mid \theta)$. The observed results might either be denoted a 'success' (e.g. the observed item contains an illegal drug, $X_i = 1$), or a 'failure' (e.g. the observed item does not contain an illegal drug, $X_i = 0$). The sample is a random sample from a binomial or hypergeometric distribution (see Appendix A) for large or small samples, respectively, for which the sampling of units may be taken to be sampling with or without replacement.

A single sampling plan and sampling based on a sequential analysis are among the most common methods for taking observations. In a single sampling plan, a sample size n is preselected and observations $x_1, \ldots, x_n$ are made, followed by an inference about the composition of the consignment. Various methods for selecting the sample size – based on the sample information – are available and have been accepted by US Courts (Aitken and Taroni 2004; Frank *et al.* 1991; Izenman 2001).

It is worth noting that a single sampling plan can be inefficient, in the sense that the examiner will continue to inspect items from a batch until the predetermined sample size is reached, without regards to what has already been observed. Consider the following example (Berger 1988). Suppose that a consignment contains a large number of items which can be classified as 'positive' or 'negative' (e.g. containing or not containing something illegal). The aim is to test $H_0 : \theta = 0.05$ versus $H_1 : \theta = 0.15$, where θ is the proportion of positive items. Assume further that the optimal fixed sample size experiment requires the inspection of a sample of size $n = 100$ and that a partial examination of half of the sample showed that all items are negative. Then there is overwhelming evidence that H_0 is true. Vice versa, it might be the case that the sample of the predetermined size is not sufficient and that more observations are needed.

As a second example, consider a typical problem in quality control where individual units are sampled from a batch and inspected to see whether they are effective or defective. Such sample information is then used to decide whether or not to consider a batch as acceptable. A basic sampling plan could state that a fixed number n of items are to be examined, and the consignment is to be rejected if more than c observations are defective (Wetherill and Glazebrook 1986). Imagine also that a partial examination of the sample showed that more than c observations are defective: it will then be unnecessary to inspect the remaining units.

If a sequential analysis is implemented, observations are taken one at a time, with a decision being made after each observation either to cease sampling, or to take another observation. The advantage of a sequential analysis is that it allows one to gather the appropriate amount of data needed for a decision of a desired accuracy.

Many procedures for drawing inferences about θ solely based on the sample information are available and belong to the so-called 'classical methods'. The likelihood principle (see Section 3.3.2) states that for an inference about θ – after the observation of the sample – all relevant experimental information is contained in the likelihood function, that is $f(x \mid \theta)$.

There may, however, be other information relevant for the statistical investigation, namely prior information. Such information could be gathered, for example, by considering an item's physical properties such as colour or shape. Prior beliefs about θ are expressed by a probability distribution $\pi(\theta)$. As an example, assume that units are sampled from a population with replacement (e.g. as may happen with large consignments), so that a binomial distribution with parameters n (sample size) and θ can be introduced to model the number of successes. A conjugate prior distribution for a binomial distribution is the beta distribution (Section 3.3).

Under these assumptions, Aitken (1999) formalized a criterion which allows one to determine the sample size required to be $100p\%$ certain that the proportion of items in the consignment which contains an illegal substance is greater than $100\theta\%$ (when all the items in the sample are found to be of type 'positive'). If some items are found to be 'negative', then the methodology suggests the number of additional items that need to be inspected.

Therefore, a Bayesian approach can provide summaries in probabilistic terms such as (for a particular case, with $p = 0.95$ and $\theta = 0.50$) 'how big a sample should be taken for it to be said that there is a 95% probability that the proportion of units in the consignment which contain drugs is greater than 50%?'

7.2.2 Large consignments

A large consignment is taken to be one which is sufficiently large that sampling is effectively with replacement. This can be as small as 50, though in many cases it will be of the order of many thousands.

A consignment of drugs containing N units will be considered as a random sample from some super-population of units which contain drugs. Let θ $(0 < \theta < 1)$ be the proportion of units in the super-population which contain drugs.

Let n be the number of sampled units. Denote the number which are found to contain drugs by y.

Consider the criterion that the scientist wishes to be $100p\%$ certain that $100\theta_0\%$ or more of the consignment contains drugs when all units sampled contain drugs $(y = n)$. The criterion may be written mathematically as

$$P(\theta > \theta_0 \mid y, n, \alpha, \beta) = p,$$

or

$$\int_{\theta_0}^{1} \frac{\theta^{\alpha+n-1}(1-\theta)^{\beta-1}}{B(\alpha+n, \beta)} d\theta = p, \tag{7.1}$$

using a beta conjugate prior distribution $\pi(\theta) = \text{Be}(\alpha, \beta)$ and a binomial distribution $f(y \mid n, \theta) = \text{Bin}(n, \theta)$ to give a beta posterior distribution $\pi(\theta \mid y = n) = \text{Be}(\alpha + n, \beta)$.

Table 7.1 contains, for different values of α and β, and different values of n, the corresponding probabilities p satisfying Equation (7.1) for $\theta_0 = 0.5$.

This criterion motivates an answer to the question 'how many units should be inspected to satisfy this criterion?'. An alternative way of looking at (7.1) is to reverse the role of the parameters and solve for n. Given specified values for θ and p and values for α and β chosen from prior beliefs, the appropriate value of n to solve (7.1) may be found by trial and error.

Table 7.1 Probability that the proportion of drugs in a large consignment is greater than 50% for various sample sizes n and prior parameters α and β, $P(\theta > 0.5 \mid y, n, \alpha, \beta)$. Note that $y = n$. (Adapted from Aitken 1999.)

		n			
α	β	2	3	4	5
1	1	0.88	0.94	0.97	0.98
0.5	0.5	0.92	0.97	0.98	0.99
0.065	0.935	0.78	0.89	0.95	0.97

Table 7.2 The sample size required to be $100p\%$ certain that the proportion of units in the consignment which contain drugs are greater than θ, when all the units inspected are found to contain drugs. The prior parameters $\alpha = \beta = 1$. (Reproduced from Aitken 1999. Journal of Forensic Sciences **44**, 750–760.)

	p		
θ	0.90	0.95	0.99
0.5	3	4	6
0.6	4	5	9
0.7	6	8	12
0.8	10	13	20
0.9	21	28	43
0.95	44	58	89
0.99	229	298	458

The dependency of the sample size on the values of p and θ is illustrated in Table 7.2 where the prior parameters α and β are set equal to 1. Consider $p = 0.90$, 0.95 and 0.99 and consider values of $\theta = 0.5$, 0.6, 0.7, 0.8, 0.9, 0.95, 0.99. The sample size n (given that $y = n$) required to be $100p\%$ certain that θ is greater than the specified value, say θ_0, is then given by evaluating (7.1) at $\alpha = \beta = 1$

$$\int_{\theta_0}^{1} \frac{\theta^n}{B(1+n,1)} d\theta = 1 - \theta_0^{n+1} = p.$$

Rearranging into a form that will determine n, that is

$$\log(1-p) = (n+1)\log(\theta_0),$$

the value of n is thus given by the smallest integer greater than $[\log(1-p)/\log(\theta_0)] - 1$.

For large consignments, of whatever size, one needs to examine only 4 units, in the first instance. If all are found to contain drugs, there is a 95% probability that at least 50% of the consignment contains drugs; then, the criterion is satisfied. This sample size is not large. However, there is not very much information gained about the exact value of θ. It is only determined that there is a probability of 0.95 that $\theta > 0.5$. This is a wide interval (from 0.5 to 1) within which the true proportion may lie.

Obviously, when considering the results in Table 7.2, the consignment size has to be taken into account, in the sense that the sample size should be small enough with respect to the size of the consignment. Thus, for the last row in particular to be useful, the size of the consignment from which the sample is to be taken will have to be of the order of several tens of thousands.

This procedure cannot be considered a single sampling plan, where the optimal sample size is determined before sampling on the basis of the desired values for p and for θ. The optimal sizes given in Table 7.2 are conditioned on the assumption that all items will be found to be positive. However, when negative items are found, the same methodology can be extended to allow for this.

Example 7.2.1 *(Inspection of pills suspected to contain drugs). Imagine that the prior distribution for the proportion θ of illicit drugs is given by $\pi(\theta) = Be(1, 1)$ and that a sample of size $n = 4$ from a consignment is inspected in order to be 95% certain that 50% or more of the consignment contains drugs. However, one of the four selected units is found not to contain drugs. How many further items should therefore be inspected? Given the specified value for θ ($\theta_0 = 0.5$), an appropriate value of the optimal sample size may be found by solving:*

$$\int_{0.5}^{1} \theta^{\alpha+y-1}(1-\theta)^{\beta+n-y-1} d\theta / B(\alpha+y, \alpha+\beta+n) = p. \qquad (7.2)$$

In particular, it can be shown (by trial and error) that if the next three successive inspected items are all found to be positive, so that six out seven items contain drugs, then

$$P(\theta > 0.5 \mid y = 6, n = 7, \alpha = 1, \beta = 1) = 0.96.$$

Therefore, three additional items should be inspected. If they all contain drugs, then it can be shown that the probability that $\theta > 0.5$, given that six out of seven contain drugs, is 0.96 and the criterion is satisfied. Conversely, if any of the three items does not contain drugs, further items need to be inspected, the number of which can be found following the same criterion.

There may be situations in which different choices of α and β may be required. For instance, there may be substantial prior beliefs about the proportion of the

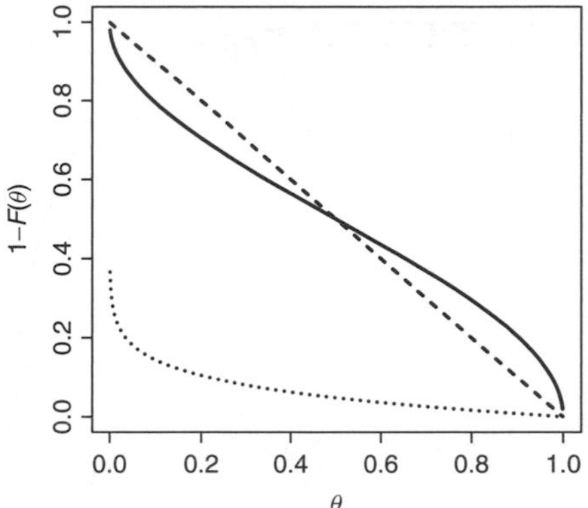

Figure 7.1 The prior probability $1 - F(\theta)$ that the proportion of units in a consignment is greater than θ, for various choices of α and β: $\alpha = \beta = 1$ (dashed curve), $\alpha = \beta = 0.5$ (solid), $\alpha = 0.065$, $\beta = 0.935$, (dotted). (Reproduced from Aitken 1999. Journal of Forensic Sciences **44**, 750–760.)

consignment which contains drugs. Such beliefs may arise, for example, from relevant, previous experience with similar consignments. In such cases, use can be made of various properties of the beta distribution (as presented in Section 4.2.1) so as to assist in choosing values for α and β. One can reproduce, for instance, results in Table 7.1 for different values of α and β, or different values of θ_0, in a straightforward manner using R routines as given at the end of this chapter.

Variation in the prior beliefs, expressed through variation in the values of α and β may have little influence on the conclusions, once some data have been observed. Figure 7.1 illustrates the prior probability that the true proportion of illegal items in a consignment is greater than a value θ, for $0 < \theta < 1$ for three choices of α and β. Figure 7.2 illustrates the posterior probability that the true proportion of illegal items in a consignment is greater than θ, for those choices of α and β, once four items have been examined and all found to be illegal. Despite the substantial difference in the prior probability curves, the respective posterior probability curves are very close.

7.2.3 Small consignments

Suppose now that the size N of the consignment is small. A sample of n units from the consignment is examined and $y(\leq n)$ units are found to contain illicit drugs. As before, let θ, satisfying $(0 < \theta < 1)$, be the proportion of units in the super-population which contains illicit drugs. The probability distribution of y, given n and θ, may be taken to be binomial. For each item, independently of the others in the consignment, the probability it contains drugs is taken to be equal to θ.

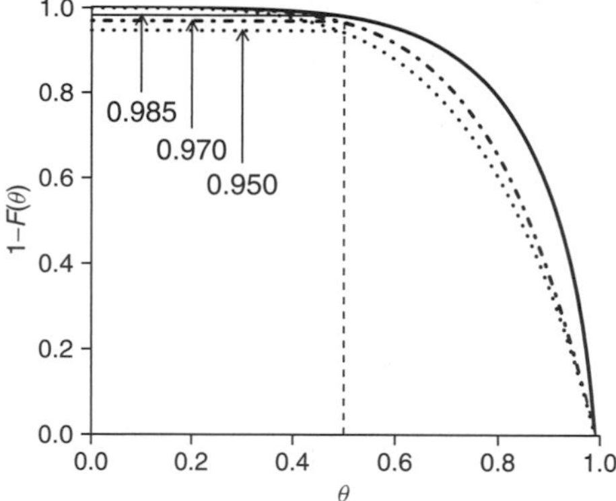

Figure 7.2 The posterior probability $1 - F(\theta)$ that the proportion of units in a consignment is greater than θ, for various choices of α and β: $\alpha = \beta = 1$, (dashed curve), $\alpha = \beta = 0.5$, (solid), $\alpha = 0.065$, $\beta = 0.935$, (dotted), after observation of four units all found to be illegal. The corresponding probabilities that at least 50% of the consignment contains illegal units are 0.985 ($\alpha = \beta = 0.5$), 0.970 ($\alpha = \beta = 1$), 0.950 ($\alpha = 0.065$, $\beta = 0.935$). (Reproduced from Aitken 1999. Journal of Forensic Sciences **44**, 750–760.)

With a $Be(\alpha, \beta)$ prior distribution for θ, the posterior distribution is another beta distribution with parameters $(\alpha + y)$ and $(\beta + n - y)$.

Since the consignment size is small, a better representation of the variability of the number of units in the non-inspected consignment which contain drugs is obtained by considering a probability distribution for this number, Z say, explicitly. Let there be m units in the remainder of the consignment (such that $n + m = N$) which have not been inspected. Then Z (unknown and $\leq m$) is the number of items in this remainder which contain drugs. Given θ, the distribution of $(Z \mid m, \theta)$, like that of $(Y \mid n, \theta)$, is binomial. However, θ has a beta distribution, and the distribution of $(Z \mid m, \theta)$ and the distribution of $(\theta \mid n, y, \alpha, \beta)$ can be combined to give a Bayesian predictive distribution for $(Z \mid n, m, y, \alpha, \beta)$, also known as a beta-binomial distribution (see Section 3.3.4 and Appendix A):

$$P(Z = z \mid n, m, y, \alpha, \beta) = \frac{\Gamma(n + \alpha + \beta)\binom{m}{z}\Gamma(z + y + \alpha)\Gamma(n + m - y - z + \beta)}{\Gamma(y + \alpha)\Gamma(n - y + \beta)\Gamma(n + m + \alpha + \beta)},$$
$$(z = 0, 1, \ldots, m). \quad (7.3)$$

Consider the beta-binomial distribution (7.3) with $\alpha = \beta = 1$. It can be shown (Aitken 1999) that

$$P(Z = z \mid n, m, y, 1, 1) = \frac{(n + 1)\binom{n}{y}\binom{m}{z}}{(n + m + 1)\binom{n+m}{y+z}}, \quad (7.4)$$

for $z = 0, 1, \ldots, m$.

Example 7.2.2 *(Inspection of pills suspected to contain drugs – continued (Aitken and Taroni 2004, p. 189)). For sake of illustration, consider a consignment of size N = 10, where five units are inspected and all five are found to contain drugs ($y = n = 5$). Assume the prior distribution for θ, the proportion of the consignment that contains illicit drugs, is $\pi(\theta) = Be(1,1)$. For the proportion of units in the consignment which contain drugs to be at least 0.7 ($\theta \geq 0.7$), it is necessary for the number of units Z in the five units not inspected to be at least 2 ($Z \geq 2$). The beta-binomial probability (7.4) gives the result*

$$P(Z \geq 2 \mid 5,5,5,1,1,) = \sum_{z=2}^{5} \frac{6\binom{5}{5}\binom{5}{z}}{11\binom{10}{5+z}} = 0.985.$$

The beta-binomial distribution assigns a probability of 0.985 to the event that $\theta \geq 0.7$.

As with large consignments, values for α and β may be chosen subjectively so as to represent prior beliefs before inspection about the proportion of the items in the consignment (considered as a random sample from the super-population) which contains drugs.

General results can also be obtained. The problem consists in choosing n, the value such that, given m, α, and β (and possible values for y, consequential on the choice of n and the outcome of the inspection), a value for z can be determined to satisfy some probabilistic criterion, e.g. the value z_0 such that $P(Z \geq z_0 \mid n, m, y, \alpha, \beta) = p$.

Example 7.2.3 *(Inspection of pills suspected to contain drugs – continued). Consider a consignment of size N = 30. If 6 pills are inspected and all of them contain drugs, then there is a probability of* 0.9 *that the number of pills that contain drugs in the remainder (24) of the consignment is at least* 17, $P(Z \geq 17 \mid 6,24,6,1,1,) = 0.9$.

If 6 *pills are inspected and one or two do not contain drugs then this number drops from* 17 *to* 13 *to* 9*:*

$$P(Z \geq 13 \mid 6,24,5,1,1,) = 0.9$$
$$P(Z \geq 9 \mid 6,24,4,1,1,) = 0.9.$$

An extension to sampling with a categorical response in which there may be more than two possible responses (e.g., with pills, the responses may be LSD, ecstasy, and licit) is given in Mavridis and Aitken (2009).

Besides prior information, sample information can also be combined with yet another source of evidence, notably quantified possible consequences of decisions, called the decision loss. In this context, decisions can consist in accepting or rejecting a consignment or, alternatively, a hypothesis about the composition of a consignment. These ideas are addressed Section 7.4.

7.3 GRAPHICAL MODELS FOR SAMPLING INSPECTION

7.3.1 Preliminaries

Bayesian approaches using beta and beta-binomial distributions, outlined in Section 7.2, have been implemented in computerized formats; scripts for R (R 2003) at the end of the chapter and Excel® spreadsheets are available[1]. These ready-to use computerized implementations greatly facilitate the practical application of thorough mathematical and statistical concepts.

The same issues can be framed within graphical models, notably Bayesian networks, which allow for (i) a flexible analysis of sampling issues with the user being able to interact directly with the respective model, (ii) an explicit and visual representation of underlying modelling assumptions, and (iii) calculation of posterior probability distributions for a consignment's true proportion of positives. Moreover, Bayesian networks also support likelihood ratio calculations under user-specified propositions as well as pre-assessment and case evaluations that account for specific customer requirements, such as the handling of competing prior beliefs (Biedermann *et al.* 2007b).

7.3.2 Bayesian network for sampling from large consignments

Figure 7.3 shows a Bayesian network useable for sampling from large consignments (Biedermann *et al.* 2007b). The proportion θ of positives in a consignment is modelled here with a discrete[2] chance node *Prop* with intervals 0–0.05, $0.05–0.1, \ldots, 0.95–1$. The probabilities assigned to the intervals of *Prop* are determined by a beta distribution whose parameters (α and β) are provided by the nodes a and b (parents of *Prop*). The states of nodes a and b are set to the numbers $0.5, 1, 2, \ldots, 10$. The choice of values for these states is a subjective one. Other values (> 0) can be defined as required. The value of 0.5 is included because the nodes a and b instantiated to this value allow the node *Prop* to have a Be(0.5,0.5) distribution which is sometimes used to represent prior beliefs according

[1]See, for example, homepage of David Lucy at (last accessed February 2009): `http://www.maths.lancs.ac.uk/~dlucy/computing.html`

[2]Currently, Bayesian networks allow genuine continuous chance nodes only for variables with a Gaussian (Normal) distribution function. One possibility of avoiding this restriction is to use a discrete chance node whose states represent disjoint intervals between 0 and 1. This allows for acceptable approximations of continuous variables.

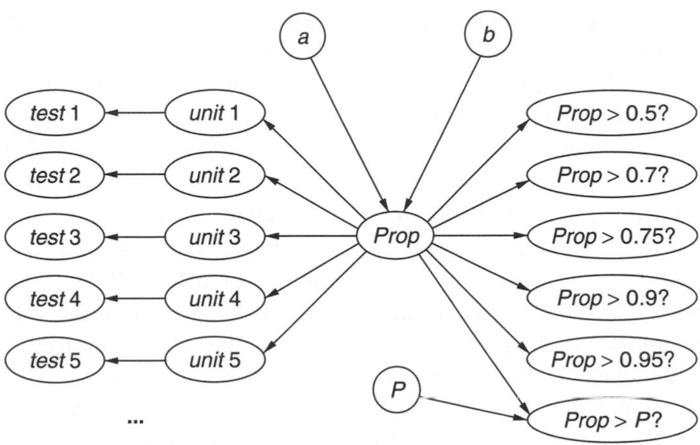

Figure 7.3 A Bayesian network for sampling from large consignments. The definitions of the nodes are as given in Table 7.3. (Reproduced from Biedermann *et al.* 2008. Law, Probability and Risk, **7**, 35–60.)

to which either no (or nearly none of the) items or all (or nearly all) items are positive.

A particular aspect of the Bayesian network shown in Figure 7.3 consists in the way in which the sampling procedure is modelled. Instead of a variable y ($y \leq n$) for the overall number of positive units in the sample of size n (Section 7.2.2), separate nodes are used to represent the target characteristic ('positive' or 'negative') of each item of a sample. In addition, a distinction is made between the true, but unknown, condition of an item (i.e. containing or not containing an illegal substance) and what is observed in the course of an experiment designed to 'detect' the presence or absence of that target characteristic. Note that in Section 7.2.2, the observation of a 'positive' (measurement) was equated with the examined unit being truly positive, which is a simplification of the underlying states of reality. Such distinctions are advocated, for example, in the context of DNA profiling analyses (Thompson *et al.* 2003), but are also challenged in other forensic disciplines (Saks and Koehler 2005).

The model shown in Figure 7.3 has the following structure. Binary nodes labelled *unit n* with states 'positive' and 'negative' represent propositions according to which the n-th item may or may not contain illegal substance. The result of a diagnostic test applied to the n-th item is modelled by a binary node *test n* with states 'positive' and 'negative'. The outcome of a test depends directly on the presence or absence of the respective characteristic. Directed edges are thus adopted between the nodes *test n* and *unit n*. Five pairs of nodes *unit n* and *test n* are incorporated in the current model. For routine use, more trials may be needed and additional nodes can be added appropriately.

There may be circumstances casting doubt on the result of a test. Possible reasons for this are the condition of the sample or erroneous experimental settings. A test may not always give a 'positive result' when the item is truly positive, or may give a 'positive result' when in fact the item is not positive. Generally, two probabilities can be used to describe the accuracy of a test: the probability of a test being positive when the item is truly positive, $P(\textit{test } n = \textit{positive} \mid \textit{unit } n = \textit{positive})$, and the probability of a test being negative when the item is actually negative, $P(\textit{test } n = \textit{negative} \mid \textit{unit } n = \textit{negative})$. Sometimes, these two values are referred to as the sensitivity and specificity of a test (Balding 2005; Kaye 1987b; Lindley 2006; Robertson and Vignaux 1995, for example). They can be used to complete the probability table of the nodes *test n*. A test is taken to indicate the presence or absence of a unit's target characteristic with certainty only if one assumes that $P(\textit{test } n = \textit{positive} \mid \textit{unit } n = \textit{positive}) = P(\textit{test } n = \textit{negative} \mid \textit{unit } n = \textit{negative}) = 1$. For the purpose of the current discussion, a hypothetical value of 0.99 is chosen for both of these probabilities.

The probability an individual item contains or does not contain an illegal substance depends directly on the proportion of items in the consignment that contain illegal substances. Directed edges are thus drawn to the node *unit n* from the node *Prop*. The probability tables of the nodes *unit n* can be completed, for example, through the expression `Distribution(Prop,1-Prop)` (Hugin syntax).

Besides, the model also contains auxilary nodes, namely P, $\textit{Prop} > P$?, $\textit{Prop} > 0.5$?, $\textit{Prop} > 0.7$?, $\textit{Prop} > 0.75$?, $\textit{Prop} > 0.9$? and $\textit{Prop} > 0.95$?. These nodes define a substructure from which cumulative probabilities may be evaluated. The definitions of these variables are given in Table 7.3.

Table 7.3 Definitions of nodes used in the Bayesian network shown in Figure 7.3.

Node	Definition	States
a, b	parameters α and β of the beta distribution defined for the node *Prop*	$0.5, 1, 2, \ldots, 10$
Prop	proportion (θ) of positives in the consignment	0–0.05, ..., 0.95–1
P	lower limit for evaluating cumulative probabilities of the proportion of positives in the consignment	$0, 0.05, \ldots, 0.95, 1$
$\textit{Prop} > P$?	is the proportion of the positives in the consignment greater than P?	*yes*, *no*
$\textit{Prop} > 0.5$ $(0.7, \ldots)$	is the proportion greater than 0.5 $(0.7, \ldots)$?	*yes*, *no*
test 1 $(2, \ldots)$	outcome of test conducted in order to determine the characteristic of item 1 $(2, \ldots)$	*positive*, *negative*
unit 1 $(2, \ldots)$	true (but unknown) characteristic of item 1 $(2, \ldots)$	*positive*, *negative*

Example 7.3.1 *(Inspection of pills suspected to contain drugs – continued). Consider again the sampling scenario introduced earlier in Example 7.2.1, but assume that all* $n = 4$ *sampled units are found to be positive. In such a situation, following the discussion in Section 7.2.1, there is – assuming a uniform prior probability distribution for* θ *– a probability greater than 95% that the proportion of positive items* θ *is greater than 0.5.*

Figure 7.4 (i) depicts the Bayesian network described above with nodes expanded and instantiations made at the relevant nodes. The parameters for the beta distribution of the node Prop are set by instantiating both nodes a and b to 1*. The nodes unit 1 to unit 4 are set to 'positive'. This represents the observation of the four items found to be positive. Instantiation of nodes 'unit n' rather than nodes 'test n' follows from the assumption that the determination of the characteristics of a sampled item is made without error. The node Prop* > 0.5? *displays the target probability* $P(\theta > 0.5 \mid n = y = 4, \alpha = 1, \beta = 1)$*. The value* 0.97 *agrees with what has been found earlier in Section 7.2.1. Notice also that the node Prop shows the updated (posterior) probability distribution for the proportion of positives in the consignment. As may be read from the graph, the result is that – compared to the assumed uniform prior – higher proportions now are more probable. Cumulative probabilities for various intervals of proportions (other than* >0.5*), are displayed at the far right-hand side.*

The proposed Bayesian network readily allows the examination of a setting in which the determination of the analytical characteristics cannot be assumed to be error-free. This is shown in Figure 7.4 (ii). Here, the nodes 'test n' are instantiated instead of the nodes 'unit n'. By defining a value of 0.99 *for the test's sensitivity and specificity, each observation of a positive test result provides a likelihood ratio of* 99 *for the proposition according to which the respective unit is in fact positive. The effect of this uncertainty on the value of the target cumulative probability is weak as it changes the result by less than* 0.01*.*

In a more general case, one could enter one observation at a time and subsequently observe the changes in the probability distributions for the nodes of interest. Notice further that one need not only consider the BN for evaluating findings that have actually been obtained. One may also evaluate the probability with which future trials, given previous observations, can be expected to yield positive and negative testing results, respectively. This is illustrated in Figure 7.4 where a probability of approximately 0.83 *is indicated for the fifth item being positive.*

Besides considering the probability of future trials resulting in positive or negative findings, one may also evaluate the information that future findings can be expected to provide. More can be learned about such a question by instantiating, for example, the node test 5 *(not shown in Figure 7.4). If a*

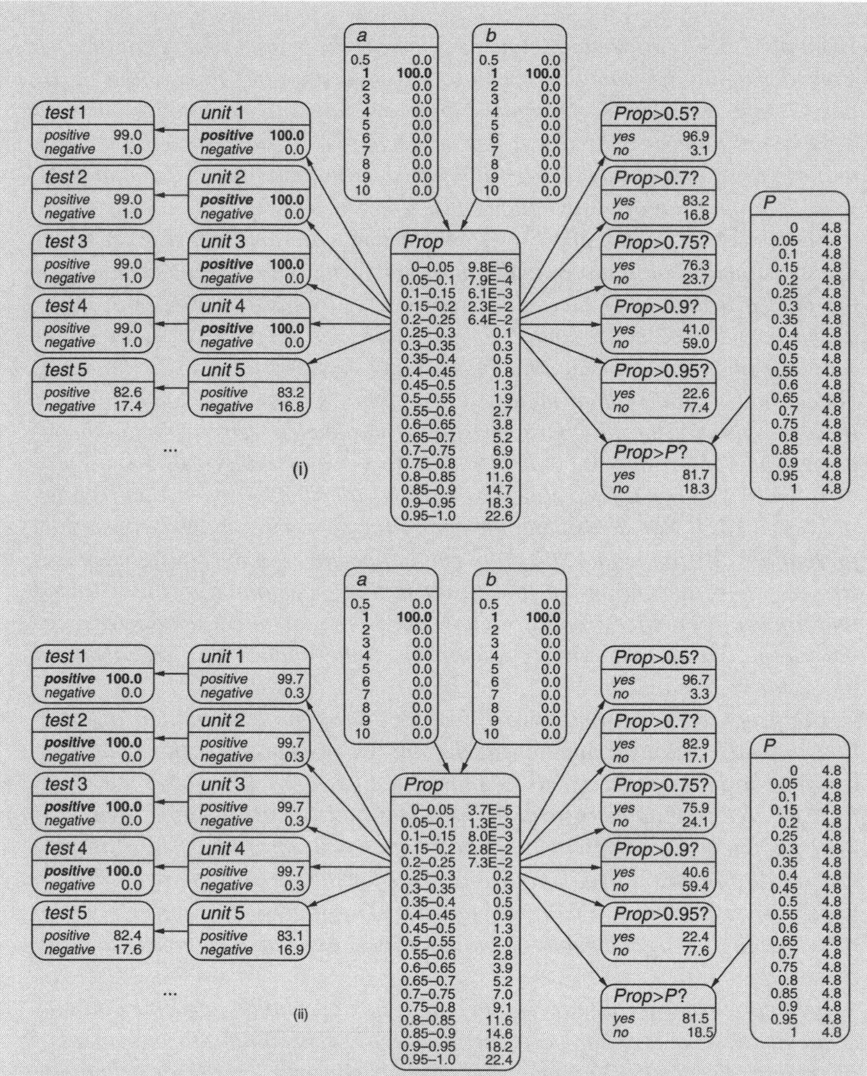

Figure 7.4 A Bayesian network for sampling from large consignments (Example 7.3.1). The definitions of the nodes are given in Table 7.3. Figure (i) displays the analysis of a setting assuming a uniform prior distribution for the parameter θ (node *Prop*) and four inspected items, all of which are found to be positive (assuming error-free analyses). Figure (ii) shows an evaluation of the scenario allowing for error in the analysis of the sampled items. Instantiations are shown in bold. (Reproduced from Biedermann *et al.* 2008. Law, Probability and Risk, **7**, 35–60.)

positive result is obtained, one can find that the probability of the proportion being greater than 0.5 *would increase a further two percentage points to approximately* 0.98. *In the case of a negative finding, this probability would decrease by approximatively six percentage points.*

7.3.3 Bayesian network for sampling from small consignments

Figure 7.5 depicts a Bayesian network for evaluating sampling scenarios that involve small consignments involving, typically, less than 50 units (Biedermann *et al.* 2007b). The probabilistic architecture underlying this network follows the Bayesian procedure described in Section 7.2.3. The target node of the Bayesian network is Z, the number of positive units among those not analyzed, and Z follows a binomial distribution. The node definitions are given in Table 7.4, except for the variables *Prop*, a and b which are the same as those given earlier in Table 7.3.

The structure of the proposed Bayesian network is based on the following considerations. There is a variable N, the consignment size, incorporated as a discrete chance node. It is modelled as a root node because it is not thought to depend on any other variable. The variable N is allowed to cover a total of 21 states, i.e. $0, 1, \ldots, 20$. A lower case letter[3] $n = 0, 1, \ldots, 20$ is used here to denote a particular instance of N. This Bayesian network enables its user to analyse scenarios involving up to 20 units.

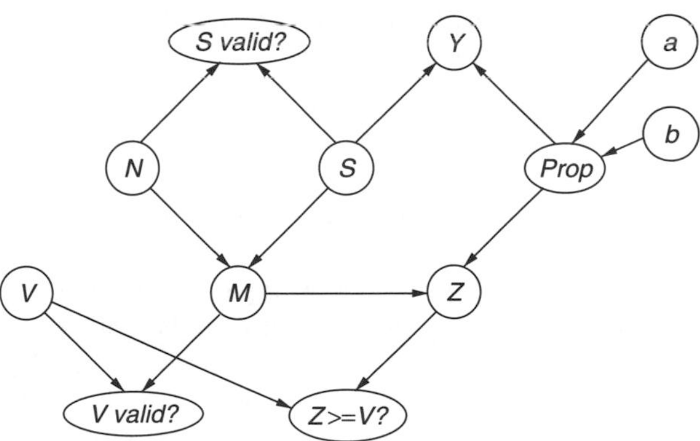

Figure 7.5 A Bayesian network for sampling from small consignments. The definitions of the nodes are given in Tables 7.3 and 7.4.

[3]Notice that this is a difference to the notation used in Section 7.2.3, where n denoted the number of inspected units.

Table 7.4 Definitions of nodes used in the Bayesian network shown in Figure 7.5.

Node	Definition	State
N	consignment size	$n = 0, 1, \ldots, 20$
S	sample size number of items inspected	$s = 0, 1, \ldots, 20$
M	number of items not inspected	$m = 0, 1, \ldots, 20$
Y	number of positive items in the sample	$y = 0, 1, \ldots, 20$
Z	number of positives among the uninspected items	$z = 0, 1, \ldots, 20$
V	lower limit for evaluating cumulative probabilities of the number of positives among the uninspected items	$v = 0, 1, \ldots, 20$
$Z >= V?$	are there at least V positives among the uninspected items?	*yes*, *no*
S *valid*?	constraint on S	*true*, *false*
V *valid*?	constraint on V	*true*, *false*

S, the number of inspected items, is defined in the same way as N, that is in terms of a numbered discrete chance node with states $s = 0, 1, \ldots, 20$. The prior probabilities to be specified for N and S are not a primary concern for the analyses considered here. Their specification is a technical matter necessary to run the model. When using the Bayesian network, the nodes N and S will be instantiated. The prior probabilities initially assigned to nodes N and S then become irrelevant. The rules for construction of Bayesian networks solely require that $\sum_n P(N = n) = \sum_s P(S = s) = 1$.

The Bayesian network also contains a substructure that assures that the instantiations that may be made at the nodes N and S satisfy the constraint $n \geq s$ $(n, s = 0, 1, \ldots, 20)$. That is, one cannot inspect more units (S) than there are units in the consignment (N). This constraint is implemented through the definition of a Boolean node S *valid*?. This node is a direct descendant of both N and S. The node table is completed as follows:

$$P(S \ \mathit{valid?} = \mathit{true} \mid N = n, S = s) = \begin{cases} 1, & n \geq s, \\ 0, & n < s, \end{cases} \quad (n, s = 0, 1, \ldots, 20).$$

In the Hugin environment, for instance, an expression of the form `N>=S` can be used to define this node table.

Knowledge about the consignment size N together with information about the number of inspected items S allows one to determine the state of a variable M, the number of units in the consignment that are not inspected. In analogy to N and S, M is defined as a discrete chance node with states $m = 0, 1, \ldots, 20$.

The node Y, with states $y = 0, 1, \ldots, 20$, represents the number of positives among S, the number of inspected items. Y is a variable with a binomial distribution whose arguments are provided by S and *Prop*. The node of primary interest, Z, is then defined analogously. It accounts for the number of positives among M, the units in the consignment that are not inspected. Following the discussion in Section 7.2.3, the distribution of Z is binomial with parameters that are given here by the nodes *Prop* and M. In addition to the node Z, there is also a substructure that is designed to provide more general statements about Z, i.e. a Boolean summary node, denoted '$Z \geq V$?', provides a probability for the event that the number of positives among the units not inspected (node Z) is at least, or greater than, V, a user-specified integer number. Here, the node V is a discrete chance node with states $v = 0, 1, \ldots, 20$. In analogy to the node S *valid*?, the node V *valid*? ensures illogical queries of the kind $v > m$ are not considered.

Example 7.3.2 *(Inspection of pills suspected to contain drugs – continued). Consider again a case of sampling from a small consignment as described earlier in Example 7.2.2. It was found, in that example, that there was a probability of* 0.985 *that there were at least* 2 *positive items among the* 5 *non-inspected units of a consignment of size* 10, *when in a sample of* 5 *inspected items all were found to be positive and a uniform beta prior distribution was assumed for* θ, the proportion of positives in the consignment.

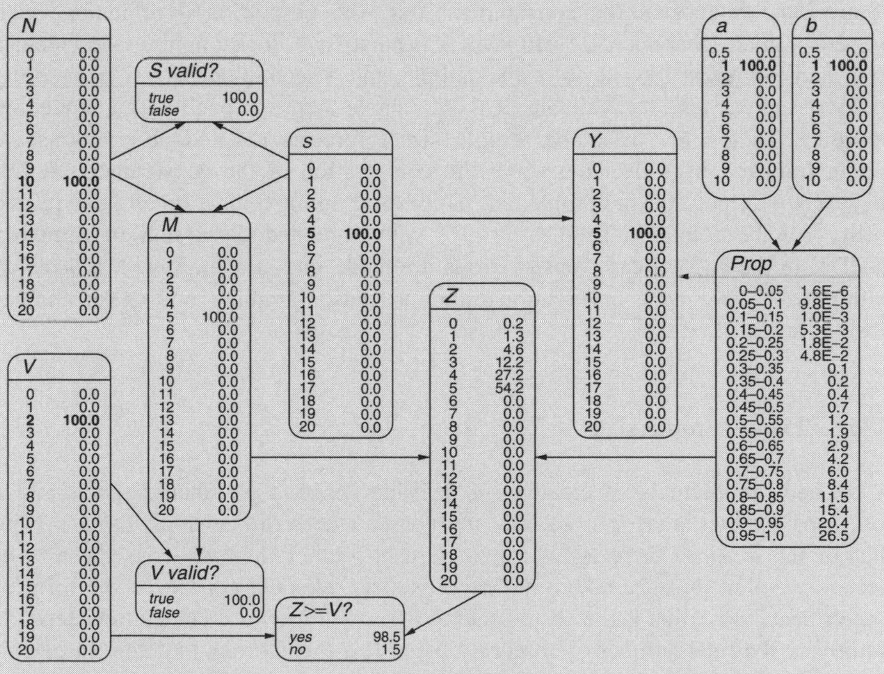

Figure 7.6 Bayesian network for the sampling problem as given by Example 7.3.2. Instantiations are shown in bold. Definitions of nodes are given in Tables 7.3 and 7.4.

This scenario is represented in Figure 7.6. The nodes representing the consignment size (N), the sample size (S), the number of observed positives (Y) and the parameters for the beta prior distribution (a and b) are instantiated. The node Z displays the probabilities for there being $z = 0, 1, 2, \ldots, m$ positives among the $m = 5$ non-inspected items. The node $Z >= V$? displays the probability that there are at least 2 positive units among the m non-inspected items. This value, 0.985, is in agreement with the result found above.

7.4 SAMPLING INSPECTION UNDER A DECISION-THEORETIC APPROACH

Suppose that the customer of a forensic laboratory, without knowing the outcome of an experiment, must make a decision, the consequences of which depend on the outcome of that experiment. Typically, the experiment is one that consists in the examination of items from a consignment, while it is of interest to make inference about some feature of the consignment (e.g. the proportion θ of items which present a given characteristic). In such a context, two decisions must be taken: a first decision about the size n of the sample, and a second decision d concerning θ. Two approaches are available: a fixed sample size approach and a sequential approach. In the first case the sample size is preselected and observations are made, followed by a decision about the composition of the consignment. A full Bayesian treatment for the problem at hand based on maximization of the expected utility can be found in Lindley (1997) with a related discussion in Bernardo (1997). In the second case, observations are made one at a time, with a decision being made after each observation either to cease sampling or to take another observation.

7.4.1 Fixed sample size

A decision problem is specified by a decision space, a parametric space, and a loss function that takes into account the consequences of decisions, $L(d, \theta)$. Note that in this setting, there is a further element – other than the consequences of decisions – that must be taken into account. That element appears in the form of experimental cost, that is the cost of taking observations, whose magnitude depends mainly on the total number of observations, n. It is then correct to define an overall loss, $L^0(d, \theta, n)$ that takes both of these aspects into account. In particular, it will be assumed that the overall loss is given by the sum of the loss function and a component of cost:

$$L^0(d, \theta, n) = L(d, \theta) + C(n). \tag{7.5}$$

Suppose that $x = (x_1, \ldots, x_n)$ have been observed, and a posterior density $\pi^n = \pi(\theta \mid x)$ becomes available. The expectation over θ of the overall loss function can be calculated and then minimized over d to select the optimal decision. Thus, the *Bayesian posterior expected loss* of decision d at time n (i.e. having observed n units), written $\bar{L}(d, \pi(\theta \mid x), n)$, is defined as

$$\begin{aligned}\bar{L}(d, \pi(\theta \mid x), n) &= E^{\pi(\theta|x)}\left[L^0(d, \theta, n)\right] \\ &= \int_\Theta L(d, \theta)\pi(\theta \mid x)d\theta + nc.\end{aligned} \tag{7.6}$$

The *Bayes risk* at time n, written $r(\pi^n, n)$, is defined to be the minimum value of the posterior expected loss over d:

$$r\left(\pi^n, n\right) = \min_{d \in \mathcal{D}} \bar{L}(d, \pi(\theta \mid x), n). \tag{7.7}$$

The choice that satisfies this criterion, denoted d^{π^n}, is called the *Bayes decision* (defined in Section 3.4.1) and represents the decision to be taken about θ if sampling has stopped after taking n observations. The objective thus is to select the decision whose associated expected loss is equal to the Bayes risk: this decision is said to be optimal because no lower risk can be attained with a different decision. The optimal sample size is the value greater than n, say n^*, which minimizes the Bayes risk:

$$n^* = \min_{n \in N} r(\pi^n, n). \tag{7.8}$$

For example, consider a random sample $(x_1, \ldots, x_n)$ from a Normal distribution, $X \sim N(\theta, \sigma^2)$, with known variance σ^2, and suppose it is of interest to make an inference about the unknown mean θ. Assume a Normal conjugate prior distribution $N(\mu, \tau^2)$ for *0*. The posterior distribution will then be Normal $N(\mu(x), \tau^2(x))$, with mean $\mu(x)$ and variance $\tau^2(x)$ as in (4.16) and (4.17). Assume a quadratic loss function $L(d, \theta) = k(d - \theta)^2$, that each observation has a constant observational cost c and that the total cost is proportional to the number of observations, so that $C(n) = nc$. With a quadratic loss function, the Bayes decision d^{π^n} is the posterior mean $\mu(x)$ (see Section 3.4.2). Then, the Bayes risk at time n is:

$$\begin{aligned}r(\pi^n, n) &= \min_{d \in \mathcal{D}} E^{\pi(\theta|x)}\left[L(d, \theta) + nc\right] = E^{\pi(\theta|x)}\left[L(d^{\pi^n}, \theta)\right] + nc \\ &= E^{\pi(\theta|x)}\left[k(\mu(x) - \theta)^2\right] + nc \\ &= k\tau^2(x) + nc.\end{aligned}$$

To find the minimum, differentiate with respect to n and set the result equal to zero:

$$\frac{\partial}{\partial n}\left[\frac{k\sigma^2\tau^2}{\sigma^2 + n\tau^2} + nc\right] = -\frac{k\sigma^2\tau^4}{(\sigma^2 + n\tau^2)^2} + c = 0.$$

Solving for n gives:

$$n^* = \frac{\sqrt{k}}{\sqrt{c}}\sigma - \frac{\sigma^2}{\tau^2}. \tag{7.9}$$

Example 7.4.1 *(Alcohol concentration in blood – continued). Imagine, as in Example 4.4.1, that one is interested in the estimation of the alcohol concentration in a blood sample. The laboratory cost is fixed at €50 for each analysis. Assume, as in Example 4.4.1, the standard deviation σ for the available measuring apparatus to be known and equal to* 0.0463, *and the standard deviation τ for the prior density of the unknown mean to be equal to* 0.3. *A quadratic loss $k(d-\theta)^2$ is chosen.*

The optimal sample size has been computed, according to (7.9), for different k-values, in particular $k = (10000, 50000, 100000)$, with results $n^* = 1, 2, 3$, respectively. Note that a considerable increase in the loss does not correspond to a substantial change in the optimal sample size. This is essentially explained by the precision of the laboratory measurements.

Once the optimal sample size has been chosen, then the Bayes decision about the unknown blood alcohol concentration is given by the posterior mean.

Consider the scenario described in Section 7.2.2, and suppose it is of interest to test the hypothesis $H_0 : \theta > \theta_0$ against $H_1 : \theta \leq \theta_0$. The parametric space is therefore given by $\Theta = \Theta_0 \cup \Theta_1$, where $\Theta_0 = (\theta_0, 1]$ and $\Theta_1 = [0, \theta_0]$. A beta conjugate prior distribution $\text{Be}(\alpha, \beta)$ and a binomial distribution $\text{Bin}(n, \theta)$ are taken to give a beta posterior distribution $\pi(\theta \mid x) = \text{Be}(\alpha + y, \beta + n - y)$. The decision space is $\mathcal{D} = \{d_0, d_1\}$, with d_0 denoting acceptance of H_0, and d_1 acceptance of H_1. A '0–k_i' loss function is chosen, that is

$$L(d_0, \theta) = \begin{cases} 0 & \theta \in \Theta_0 \\ k_0 & \theta \in \Theta_1 \end{cases} \quad ; \quad L(d_1, \theta) = \begin{cases} 0 & \theta \in \Theta_1 \\ k_1 & \theta \in \Theta_0 \end{cases}, \tag{7.10}$$

and the overall loss function $L^0(d, \theta, n)$ is taken to be as in (7.5)

$$L^0(d, \theta, n) = L(d, \theta) + C(n), \tag{7.11}$$

where the total cost is supposed to be proportional to the number of observations and the observational cost is assumed constant, that is $C(n) = nc$. The posterior expected losses are easily obtained

$$\begin{aligned}
\bar{L}(d_0, \pi(\theta \mid x), n) &= \int_{\Theta} L(d_0, \theta)\pi(\theta \mid x)d\theta + nc \\
&= \int_{\Theta_1} k_0 \pi(\theta \mid x)d\theta + nc = k_0 P^{\pi(\theta|x)}(\theta \leq \theta_0) + nc,
\end{aligned}$$

and

$$\bar{L}(d_1, \pi(\theta \mid x), n) = \int_{\Theta} L(d_1, \theta)\pi(\theta \mid x)d\theta + nc$$
$$= \int_{\Theta_0} k_1 \pi(\theta \mid x)d\theta = k_1 P^{\pi(\theta|x)}(\theta > \theta_0) + nc.$$

The Bayes risk (7.7) is therefore

$$r(\pi^n, n) = \min_{d_0, d_1} \left\{ \bar{L}(d_0, \pi(\theta \mid x), n), \bar{L}(d_1, \pi(\theta \mid x), n) \right\} \tag{7.12}$$

and the optimal sample size n^* is given by the minimum of (7.12) with respect to n, as in (7.8).

Example 7.4.2 *(Inspection of pills suspected to contain drugs – continued). A large consignment of pills is seized and it is desired to analyze a sample to test $H_0 : \theta > 0.5$ against $H_1 : \theta \leq 0.5$. Choose a uniform prior $\pi(\theta) = Be(1, 1)$.*

A '0–k_i' loss function is used. This example is a good place at which to introduce consideration of the quantification of a loss function in monetary terms. There are two components to the loss function in this example. The first, k_0, is the amount of compensation which may need to be allocated to an individual found to be guilty but who is truly innocent and subsequently exonerated. The second component, k_1, may be interpreted as the monetary value that would have been confiscated by the government as a penalty if the individual had not, incorrectly, been found not guilty. For the purpose of illustration in this example, k_0 and k_1 have been chosen equal with value €100,000. There is also a laboratory cost of €50 for each item analyzed.

The optimal sample size is found by minimizing the Bayes risk with respect to n:

$$r(\pi^n, n) = \min_{d_0, d_1}\{100000P(\theta \leq 0.5 \mid n, y, \alpha, \beta) + 50n, 100000P(\theta > 0.5 \mid n, y, \alpha, \beta) + 50n\}. \tag{7.13}$$

Assuming that all items are found to be positive, the optimal sample size is $n^* = 9$. The size of the consignment has been assumed equal to 100, so the Bayes risk has been computed for increasing values of n ($n = 1, \ldots, 100$), and then minimized over d_0 and d_1. The Bayes risk decreases until $n = 9$, and then increases as the cost of analysis of an item is the dominant cost.

The effect on the sample size of variation in the values of k_0 and k_1, whilst still retaining equality, is shown in Table 7.5. In the presence of negative

Table 7.5 Optimal sample size for a symmetric loss. A large consignment of pills is seized and it is desired to analyze a sample to test $H_0 : \theta > 0.5$ against $H_1 : \theta \leq 0.5$, given $\pi(\theta) = \text{Be}(1, 1)$, and a '$0-k_i$' loss function with $k_0 = k_1 = k$. The laboratory cost is €50.

Loss k	optimal size n^*
10000	6
50000	8
100000	9

items, the same methodology can be extended to allow for units which do not contain drugs. The minimization can be carried out using, for example, the R routines given at the end of the chapter, where provision is made for the situation where $k_0 \neq k_1$.

7.4.2 Sequential analysis

Consider a consignment of size N and assume observations can be taken sequentially. After observing $(x_1, \ldots, x_n)$ at most $m = N - n$ additional observations can be taken. The sequential decision procedure incorporates two components: the stopping rule s^π, that defines when to stop sampling, and the decision rule d^π, that defines the action to be taken if the sampling has been stopped. Within a sequential sampling procedure, denoted $\mathbf{d} = (s^\pi, d^\pi)$, the decision problem is rephrased after each observation as starting at that point. The procedure works by making a comparison between the Bayes risk that can be attained by stopping sampling and taking a decision immediately, $r_0(\pi^n, n)$, and the smallest Bayes risk that can be attained by taking more observations (up to m), $r_m(\pi^n, n)$.

Imagine a client who, having observed $(x_1, \ldots, x_n)$, must decide whether to take a decision about the quantity of interest θ without further observations, or to instruct the scientist to continue and observe X_{n+1}, whose outcome is unknown. The Bayes risk of an immediate decision, $r_0(\pi^n, n)$, is to be compared with the Bayes risk that one can expect by observing X_{n+1} (in which case at most $(m-1)$ observations can be taken), namely

$$E^X\left[r_{m-1}\left(\pi^n(\theta \mid X_{n+1}), n+1\right)\right], \tag{7.14}$$

where the expectation is taken over X with respect to the marginal density of X, $m(x)$. The smallest Bayes risk that can be attained, $r_m(\pi^n, n)$, is given by

$$r_m(\pi^n, n) = \min\left\{r_0(\pi^n, n), E^X\left[r_{m-1}\left(\pi^n(\theta \mid X_{n+1}), n+1\right)\right]\right\}. \qquad (7.15)$$

The optimal sampling inspection procedure is to cease when the overall risk is minimized. Therefore, the optimal course of action is to stop sampling and make a decision when $r_0(\pi^n, n) = r_m(\pi^n, n)$. The procedure works iteratively as follows. At stage 0, compare the Bayes risk of an immediate decision, $r_0(\pi, 0)$ (no observation taken) with the smallest Bayes risk that can be attained by inspecting at most m items (at stage 0, $m = N$), $r_m(\pi, 0)$. If $r_0(\pi, 0)$ is smaller than $r_m(\pi, 0)$ make no observations and take a decision immediately, otherwise proceed to stage 1. At stage 1, once x_1 has been observed, compare the Bayes risk of an immediate decision, $r_0(\pi^1, 1)$, with the smallest Bayes risk that can be attained by inspecting at most $(m-1)$ items, $r_{m-1}(\pi^1, 1)$; continue sampling if the latter is smaller, and so on.

Assume a linear loss function as in (7.5) with $C(n) = nc$. The Bayes risk of an immediate decision, is given by

$$r_0(\pi^n, n) = \rho_0(\pi^n) + nc,$$

where $\rho_0(\pi^n) = E^{\pi(\theta|x)}\left[L(d^{\pi^n}, \theta)\right]$ denotes the posterior Bayes decision risk. It follows from (7.15) that

$$\rho_m(\pi^n) = \min\left\{\rho_0(\pi^n), E^X\left[\rho_{m-1}\left(\pi^n(\theta \mid X_{n+1})\right)\right] + c\right\}. \qquad (7.16)$$

The optimal course of action is to stop sampling and make a decision at the point of the first n observations for which

$$\rho_0(\pi^n) = \rho_m(\pi^n). \qquad (7.17)$$

The difficulty of this approach can be effectively illustrated through a simple example. Suppose it is desired to test $H_0 : \theta = \theta_0$ against $H_1 : \theta = \theta_1$. The parameter space thus has two points: $\Theta = \{\theta_0, \theta_1\}$. Let π_i denote the prior probability that H_i is true. The decision space is $\mathcal{D} = \{d_0, d_1\}$, with d_i denoting acceptance of H_i, $i = 0, 1$. Consider a symmetric '0–k_i' loss function. The loss is zero if the decision is correct, a positive value $k = k_0 = k_1$ applies for decisions that are not correct. The experimental cost for each inspected unit is c. Assume observations are sampled sequentially and that the probability distribution for the outcome of an inspection is taken to be binomial, $\text{Bin}(1, \theta)$. At stage 0, the decision maker must evaluate the opportunity of taking a decision immediately with no cost of inspection, or observing one unit. The expected losses of an immediate decision are respectively:

$$\bar{L}(d_0, \pi) = E^\pi\left[L(d_0, \theta)\right] = k(1 - \pi_0),$$
$$\bar{L}(d_1, \pi) = E^\pi\left[L(d_1, \theta)\right] = k\pi_0.$$

Therefore, the Bayes risk of an immediate decision, Equation (7.7), is

$$r_0(\pi, 0) = \rho_0(\pi) = \min\left(k(1-\pi_0), k\pi_0\right) = \begin{cases} k\pi_0 & 0 \le \pi_0 \le 1/2 \\ k(1-\pi_0) & 1/2 < \pi_0 \le 1 \end{cases}. \tag{7.18}$$

This should be compared with the Bayes risk expected with an observation of X_1, that is

$$E^X\left[\rho_{m-1}\left(\pi(\theta_0 \mid X_1)\right)\right] + c = \rho_{m-1}\left(\pi(\theta_0 \mid 0)\right) m(0) + \rho_{m-1}\left(\pi(\theta_0 \mid 1)\right) m(1) + c. \tag{7.19}$$

It is necessary to compute the posterior Bayes decision risk $\rho_{m-1}\left(\pi(\theta \mid x)\right)$, the marginal density $m(x)$, and the posterior distribution $\pi(\theta \mid x)$ at $x = 0, 1$ to solve (7.19).

The marginal density $m(x)$ at $x = 0, 1$ is equal to

$$m(x) = E^\pi\left[f(x \mid \theta)\right] = f(x \mid \theta_0)\pi_0 + f(x \mid \theta_1)\pi_1 = \begin{cases} (1-\theta_0)\pi_0 + (1-\theta_1)\pi_1 & \text{if} \quad x = 0 \\ \theta_0\pi_0 + \theta_1\pi_1 & \text{if} \quad x = 1 \end{cases}. \tag{7.20}$$

The posterior distribution $\pi(\theta_0 \mid x)$ at $x = 0, 1$ is determined by

$$\pi(\theta_0 \mid x) = \frac{\pi_0 f(x \mid \theta_0)}{m(x)} = \frac{\pi_0\theta_0^x(1-\theta_0)^{1-x}}{m(x)} = \begin{cases} \frac{\pi_0(1-\theta_0)}{\pi_0(1-\theta_0)+\pi_1(1-\theta_1)} & \text{if} \quad x = 0 \\ \frac{\pi_0\theta_0}{\pi_0\theta_0+\pi_1\theta_1} & \text{if} \quad x = 1 \end{cases}. \tag{7.21}$$

Finally, the posterior expected losses are

$$\rho_{m-1}\left(\pi(\theta_0 \mid 0)\right) = \min\left\{k\pi(\theta_0 \mid 0), k(1 - \pi(\theta_0 \mid 0))\right\}$$
$$\rho_{m-1}\left(\pi(\theta_0 \mid 1)\right) = \min\left\{k\pi(\theta_0 \mid 1), k(1 - \pi(\theta_0 \mid 1))\right\}.$$

Now the decision maker has all the structure necessary to implement a sequential decision procedure. Consider $k = 5$, $\theta_0 = 3/4$, $\theta_1 = 1/4$, $c = 1$. The Bayes risk of an immediate decision, Equation (7.18), is simply

$$\rho_0(\pi) = \begin{cases} 5\pi_0 & 0 \le \pi_0 \le 1/2 \\ 5(1-\pi_0) & 1/2 < \pi_0 \le 1 \end{cases}.$$

In the same way, from (7.20) and (7.21), the marginal density $m(x)$ and the posterior distribution $\pi(\theta_0 \mid x)$ are computed at $x = 0, 1$, to be

$$m(x) = \begin{cases} \frac{3-2\pi_0}{4} & \text{if} \quad x = 0 \\ \\ \frac{1+2\pi_0}{4} & \text{if} \quad x = 1 \end{cases},$$

and

$$\pi(\theta_0 \mid x) = \begin{cases} \frac{\pi_0}{3-2\pi_0} & \text{if} \quad x = 0 \\ \\ \frac{3\pi_0}{1+2\pi_0} & \text{if} \quad x = 1 \end{cases}.$$

Therefore, one can compute

$$\begin{aligned} \rho_{m-1}\left(\pi(\theta_0 \mid 0)\right) &= \min\left\{5\pi(3/4 \mid 0), 5\left(1 - \pi(3/4 \mid 0)\right)\right\} \\ &= \begin{cases} 5\pi\left(\frac{3}{4} \mid 0\right) & \text{if} \quad \pi\left(\frac{3}{4} \mid 0\right) \leq 1/2 \\ \\ 5\left(1 - \pi\left(\frac{3}{4} \mid 0\right)\right) & \text{if} \quad \pi\left(\frac{3}{4} \mid 0\right) > 1/2 \end{cases} \\ &= \begin{cases} \frac{5\pi_0}{3-2\pi_0} & \text{if}\, \pi_0 \leq \frac{3}{4} \\ \\ \frac{5(3-3\pi_0)}{3-2\pi_0} & \text{if}\, \pi_0 > \frac{3}{4} \end{cases}, \end{aligned}$$

and similarly

$$\begin{aligned} \rho_{m-1}\left(\pi(\theta_0 \mid 1)\right) &= \min\left\{5\pi(3/4 \mid 1), 5\left(1 - \pi(3/4 \mid 1)\right)\right\} \\ &= \begin{cases} 5\pi\left(\frac{3}{4} \mid 1\right) & \text{if}\, \pi\left(\frac{3}{4} \mid 1\right) \leq 1/2 \\ \\ 5\left(1 - \pi\left(\frac{3}{4} \mid 1\right)\right) & \text{if}\, \pi\left(\frac{3}{4} \mid 1\right) > 1/2 \end{cases} \\ &= \begin{cases} \frac{15\pi_0}{1+2\pi_0} & \text{if}\, \pi_0 \leq \frac{1}{4} \\ \\ \frac{5(1-\pi_0)}{1+2\pi_0} & \text{if}\, \pi_0 > \frac{1}{4} \end{cases}. \end{aligned}$$

Finally, considering separately the intervals $(0, 1/4]$, $(1/4, 3/4]$, $(3/4, 1]$, the Bayes risk expected with an observation X_1 is given by

$$\begin{aligned} E^X\left[\rho_{m-1}\left(\pi(\theta \mid X_1)\right)\right] &= \rho_{m-1}\left(\pi(\theta_0 \mid 0)\right) m(0) + \rho_{m-1}\left(\pi(\theta_0 \mid 1)\right) m(1) \\ &= \begin{cases} \left(\frac{5\pi_0}{3-2\pi_0}\right) \frac{3-2\pi_0}{4} + \frac{15\pi_0}{1+2\pi_0} \frac{1+2\pi_0}{4} & \text{if}\, \pi_0 \leq \frac{1}{4} \\ \\ \frac{5\pi_0}{3-2\pi_0} \frac{3-2\pi_0}{4} + \frac{5(1-\pi_0)}{1+2\pi_0} \frac{1+2\pi_0}{4} & \text{if}\, \frac{1}{4} < \pi_0 \leq \frac{3}{4} \\ \\ \frac{5(3-3\pi_0)}{3-2\pi_0} \frac{3-2\pi_0}{4} + \frac{5(1-\pi_0)}{1+2\pi_0} \frac{1+2\pi_0}{4} & \text{if}\, \pi_0 > \frac{3}{4} \end{cases} \\ &= \begin{cases} 5\pi_0 & \text{if}\, \pi_0 \leq \frac{1}{4} \\ \\ \frac{5}{4} & \text{if}\, \frac{1}{4} < \pi_0 \leq \frac{3}{4} \\ \\ 5(1 - \pi_0) & \text{if}\, \pi_0 > \frac{3}{4} \end{cases}. \end{aligned}$$

Then, the smallest Bayes risk that can be attained is

$$\rho_m(\pi) = \min\left\{\rho_0(\pi), E^X\left[\rho_{m-1}\left(\pi(\theta \mid X_1)\right)\right] + c\right\}$$
$$= \begin{cases} 5\pi_0 & \text{if} \quad \pi_0 \le \frac{9}{20} \\ \frac{9}{4} & \text{if} \quad \frac{9}{20} < \pi_0 \le \frac{11}{20} \\ 5(1-\pi_0) & \text{if} \quad \pi_0 > \frac{11}{20} \end{cases} .$$

Let $\pi_0 = 0.6$, for example. Then, $\rho_0(\pi) = 5 \times 0.4 = 2$, and $\rho_m(\pi) = 5 \times 0.4 = 2$, and the optimal sequential procedure is to take a decision immediately (no sampling). Conversely, imagine $\pi_0 = 0.5$, then $\rho_0(\pi) = 5 \times 0.5 = 2.5$, and $\rho_m(\pi) = \frac{9}{4}$. The optimal sequential decision procedure is to observe X_1. Once x_1 has been observed, one must compare $\rho_0(\pi^1)$ with $\rho_{m-1}(\pi^1)$, and so on. However, computations soon become unfeasible.

7.4.3 Sequential probability ratio test

The most commonly used sequential procedure is the sequential probability ratio test (SPRT), introduced by Wald in the 1940s (Wald 1947). The SPRT is designed to test a simple null hypothesis against a simple alternative hypothesis. Such testing scenarios may not meet the requirements in some cases, in particular in the area of forensic science. The procedure will thus be generalized to cases of interest that involve composite hypotheses (Bozza and Taroni 2009; Bozza *et al.* 2008a).

SPRT for testing simple hypotheses

Consider initially a pair of simple hypotheses about the composition of a consignment: $H_0 : \theta = \theta_0$ and $H_1 : \theta = \theta_1$. Assume a sample of items can be inspected sequentially at a constant cost c per observation. Formally, the parameter space is $\Theta = \{\theta_0, \theta_1\}$, so that a prior π can be specified by $\pi_0 = P(\theta = \theta_0)$; $\pi_1 = P(\theta = \theta_1)$. The decision space, too, has only two elements, $\mathcal{D} = \{d_0, d_1\}$:

d_0: the proportion of 'positive' items is θ_0;
d_1: the proportion of 'positive' items is θ_1.

Consider a '0–k_i' loss function as in (7.10), and assume that $L^0(d,\theta,n) = L(d,\theta) + nc$, where n is the sample size. As outlined in Section 7.4.2 the optimal decision procedure consists in stopping sampling for the first n observations at which the Bayes risk of an immediate decision, $\rho_0(\pi^n)$, equals the smallest Bayes risk that can be attained among procedures that involve at least one and at most m observations, $\rho_m(\pi^n)$.

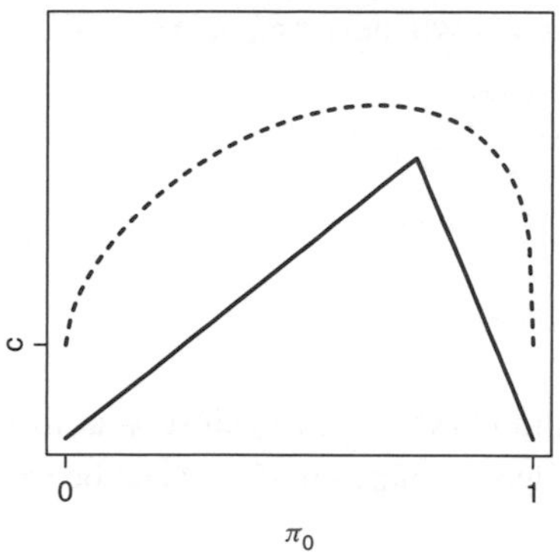

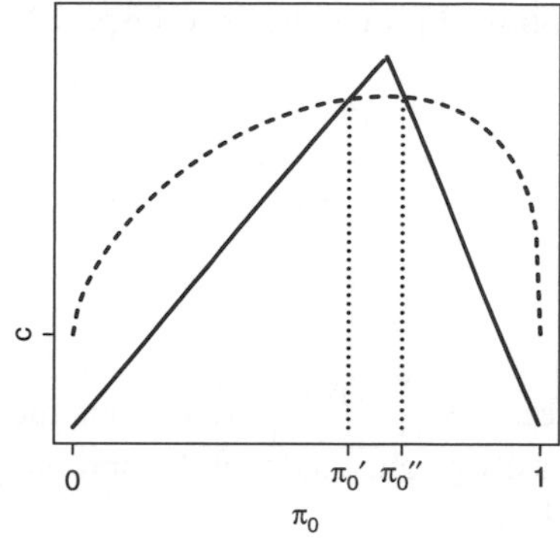

Figure 7.7 Bayes risk $\rho_0(\pi)$ associated with an immediate decision, solid line; smallest Bayes risk $\rho_m(\pi)$ associated with procedures involving at least one observation and at most m, dashed line. Case 1: $\rho_0(\pi) \leq \rho_m(\pi)$ for all π_0 (left); Case 2: $\rho_0(\pi) > \rho_m(\pi)$ for some π_0; π_0', $\pi_0 < \pi_0''$ (right).

At stage 0, $\rho_0(\pi)$ is compared with $\rho_m(\pi)$, that is a decision is to be taken whether to make or not to make an observation. The Bayes risk at stage 0, Equation (7.18), is a piecewise linear function, with $\rho_0(0) = \rho_0(1) = 0$, see Figure 7.7. As far as $\rho_m(\pi)$ is concerned, it can be demonstrated that it is a concave continuous function on the interval $0 \leq \pi_0 \leq 1$ (Berger 1988). Moreover, since it is computed for decision procedures involving at least one observation, then the sampling cost is at least c, so $\rho_m(0) = \rho_m(1) = c$, and $\rho_m(\pi_0) \geq c$ for all values of π_0. Two cases need to be distinguished:

1. $\rho_0(\pi) \leq \rho_m(\pi)$ for all π_0 (Figure 7.7, left).
   In this situation the Bayes procedure is to take no observations, since no lower risk can be attained.
2. $\rho_0(\pi) > \rho_m(\pi)$ for some π_0 (Figure 7.7, right).
   There will be some values π_0 for which it is optimal to terminate sampling. Generally, these values are close to 0 or 1 because the reduced margin of uncertainty does not suggest the inspection of more items. In particular, there exist two bounds π_0' and π_0'' such that for any $\pi_0 < \pi_0'$ or $\pi_0 > \pi_0''$, then $\rho_0(\pi) < \rho_m(\pi)$, and it is preferable to stop sampling. For other values of π_0, in particular $\pi_0' \leq \pi_0 \leq \pi_0''$, then $\rho_0(\pi) > \rho_m(\pi)$, and it is worthwhile to analyze more units because the overall risk is smaller.

Suppose, for instance, that the prior probability π_0 lies in the interval $\pi_0' \leq \pi_0 \leq \pi_0''$, so that it is worthwhile to take at least one observation. Suppose n observations $(x_1, \ldots, x_n)$, denoted x, are taken, and the posterior probability is $\pi_0^n = \pi(\theta_0 \mid x)$.

The optimal procedure is to inspect further units whenever $\pi_0' < \pi_0^n < \pi_0''$, that is

$$\pi_0' < \frac{\prod_{i=1}^n f(x_i \mid \theta_0)\pi_0}{\prod_{i=1}^n f(x_i \mid \theta_0)\pi_0 + \prod_{i=1}^n f(x_i \mid \theta_1)\pi_1} < \pi_0''. \tag{7.22}$$

The posterior distribution π_0^n can be expressed as

$$\frac{1}{1 + \frac{\pi_1}{\pi_0} L_n},$$

where $L_n = \frac{\prod_{i=1}^n f(x_i | \theta_1)}{\prod_{i=1}^n f(x_i | \theta_0)}$ represents the likelihood ratio of θ_1 against θ_0 at time n. It can easily be checked that the optimal procedure obtained in (7.22) can be rewritten as follows:

$$A = \frac{\pi_0(1 - \pi_0'')}{\pi_1 \pi_0''} < L_n = \frac{\prod_{i=1}^n f(x_i \mid \theta_1)}{\prod_{i=1}^n f(x_i \mid \theta_0)} < \frac{\pi_0(1 - \pi_0')}{\pi_1 \pi_0'} = B. \tag{7.23}$$

The optimal Bayes procedure is to continue with the inspection of items whenever the last relation, (7.23), is satisfied. That is:

- if $L_n \leq A$, *stop sampling and decide* d_0;
- if $L_n \geq B$, *stop sampling and decide* d_1;
- if $A < L_n < B$, *take another observation*.

The procedure can be reformulated as follows. Define the random variable

$$Z_i = \log\left(\frac{f(x_i \mid \theta_1)}{f(x_i \mid \theta_0)}\right).$$

Then, relation (7.23) becomes:

$$l_A < S_n = \log(L_n) = \sum_{i=1}^n Z_i =< l_B, \tag{7.24}$$

where $l_A = \log A$ and $l_B = \log B$. The sequential probability ratio test then works as follows:

- if $S_n \leq l_A$, *stop sampling and decide* d_0;
- if $S_n \geq l_B$, *stop sampling and decide* d_1;
- if $l_A < S_n < l_B$, *take another observation*.

The problem can then be rephrased as that of choosing l_A and l_B such as to minimize the Bayes risk. Let:

$$\alpha_0 = P_{\theta_0}(\text{deciding}\, d_1) = P_{\theta_0}\,(S_n \geq l_B)$$
$$\alpha_1 = P_{\theta_1}(\text{deciding}\, d_0) = P_{\theta_1}\,(S_n \leq l_A)\,.$$

Let $E_{\theta_0} n^*$ and $E_{\theta_1} n^*$ denote the expected stopping times under θ_0 and θ_1, respectively, where $n^* = \min\{n : S_n \leq l_A \text{ or } S_n \geq l_B\}$ denotes the unknown stopping time. The optimal sequential procedure **d** is obtained by minimizing the Bayes risk that is equal to

$$r(\pi, \mathbf{d}) = \pi_0\left(\alpha_0 k_1 + cE_{\theta_0} n^*\right) + \pi_1\left(\alpha_1 k_0 + cE_{\theta_1} n^*\right). \tag{7.25}$$

The problem reduces to the calculation of α_0, α_1, $E_{\theta_0}n^*$, $E_{\theta_1}n^*$, and the subsequent minimization of (7.25) over l_A and l_B. Reasonably accurate approximations exist and simplify the calculation considerably (Berger 1988).

SPRT for testing composite hypotheses

The aim is to test $H_0 : \theta \leq \theta_0$ versus $H_1 : \theta \geq \theta_1$, a setting for which the SPRT needs some convenient generalization.

In the current sequential decision problem, the decision space is $\mathcal{D} = \{d_0, d_1\}$:

d_0: the proportion θ of 'positive' units is lower than θ_0;
d_1: the proportion θ of 'positive' units is greater than θ_1,

while the parameter space is $\Theta = \Theta_0 \cup \Theta_1$, with $\Theta_0 = [0, \theta_0]$ and $\Theta_1 = [\theta_1, 1]$. The observational cost is still taken to be constant and equal to c, and the loss function introduced above also remains unchanged, that is $L^0(d, \theta, n) = L(d, \theta) + nc$, with $L(d, \theta)$ being a '0–k_i' loss function.

Here, θ_0 and θ_1 represent the boundaries of regions between which it is important to distinguish, while the SPRT is designed to test simple hypotheses. In this specific case, all possible values of θ may need to be considered. Hence, it is important to investigate the error probabilities and the expected sample sizes of an SPRT for all values of θ. Let:

$$\beta(\theta) = P_\theta(\text{deciding}\, d_1) = P_\theta(S_N \geq l_B)$$
$$\alpha(\theta) = P_\theta(\text{deciding}\, d_0) = P_\theta(S_N \leq l_A),$$

and let $E_\theta(n^*)$ be the expected stopping time for an arbitrary value of θ. In analogy to the SPRT designed for testing pairs of simple hypotheses, the problem can be rephrased as that of choosing l_A and l_B such as to minimize the Bayes risk of the sequential procedure, which is equal to:

$$r(\pi, \mathbf{d}) = \int_{\Theta_0} \left[\beta(\theta)k_1 + cE_\theta n^*\right]\pi(\theta)d\theta + \int_{\Theta_1} \left[\alpha(\theta)k_0 + cE_\theta n^*\right]\pi(\theta)d\theta. \quad (7.26)$$

Error probabilities and expected sample sizes need to be approximated. The basic tool that is used in approximating $\beta(\theta)$ and $E_\theta n^*$ is the *fundamental identity of sequential analysis*, first established by Wald (DeGroot 1970; Wald 1947). Suppose that $Z_1, Z_2, \ldots$ is a sequence of independent and identically distributed random variables such that $E_\theta(Z_i) = \mu_\theta$ and $E_\theta(Z_i - \mu_\theta)^2 = \sigma_\theta^2$. Then, under general conditions, for any sequential procedure for which the moment generating function $M_\theta(t)$[4] exists for t in a neighbourhood of the origin, the Wald approximations to

[4]Let X be a random variable with cumulative distribution function $F(x)$. The moment generating function (*mgf*) of X, denoted by $M_X(t)$ is

$$M_X(t) = Ee^{tX}.$$

As the name suggests, the moment generating functions can be used to calculate moments. However, the main use is for help in the characterization of a distribution (Casella and Berger 2002).

the error probability $\beta(\theta)$ and to the expected number of observations $E_\theta n^*$ can be obtained. In particular, it can be shown that:

$$\beta(\theta) \cong \tilde{\beta}(\theta) = \begin{cases} \frac{1-\exp(t_\theta l_A)}{\exp(t_\theta l_B)-\exp(t_\theta l_A)} & \text{if } \mu_\theta \neq 0 \\ \frac{-l_A}{l_B - l_A} & \text{if } \mu_\theta = 0 \end{cases}, \tag{7.27}$$

with t_θ such that $M_\theta(t_\theta) = 1$ $(t_\theta \neq 0)$, and

$$E_\theta n^* \cong \tilde{E}_\theta n^* = \begin{cases} \frac{l_A + (l_B - l_A)\tilde{\beta}(\theta)}{\mu_\theta} & \text{if } \mu_\theta \neq 0 \\ \frac{-l_A l_B}{\sigma_\theta^2} & \text{if } \mu_\theta = 0 \end{cases}. \tag{7.28}$$

The Bayes risk of the sequential procedure will be that risk which is the minimum value of

$$r(\pi, \mathbf{d}) \cong \int_{\Theta_0} \left[\tilde{\beta}(\theta)k_1 + c\tilde{E}_\theta n^*\right] \pi(\theta) d\theta + \int_{\Theta_1} \left[\tilde{\alpha}(\theta)k_0 + c\tilde{E}_\theta n^*\right] \pi(\theta) d\theta, \tag{7.29}$$

and where minimization is over l_A and l_B.

Example 7.4.3 *(Inspection of pills suspected to contain drugs – continued). A consignment of individual tablets is seized and at least a certain proportion of it is suspected to contain an illegal substance (e.g. a substance belonging to the general class of amphetamines). Suppose that the positions taken by the prosecution and the defence are such that the following pair of competing propositions can be formulated:*

$H_p : \theta \in [\theta_p, 1]$. *The unknown proportion θ of pills in the consignment that contain an illegal substance is greater than θ_p;*

$H_d : \theta \in [0, \theta_d]$. *The unknown proportion θ of pills in the consignment that contain an illegal substance is lower than $\theta_d, \theta_d < \theta_p$.*

The subscripts p and d are used here – in accordance with a predominant part of literature on forensic statistics – as an indicator of the proposition forwarded by, respectively, the prosecution and the defence.

Notice that θ may be considered as a threshold, for a court of justice, to declare a guilty judgement. The defence might not sustain the hypothesis of a proportion of positive items in the consignment as a sign of guilt. It could argue that the presence of positive elements is an example of external contamination for which the suspect is not responsible. For these reasons, the boundary θ_d of the region for H_d may be much lower than the boundary θ_p

for H_p. Here, a boundary of 0.65 *for both hypotheses is considered for the sake of illustration only.*

When it is decided that individual tablets will be sampled and inspected, then that process will end, at some point, with a decision which reflects the acceptance or rejection of a target hypothesis. In particular, the decision space is $\mathcal{D} = \{d_p, d_d\}$:

– d_p: the proportion θ of pills with illegal content is greater than θ_p;
– d_d: the proportion θ of pills with illegal content is lower than θ_d.

Each item can be inspected separately at a constant laboratory cost c of €100. The consequences of a wrong decision are evaluated and quantified in terms of €100,000 and are assumed symmetric, that is $k_p = k_d =$ €100,000, as in Example 7.4.2. Here, k_p is the loss associated with d_p when H_d holds whereas k_d is the loss associated with d_d when H_p holds.

Generally, the physical aspect of the seized units, such as colour, size or shape, may be used to shape a prior distribution of θ. For the purpose of the current discussion, this is not investigated in further detail. As merely a technical convention, a uniform beta distribution (with parameters $\alpha = \beta = 1$) is retained.

Observations can be considered as draws from a binomial distribution $Bin(1, \theta)$. Then, one has

$$\begin{aligned} Z_i &= \log \frac{f(x_i \mid \theta_p)}{f(x_i \mid \theta_d)} = X_i \log \frac{\theta_p}{\theta_d} + (1 - X_i) \log \left(\frac{1-\theta_p}{1-\theta_d} \right) \\ &= \log \left(\frac{\theta_p(1-\theta_d)}{\theta_d(1-\theta_p)} \right) X_i + \log \left(\frac{1-\theta_p}{1-\theta_d} \right) \end{aligned}$$

and

$$S_n = \sum_{i=1}^{n} Z_i = \log \left(\frac{\theta_p(1-\theta_d)}{\theta_d(1-\theta_p)} \right) \sum_{i=1}^{n} X_i + n \log \left(\frac{1-\theta_p}{1-\theta_d} \right). \quad (7.30)$$

Note that Z_i can be rewritten, for short, as $Z_i = f X_i + g$, where $f = \left(\log \frac{\theta_p(1-\theta_d)}{\theta_d(1-\theta_p)} \right)$ and $g = \log \left(\frac{1-\theta_p}{1-\theta_d} \right)$. The quantities μ_θ and σ_θ^2 that are needed for the Wald approximations, are then given by

$$\begin{aligned} \mu_\theta &= E_\theta\left(Z_i\right) = f\theta + g \\ \sigma_\theta^2 &= E_\theta\left(Z_i - \mu_\theta\right)^2 \\ &= E_\theta\left(f^2(X_i - \theta)^2\right) = f^2\theta(1-\theta). \end{aligned}$$

A standard calculation (Casella and Berger 2002) shows that the moment generating function of Z_i, that is also necessary for the Wald approximations, is

$$M_\theta(t) = E_\theta\left[e^{tZ_i}\right] = e^{gt}\left[\theta e^{ft} + (1-\theta)\right].$$

The Bayes risk is calculated using approximations as in (7.27) and (7.28) (Bozza 2008a).

The numerical minimization of the approximated Bayes risk, (7.29), gives $l_A = -0.2$ and $l_B = 0.2$. According to the proposed procedure, individual units are sampled and inspected sequentially, with S_n calculated according to Equation (7.30) at each iteration. This process ends as soon as $S_n \leq -0.2$ or $S_n \geq 0.2$. In the current scenario a final sample size of $n = 13$ is obtained if all items are found to be positive. This sequential decision procedure is illustrated in Table 7.6.

Table 7.6 Sequential decision procedure when all inspected units are found to be positive with S_n as given in (7.30) (Example 7.4.3: $c = 100$, $k_p = k_d = 100000$).

Item	Characteristic	S_n	Decision
1	Positive	0.015	take another observation
2	Positive	0.031	take another observation
3	Positive	0.046	take another observation
4	Positive	0.062	take another observation
5	Positive	0.077	take another observation
6	Positive	0.093	take another observation
7	Positive	0.108	take another observation
8	Positive	0.124	take another observation
9	Positive	0.139	take another observation
10	Positive	0.155	take another observation
11	Positive	0.170	take another observation
12	Positive	0.186	take another observation
13	Positive	0.201	stop and decide d_p

The optimal number of observations depends on the outcome of the experiment, the laboratory cost and the assessment of the loss function.

Consider the same case study with unchanged overall loss, but with the second item found to be negative. Table 7.7 shows that this results in a final sample size increased by three items from that in Table 7.6.

Table 7.7 Sequential decision procedure when the second unit inspected is negative with S_n as given in (7.30) ($c = 100$, $k_p = k_d = 100000$).

Item	Characteristic	S_n	Decision
1	Positive	0.015	take another observation
2	Negative	−0.012	take another observation
3	Positive	0.002	take another observation
4	Positive	0.018	take another observation
5	Positive	0.033	take another observation
6	Positive	0.049	take another observation
7	Positive	0.064	take another observation
8	Positive	0.080	take another observation
9	Positive	0.095	take another observation
10	Positive	0.111	take another observation
11	Positive	0.126	take another observation
12	Positive	0.142	take another observation
13	Positive	0.157	take another observation
14	Positive	0.173	take another observation
15	Positive	0.188	take another observation
16	Positive	0.204	stop and decide d_p

The laboratory cost has an influence on the sample size in the sense that the higher the laboratory cost, the greater the Bayes risk. This will result in a smaller final sample size with the aim of minimizing the Bayes risk. Thus, consider a laboratory cost of €200 instead of €100, while the loss function remains unchanged. The minimization of the Bayes risk gives $l_A = -0.15$ *and* $l_B = 0.15$*. Then, Table 7.6 shows that, if all items are found to contain an illegal substance, the final sample size will be* $n = 10$*. This contrasts with* $n = 13$ *which was found for* $c =$ *€100. The higher cost of experimentation thus tends to reduce the sample size. Conversely, the lower cost allows one to examine more units to counterbalance the loss.*

The loss function has an inverse influence on the sample size. The more severe the consequences of a wrong decision, the greater will be the number of items that needs to be inspected, and vice versa. Suppose, for instance, that the loss function is quantified to be $k_p = k_d = 50000$*, while the laboratory cost remains unchanged at its initial value of €100. Then the minimization of the Bayes risk gives* $l_A = -0.15$ *and* $l_B = 0.15$*. The interval* $[l_A, l_B]$ *thus is reduced and fewer observations will be necessary to reach a decision. In particular, it can be found that, if all items are positive, it will be sufficient to inspect a sample of size* $n = 10$ *(see Table 7.6).*

Different assessments of the loss functions, or of laboratory costs, do not affect the decision but do affect the required amount of observations.

Finally, consider a comparison between the fixed and the sequential procedures. If all items are found to be positive, both procedures end with the same optimal size (given an identical loss function and cost). In particular, the result of the fixed sampling procedure is that $n = 13$ *is the optimal sample size to test* $H_0 > \theta_0$ *against* $H_1 \leq \theta_0$ *given* $k_0 = k_1 = 100000$ *and a laboratory cost of €100 (it is sufficient to minimize the Bayes risk in Equation (7.13) in correspondence with the boundary region* $\theta_0 = 0.65$ *and a laboratory cost of €100). This output is the same as that given by the sequential procedure developed in this example. However, whenever some items are found to be negative, the sequential procedure allows for a lower number of observations. Imagine one item is found to be negative. It has been shown for such a setting that the sequential procedure ends with an optimal sample size of* $n = 16$*. The fixed sampling procedure with the updated parameters* $\alpha^* = 13$ *and* $\beta^* = 2$ *gives an optimal sample size of* $n = 5$ *(it is sufficient to minimize the Bayes risk in Equation (7.13) for* $H_0 : \theta \leq 0.65$*). Therefore, the total number of observations required is* $n = 18$.

7.5 R CODE

A symbol '$*$', '$+$', ',' and so on at the end of a line indicates that the command continuous to the following line. The absence of such a symbol indicates the end of a command.

Example 7.2.1

Input values

```
theta0=0.5
alpha=1
beta=1
n=4
y=3
```

Determination of the optimal sample size n

```
alphap=alpha+y
betap=beta+n-y
p=pbeta(theta0,alphap,betap,lower.tail=FALSE)
while (p<0.95){
n=n+1
print('additional item is found to be positive')
y=y+1
```

```
alphap=alpha+y
betap=beta+n-y
p=pbeta(theta0,alphap,betap,lower.tail=FALSE)
}
print(paste('Observations required if all successive items
 are found to be positives: n =',n))
```

Example 7.2.2

Input values

```
N=10
n=5
y=5
alpha=1
beta=1
z0=2
```

Output

```
m=N-n
z=seq(z0,m,1)
sum((gamma(n+alpha+beta)*choose(m,z)*gamma(z+y+alpha)*
gamma(n+m-y-z+beta))/(gamma(y+alpha)*gamma(n-y+beta)*
gamma(n+m+alpha+beta)))
```

Example 7.4.1

Input values

```
c=50
sigma=0.0463
tau=0.3
k=c(10000,50000,100000)
```

Determination of the optimal sample size

```
n=ceiling(sqrt(k)*sigma/sqrt(c)-(sigma^2)/(tau^2))
for (i in 1: length(k)){
print(paste('Optimal sample size n =',n[i],'(k =',k[i],')'))
}
```

Example 7.4.2

Input values

```
c=50
theta0=0.5
k0=100000
k1=100000
alpha.in=1
beta=1
```

Determination of the optimal sample size

```
N=100
r=matrix(0,nrow=N,ncol=3)
nopt=matrix(0,nrow=length(k),ncol=1)

for (i in 1 :N){
alpha=alpha.in+i
d0=k1*pbeta(theta0,alpha,beta)+i*c
d1=k0*pbeta(theta0,alpha,beta,lower.tail=FALSE)+i*c
r[i,1]=d0
r[i,2]=d1
r[i,3]=min(d0,d1)
}
nopt=which(r[,3]==min(r[,3]))
print(paste('Optimal sample size n =',nopt))
```

8

Classification of Observations

8.1 INTRODUCTION

Forensic scientists are routinely faced with the problem of classifying an observation (e.g. an individual, an item) into one of several populations on the basis of the available measurements of some attributes. Suppose for example that some skeletal remains are recovered, and the mandible is available and suitable for inspection (Schmittbuhl *et al.* 2007). The external appearances suggest that the recovered remains belong to an adult individual from the hominoid species. Moreover, the observation of the mandibular outline suggests the possible genus of the individual. So, it might be of interest for the scientist to classify the individual on the basis of the genus. Analogously, in a forensic scenario, consider a scientist who may be called on to conduct laboratory analyses to aid in the determination of the source of a particular sample. For example, imagine a case concerning the contamination of bank notes. The scientist may be asked to classify single items (i.e. bank notes) in a group of bank notes seized during drug trafficking investigations or in a group of bank notes in general circulation. As a third example, an important decision of medicine is diagnosis, for example in the task of assigning an individual to one of two categories (diseased or not diseased) on the basis of available information (Parmigiani 2002). Note that in medicine (as well as in the legal context), diagnosis is not an end in itself, but rather a means of assisting a later decision on treatment.

A fundamental assumption throughout this chapter is that there are a finite number of populations (categories) from which the observation may have come, and that each population is characterized by a probability distribution of the measurements. Different observations belonging to different populations will yield different

Data Analysis in Forensic Science: A Bayesian Decision Perspective Franco Taroni, Silvia Bozza, Alex Biedermann, Paolo Garbolino and Colin Aitken

measurements, and this variability will be expressed in probabilistic terms. Therefore, the scientist can treat the observation as a random observation from one of these populations, the distribution of which depends on the actual population. The problem is to classify the observation in the correct population. In some cases the populations are completely specified in the sense that the probability distributions are assumed known; in other cases only the form of each distribution is specified, but parameters need to be estimated. The problems that arise whenever the probabilistic structure is not known will not be considered here (Duda *et al.* 2001; Neal 1996).

The problem of classification can be treated as a problem of testing statistical hypotheses: each hypothesis is that the probability distribution that has originated the observation has a given form. Acceptance or rejection of a specific hypothesis allows the classification or not of an observation to a given population. If only two populations are considered, then the Bayesian approach presented in Chapter 6 to compare hypotheses can be applied here.

8.2 STANDARDS OF COHERENT CLASSIFICATION

Suppose only two populations are considered, say population 1 (p_1) and population 2 (p_2). One possible approach to classification is known as the *Bayesian predictive approach* (see Press 2003 and related references).

Assume that each population is characterized by a probability distribution, with parameter θ_i (possibly vector-valued):

$$f_i(x \mid \theta_i, p_i) \qquad i = 1, 2 \tag{8.1}$$

and that there are several observations available, known to have come from each population. Suppose probability distributions (8.1) are completely specified and let $P(p_i)$ be the prior probability of population p_i, $i = 1, 2$. This reflects the prior knowledge of how likely it is to have an observation from population 1(2).

A new observation $\tilde{x}$ is available and known to come from one of these populations, but it is not known which one. The observation is to be classified. Two propositions – sometimes also called *models* – are considered:

H_1: the observation comes from population 1;
H_2: the observation comes from population 2.

The posterior probability of belonging to population i can be obtained by Bayes' theorem (see Section 2.3.1):

$$P(p_i \mid \tilde{x}) = \frac{P(p_i) f_i(\tilde{x} \mid \theta_i, p_i)}{\sum_{j=1}^{2} P(p_j) f_j(\tilde{x} \mid \theta_j, p_j)}.$$

If the posterior probability $P(p_1 \mid \tilde{x})$ is greater (lower) than $P(p_2 \mid \tilde{x})$, then proposition $H_1(H_2)$ is supported and the scientist would be naturally inclined to classify the observation in population 1(2). As outlined in Section 6.2.1, one can compute the ratio of the posterior probabilities, i.e. the *posterior odds*, that is

$$\frac{P(p_1 \mid \tilde{x})}{P(p_2 \mid \tilde{x})} = \frac{P(p_1)f_1(\tilde{x} \mid \theta_1, p_1)}{P(p_2)f_2(\tilde{x} \mid \theta_2, p_2)},$$

and classify the observation in population 1(2) if the posterior odds is greater (lower) than 1. The *Bayes factor* is the ratio of the posterior odds to the prior odds,

$$BF = \frac{P(p_1 \mid \tilde{x})}{P(p_2 \mid \tilde{x})} \Big/ \frac{P(p_1)}{P(p_2)} = \frac{f_1(\tilde{x} \mid \theta_1, p_1)}{f_2(\tilde{x} \mid \theta_2, p_2)},$$

and measures the change produced by the evidence $\tilde{x}$ in the odds when going from the prior to the posterior distribution. It can be observed that the Bayes factor in this case is just the likelihood ratio of H_1 to H_2, since the problem can be transformed to one of testing a simple hypothesis versus a simple hypothesis. Note also that if $P(p_1) = P(p_2)$ (i.e. the prior odds equal 1) the posterior odds is equivalent to the likelihood ratio.

In some cases, the form of the probability distributions are known but the values of the parameters are not known so a prior distribution must be introduced. Let $\pi_i(\theta_i)$ ($\theta_i \in \Theta_i$; and θ_i may be vector-valued, e.g. the mean and variance of a Normal distribution) denote the prior distribution that incorporates prior beliefs about the population parameters (this can also incorporate information available from previous experiments, see Section 3.3.1 and the sequential use of Bayes' theorem). Then, the predictive distribution $f_i(\tilde{x} \mid p_i)$ can be computed, that is

$$f_i(\tilde{x} \mid p_i) = \int_{\Theta_i} f_i(\tilde{x} \mid p_i, \theta_i)\pi_i(\theta_i)d\theta_i.$$

The posterior probability of each population is therefore

$$P(p_i \mid \tilde{x}) = \frac{P(p_i)f_i(\tilde{x} \mid p_i)}{\sum_{j=1}^{2} P(p_j)f_j(\tilde{x} \mid p_j)}.$$

The posterior odds is given by:

$$\frac{P(p_1 \mid \tilde{x})}{P(p_2 \mid \tilde{x})} = \frac{P(p_1)f_1(\tilde{x} \mid p_1)}{P(p_2)f_2(\tilde{x} \mid p_2)}, \tag{8.2}$$

and the Bayes factor is the ratio of the predictive distributions,

$$BF = \frac{f_1(\tilde{x} \mid p_1)}{f_2(\tilde{x} \mid p_2)}. \tag{8.3}$$

Bayesian decision theory offers another approach to address the problem of classification in this context. Populations are described by a probabilistic distribution whose structure is known. This approach incorporates the quantification of the consequences (losses) that accompany errors of classification. Let $\mathcal{D} = \{d_1, d_2\}$ denote the decision space, where $d_{1(2)}$ represents the decision of classifying the observation $\tilde{x}$ in population 1(2), while there are two possible states of nature, population 1 (p_1) or population 2 (p_2). Consider a '0–k_i' *loss function* as in Table 8.1, with k_1 representing the loss of classifying a member of population 2 as a member of population 1 and k_2 representing the loss of classifying a member of population 1 item in population 2. In a medical context (Parmigiani and Inoue 2009), letting population 1 represent healthy patients and population 2 diseased patients, k_1 will represent the loss of diagnosing a diseased patient as healthy and k_2 will represent the loss of diagnosing a healthy patient as diseased.

Table 8.1 The '0–k_i' loss function. Decision d_1, d_2: classify observation in population 1 and 2, respectively.

	Population 1	Population 2
d_1	0	k_1
d_2	k_2	0

A coherent classification procedure is the *Bayes decision* procedure since it minimizes the probability of misclassification. According to this, decision d_1 is taken (i.e. the observation is classified in population 1) if (see Section 6.2.2)

$$P(p_1 \mid \tilde{x})k_2 > P(p_2 \mid \tilde{x})k_1,$$

that is if

$$\frac{P(p_1 \mid \tilde{x})}{P(p_2 \mid \tilde{x})} > \frac{k_1}{k_2}. \tag{8.4}$$

As observed in Section 6.2.2, the larger k_1/k_2, that is the more an incorrect decision under H_2 is penalized relative to that under H_1, the larger the posterior probability of population 1 needs to be in order for H_1 to be accepted. A threshold for the interpretation of the Bayes factor can be obtained by multiplying both sides of Equation (8.4) by the prior odds $P(p_2)/P(p_1)$, that is

$$\frac{P(p_2)}{P(p_1)}\frac{P(p_1 \mid \tilde{x})}{P(p_2 \mid \tilde{x})} > \frac{k_1}{k_2}\frac{P(p_2)}{P(p_1)}.$$

Therefore, when applying a '0–k_i' loss function, the optimal decision is d_1 whenever

$$BF > \frac{k_1}{k_2}\frac{P(p_2)}{P(p_1)}. \tag{8.5}$$

Note that if $P(p_1) = P(p_2)$ the threshold reduces to k_1/k_2.

The problem of classification can be generalized to several populations, say $p_1, \ldots, p_k$. Each population is characterized by a probability distribution $f_i(x \mid \theta_i, p_i)$. Given prior probabilities for each population, say $P(p_1), \ldots, P(p_k)$, and the observation $\tilde{x}$, the posterior probability of each population can be easily computed and is

$$P(p_i \mid \tilde{x}) = \frac{P(p_i) f_i(\tilde{x} \mid \theta_i, p_i)}{\sum_{j=1}^{k} P(p_j) f_j(\tilde{x} \mid \theta_j, p_j)}. \tag{8.6}$$

Consider now a '0–k_i' loss function extended to the case of several populations, as in Table 8.2 for $k = 3$, where $k_{j|i}, i, j = 1, 2, 3, i \neq j$, denotes the loss of misclassifying an observation from population i as from population j. This loss function will be termed as '0–$k_{j|i}$'. The optimal classification procedure (Anderson 2003) is to assign an observation $\tilde{x}$ to population l if

$$\sum_{i=1; i \neq l}^{k} P(p_i) f_i(\tilde{x} \mid \theta_i, p_i) k_{l|i} < \sum_{i=1; i \neq j}^{k} P(p_i) f_i(\tilde{x} \mid \theta_i, p_i) k_{j|i} \qquad j = 1, \ldots, k \; ; \; j \neq l \tag{8.7}$$

When $k_{j|i} = 1$ for all i and j, $i \neq j$, then the optimal classification procedure reduces to the assignment of an observation $\tilde{x}$ to population l if

$$P(p_j) f_j(\tilde{x} \mid \theta_j, p_j) < P(p_l) f_l(\tilde{x} \mid \theta_l, p_l) \qquad j \neq l. \tag{8.8}$$

Table 8.2 The '0–$k_{j|i}$' loss function for three populations. Decision d_1, d_2, d_3: classify observation in population 1, 2 and 3, respectively.

	Population 1	Population 2	Population 3
d_1	0	$k_{1\|2}$	$k_{1\|3}$
d_2	$k_{2\|1}$	0	$k_{2\|3}$
d_3	$k_{3\|1}$	$k_{3\|2}$	0

8.3 COMPARING MODELS USING DISCRETE DATA

8.3.1 Binomial distribution and cocaine on bank notes

Consider two populations formed by elements (e.g. individuals, items) of two possible kinds in different proportions. These populations are referred as population 1 (p_1) and population 2 (p_2). For example, imagine the case of a population of bank notes from drug trafficking investigations (p_1) and a population of bank notes in general circulation (p_2) that will be developed in Example 8.3.1. It is known that bank notes may be contaminated with cocaine in a higher or lower proportion depending on whether they have or have not been involved in drug dealing. Imagine a scenario where some bank notes are seized on a suspect and some of them, after inspection, are found to be contaminated with cocaine. A typical question a forensic scientist may be called to answer is whether or not the bank notes have been connected with drug trafficking.

In a simplistic model, the number X_i of bank notes contaminated in samples of size $n_{1(2)}$ from the two populations can be modelled by a binomial distribution, $X_i \sim \text{Bin}(n_i, \theta_i), i = 1, 2$ where $\theta_{1(2)}$ denotes the probability that a bank note is contaminated in each of the two populations. This model is simplistic because it ignores the possibility of the correlation of levels of cocaine between adjacent notes in a bundle (see footnote 1 in Example 8.3.1).

Imagine now that some bank notes are seized on a suspect. The number seized equals n and, after inspection, $\tilde{x}$ are found to be contaminated with cocaine. The evidence E to be evaluated is the number $\tilde{x}$ of bank notes found to be contaminated out of a sample of size n. The sample size n is taken to be fixed. Consider two propositions of interest:

H_1 : the bank notes seized on the suspect have been involved in drug dealing;
H_2 : the bank notes seized on the suspect are part of general circulation.

In the simplest case the proportions θs are known: if H_1 is true the probability a single bank note is contaminated with cocaine is θ_1 and, similarly, if H_2 is true the probability a single bank note is contaminated is θ_2. The posterior odds reduces to the product of the prior odds $P(p_1)/P(p_2)$ times the value of the evidence that can be determined by considering the ratio of the binomial likelihoods (Aitken and Taroni 2004). In particular, the value V of evidence is given by

$$V = \frac{P(E \mid H_1)}{P(E \mid H_2)} = \frac{f(\tilde{x} \mid \theta_1)}{f(\tilde{x} \mid \theta_2)} = \frac{\binom{n}{\tilde{x}}\theta_1^{\tilde{x}}(1-\theta_1)^{(n-\tilde{x})}}{\binom{n}{\tilde{x}}\theta_2^{\tilde{x}}(1-\theta_2)^{(n-\tilde{x})}}, \tag{8.9}$$

and is equivalent to the Bayes factor, as discussed in the previous section. The posterior odds is obtained multiplying (8.9) by the prior odds.

However, the probabilities of finding contaminated bank notes in the two populations are generally unknown, though some knowledge may be available from previous investigations. On the basis of the prior knowledge, a conjugate beta prior density for the unknown proportions is introduced, $\theta_i \sim \text{Be}(\alpha_i, \beta_i)$. The predictive distribution $f(\tilde{x} \mid p_i)$ can be calculated as in Example 3.3.2 and is a beta-binomial distribution with parameters n, α_i and β_i,

$$f(\tilde{x} \mid p_i) = \binom{n}{\tilde{x}} \frac{\Gamma(\alpha_i + \beta_i)}{\Gamma(\alpha_i)\Gamma(\beta_i)} \frac{\Gamma(\alpha_i + \tilde{x})\Gamma(\beta_i + n - \tilde{x})}{\Gamma(\alpha_i + n + \beta_i)}. \tag{8.10}$$

The posterior odds and the Bayes factor can be obtained by substituting (8.10) in (8.2) and (8.3) respectively. Numerical difficulties that may arise (e.g. from calculating factorials of large integers) when computing the beta-binomial density can be overcome by taking the logarithm of the probability, $\log f(\tilde{x} \mid p_i)$. Therefore, the posterior odds can be computed as

$$\frac{P(p_1 \mid \tilde{x})}{P(p_2 \mid \tilde{x})} = \exp\left\{\log \frac{P(p_1)}{P(p_2)} + \log f(\tilde{x} \mid p_1) - \log f(\tilde{x} \mid p_2)\right\}, \tag{8.11}$$

and, equivalently, the Bayes factor

$$BF = \exp\left\{\log f(\tilde{x} \mid p_1) - \log f(\tilde{x} \mid p_2)\right\}. \tag{8.12}$$

Example 8.3.1 *(Contamination of bank notes). Consider the following scenario. Some bank notes, $n = 100$, are seized on a suspect and, after inspection, $\tilde{x} = 76$ are found to be contaminated with cocaine. The scientist is interested in evaluating $\tilde{x}$ under the following two propositions, say H_1, the bank notes seized on the suspect have been involved in drug dealing, and H_2, the bank notes seized on the suspect are part of the general circulation.*

Let θ_1 and θ_2 denote the probability that a bank note is contaminated with cocaine from the drug dealing and from general circulation, respectively. Parameters of the beta densities introduced to model the uncertainty about the unknown proportions θ_1 and θ_2 are chosen according to the procedure described in Section 4.2.1 that allows the prior distribution to be based on available knowledge. In particular, it is known (Besson 2003) that among a population of $n_1 = 462$ bank notes from drug trafficking investigations, a number of $x_1 = 382$ have been found to be contaminated; moreover, among a population of $n_2 = 992$ bank notes from general circulation, a total of $x_2 = 562$ have been found to be contaminated. Parameters are then chosen according to equations (4.3) and (4.4), and results in a $Be(381, 80)$ distribution for θ_1, and a $Be(561, 430)$ distribution for θ_2.

The values of the beta-binomial densities can be obtained using the R routines given at the end of the chapter, and are given by $\log f(\tilde{x} \mid p_1) = -3.639$ *and* $\log f(\tilde{x} \mid p_2) = -9.887$. *The Bayes factor can be obtained by substituting these values in Equation (8.12) and gives*

$$BF = \exp(-3.639 + 9.845) = 517,$$

in favour of the hypothesis of association with drug dealing.

According to the verbal scale in Table 6.2, a Bayes factor equal to 517 *represents moderately strong evidence to support that the bank notes seized on the suspect have been involved in drug dealing*[1]. *If the prior odds (the ratio between the prior probabilities that the recovered sample comes from population 1 or 2) is set equal to* r*, then the posterior odds is* $517r$. *Figure 8.1 shows the impact of different prior probabilities for the classification in population 1 on the posterior odds. It can easily be checked that, whenever the prior probability* $P(p_1) \geq 0.003$*, the posterior odds is greater than 1; this means that the recovered sample should coherently be classified into population 1.*

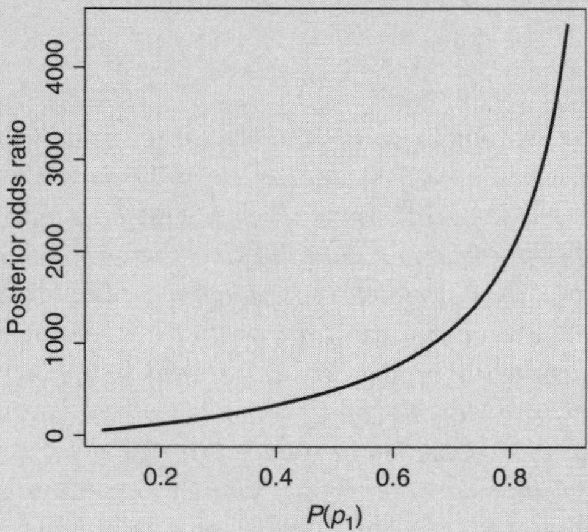

Figure 8.1 Posterior odds $P(p_1 \mid \tilde{x})/P(p_2 \mid \tilde{x})$ obtained for increasing probability values $P(p_1)$ of population 1.

The same conclusions in favour of proposition H_1 *may be obtained under a decision-oriented approach. Assuming a '*0*–*k_i*' loss function, the decision criterion is the one outlined in Equation (8.4), but the values for* k_1 *and* k_2 *may be difficult to assess. However, what really matters is the ratio of the two values. Assuming that classifying seized bank notes as having been involved in drug dealing when they have not been involved is worse than the opposite, that is classifying seized bank notes as from general circulation when they*

[1]Note that such an example is a simplification of the reality, because the use of the binomial distribution may be questioned here. One of the modelling assumptions for a binomial distribution is that all members of the sample have a constant probability of 'success', independent of other members of the sample. Models which allow for dependence amongst members of the sample are beyond the scope of this book.

have been involved in drug dealing, then the value of k_1 will be greater than k_2. However, the consequences of erroneously taking decision d_1 when H_2 is true will need to be very much greater than the consequences of taking d_2 when H_1 is true, in order to overturn a BF of nearly 500. Consider now prior probabilities are not felt to be equal, and population 2 is believed a priori much more probable, say $P(p_2) = 0.9$, then Equation (8.5) becomes

$$517 > \frac{k_1}{k_2}\frac{0.9}{0.1},$$

so that decision d_2 becomes the Bayes decision if and only if the ratio k_1/k_2 exceeds 57. Otherwise, a Bayes factor equal to 517 is still sufficiently large to confirm d_1 as the Bayes decision.

8.3.2 Poisson distributions and firearms examination

Consider the following scenario. A bullet is found at a crime scene and a suspect is apprehended with a gun. The following propositions are of interest:

- H_1: the bullet found at the crime scene was fired from the suspect's gun;
- H_2: the bullet found at the crime scene was fired from a gun other than the suspect's gun.

A statistic often used for quantifying the extent of agreement between marks (left by firearms) is that of *consecutive matching striations* or *CMS* for short. An examiner studies bullets under a macroscope and decides what is a striation and, for comparison, what striations match between the two bullets. The recovered bullet and a bullet fired from the gun which is suspected of being used in the crime (suspect gun) are compared. Observations are made of the matching striations and of the differences.

For the analysis of *CMS*, a scientist fires numerous bullets through many firearms of the same make and model. A macroscopic comparison is then made of specimens known to have been fired from the same barrel and specimens known to have been fired from different barrels, and counts are made of the numbers of matching striations. Bunch (2000) describes a model in which the only *CMS* run on a bullet which matters is the one, or more, which features the maximum *CMS* count. Two data sets can be compiled, one for pairs of bullets fired from the same gun and one for pairs of bullets fired from different guns (i.e. different make and model). Let Y be the maximum *CMS* count for a particular bullet found at a crime scene when compared with a bullet fired from a gun, known as the suspect gun. One model for the *CMS* count is a Poisson model with $P(Y = y \mid \lambda) = \text{Pn}(\lambda)$, where the parameter λ is the weighted average maximum *CMS* count. Two Poisson distributions are required, one for pairs of bullets fired from the same gun (S), Pn(λ_S) and one for pairs of bullets fired from different guns (D), Pn(λ_D). If the suspect gun is the same gun as the one that fired the bullet found at the crime scene (H_1 is true), then

$$f(y \mid \lambda_S) = \frac{\lambda_S^y}{y!} e^{-\lambda_S} \qquad y = 0, 1, \ldots$$

If the suspect gun is a different gun to the one that fired the bullet found at the crime scene (H_2 is true), then

$$f(y \mid \lambda_D) = \frac{\lambda_D^y}{y!} e^{-\lambda_D} \qquad y = 0, 1, \ldots$$

The evidence E to be considered is the observed number $\tilde{y}$ of CMS. If parameters λ_S and λ_D are known, the value of evidence is given by the following likelihood ratio:

$$V = \frac{P(E \mid H_1)}{P(E \mid H_2)} = \frac{f(\tilde{y} \mid \lambda_S)}{f(\tilde{y} \mid \lambda_D)} = \left(\frac{\lambda_S}{\lambda_D}\right)^{\tilde{y}} e^{\lambda_D - \lambda_S}.$$

The use of CMS through the likelihood ratio enables a summary of the evidence of CMS to be in a phrase of the form 'the evidence is so many times more likely if H_1 is true than if H_2 is true'. This provides a good summary of what the statistics of CMS means in the context of determining the origin of the bullet found at a crime scene.

However, if parameters λ_S and λ_D are unknown, a gamma conjugate prior distribution can be assumed, $\lambda_g \sim Ga(\alpha_g, \beta_g)$, for $g = S, D$, as discussed in Section 4.3. The predictive distribution under hypothesis H_1 is the Poisson-gamma mixture (see also Example 3.3.9), that is

$$\begin{aligned} f(\tilde{y} \mid H_1) &= \int_0^\infty \frac{e^{-\lambda_S} \lambda_S^{\tilde{y}}}{\tilde{y}!} Ga(\alpha_S, \beta_S) d\lambda_S \\ &= \frac{1}{\tilde{y}!} \frac{\beta_S^{\alpha_S}}{\Gamma(\alpha_S)} \frac{\Gamma(\alpha_S + \tilde{y})}{(\beta_S + 1)^{\alpha_S + \tilde{y}}}. \end{aligned} \tag{8.13}$$

Similarly, the predictive distribution under hypothesis H_2 is

$$f(\tilde{y} \mid H_2) = \frac{1}{\tilde{y}!} \frac{\beta_D^{\alpha_D}}{\Gamma(\alpha_D)} \frac{\Gamma(\alpha_D + \tilde{y})}{(\beta_D + 1)^{\alpha_D + \tilde{y}}}. \tag{8.14}$$

The posterior odds ratio can be obtained by substituting (8.13) and (8.14) in (8.2).

Example 8.3.2 *(Firearm assessment). Consider a firearm examination scenario in which the observed number of CMS counts is* $\tilde{y} = 5$*. At first, consider the case where parameters describing the two populations (*S *and* D*) are known from previous experiments (see Examples 4.3.1 and 4.3.2). Consider that, for bullets fired from the same gun (*S*), the weighted average maximum CMS (see Bunch 2000, for details) count is given by* $\lambda_S = 3.91$*. For bullets fired from different guns (*D*), the weighted average maximum CMS count is given by* $\lambda_D = 1.32$*. The ratio of* $f(\tilde{y} \mid \lambda_S) = 0.153$ *to* $f(\tilde{y} \mid \lambda_D) = 0.0089$ *equals 17.1 and is then the value of the evidence* $\tilde{y}$ *of the maximum CMS count. This result moderately supports the proposition that the fired bullet comes from the suspect's gun.*

Consider the same scenario but the parameters describing the two populations are unknown. Assume for λ_S *and for* λ_G *a* $Ga(125, 32)$ *and a* $Ga(7, 5)$ *prior density, respectively (Figure 8.2). Parameters have been chosen by starting from prior beliefs about the mean and the standard deviation as illustrated in Section 4.3.1. Given the predictive distributions in (8.13) and (8.14),* $f(5 \mid H_1) = 0.15$ *and* $f(5 \mid H_2) = 0.016$*, approximately. The value of the evidence is therefore* 9.06 *and represents limited evidence to support the proposition that the fired bullet comes from the suspect gun. For equal prior probabilities* $P(p_1) = P(p_2) = 0.5$*, this value is also the posterior odds, which allows one to classify* $\tilde{y}$ *into population 1 (suspect's gun). Figure 8.3 shows the influence of different prior probabilities* $P(p_1)$ *on the posterior odds. In particular, it can be seen from the right-hand figure that prior probabilities* $P(p_1) > 0.1$ *allows for classification of the observation in population* 1.

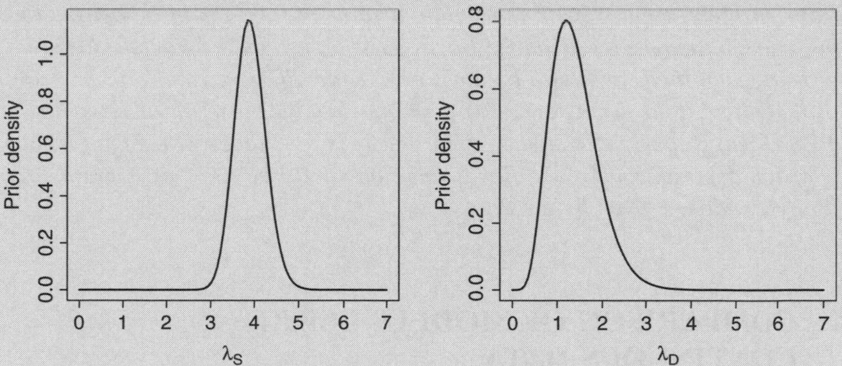

Figure 8.2 Gamma prior densities, $Ga(125, 32)$ (left) and $Ga(7, 5)$ (right) describing the prior beliefs about the mean and the standard deviation on the *CMS* from bullets fired by the same (left) and different (right) firearms.

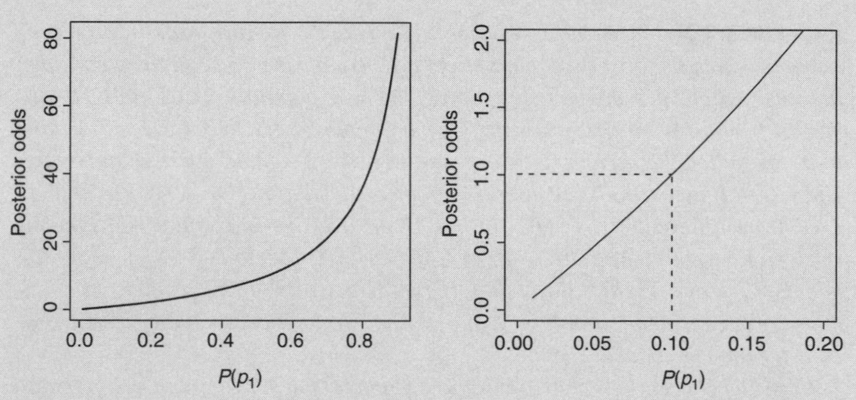

Figure 8.3 Effect of prior probability $P(p_1)$ on the posterior odds in Example 8.3.2; the section of the left-hand graph for $P(p_1) < 0.20$ is illustrated in the right-hand graph.

From a Bayesian decision perspective, and using an asymmetric loss function '0–k_i', the decision criterion is the one outlined in Equation (8.4). How can k_1 and k_2 be assessed? As already outlined, what really matters is the ratio of the two losses, k_1/k_2. At first, it is believed the ratio of the two losses k_1/k_2 should be greater than 1 since to link falsely a bullet to the suspect's gun is considered worse than to exclude falsely a link. So, given a posterior odds equal to 9.06*, and the decision criterion in (8.4), the ratio of the losses should be greater than* 9 *to change the classification decision from d_1 (the bullet has been fired from the suspect's gun) to d_2 (the bullet has been fired from a different gun). Note that while a ratio of* 9.06 *could be considered big enough to consider d_1 as the Bayes decision (no matter how the loss values are quantified), it would be not so clear for different values of the prior probabilities. Take for instance $P(p_1) = 0.3$, then the posterior odds would be* 3.88*. Then, if taking decision d_1 erroneously is four times worse than taking decision d_2 erroneously, the Bayes decision would be d_2. This confirms the limited evidence given by the data.*

8.4 COMPARISON OF MODELS USING CONTINUOUS DATA

8.4.1 Normal distribution and colour dye (case with known variance)

A scientist is interested in evaluating the measurement of colour dye concentration in ecstasy tablets. Imagine a scenario that involves the comparison between measurements (Y) on a suspect tablet and measurements (X_c) on a consignment of tablets (denoted C) for which laboratory analysis has revealed the presence of a certain kind of colour dye. One would like to evaluate the possibility that the incriminated tablet is linked to the consignment above. Two hypotheses need to be compared:

H_1 : the existence of a link to a population C;
H_2 : the absence of a link.

Colour dye concentration, say X, is a continuous measurement for which a Normal distribution is appropriate. The colour dye concentration in tablets for which laboratory analyses are available, the results of which are denoted X_c, is assumed to follow a Normal distribution, $X_c \sim N(\theta_c, \sigma_c^2)$. The evidence E to be considered is the measurement y of colour dye concentration in the suspect tablet. The likelihood ratio for evaluating the existence of a link to the consignment of tablets C can be found by evaluating the probability densities of the colour dye concentration y given the existence of a link, and given the absence of a link, to the consignment at hand. In the latter case a population of unrelated cases, denoted P, is considered, with measurements X_p, Normally distributed, $X_p \sim N(\theta_p, \sigma_p^2)$. So, the value of the evidence

$$V = \frac{P(E \mid H_1)}{P(E \mid H_2)} = \frac{f(y \mid \theta_c, \sigma_c^2)}{f(y \mid \theta_p, \sigma_p^2)} = \frac{(\sigma_c^2)^{-1/2} \exp\left\{-\frac{1}{2}\left(\frac{y-\theta_c}{\sigma_c}\right)^2\right\}}{(\sigma_p^2)^{-1/2} \exp\left\{-\frac{1}{2}\left(\frac{y-\theta_p}{\sigma_p}\right)^2\right\}}.$$

Alternatively, it may be of interest to test whether the suspect tablet is linked to the consignment of a specific producer (Z), where measurements, denoted X_z, are Normally distributed, $X_z \sim N(\theta_z, \sigma_z^2)$.

Consider a series of independent measurements $y_1, y_2, \ldots, y_n$ on the n suspect tablets. As was outlined in Section 6.4.3, the probability densities of the sample are $f(y_1, \ldots, y_n \mid H_1) = N(\bar{y} \mid \theta_c, \frac{\sigma_c^2}{n})$ and $f(y_1, \ldots, y_n \mid H_2) = N(\bar{y} \mid \theta_p, \frac{\sigma_p^2}{n})$ with $\bar{y} = \sum_{i=1}^{n} y_i / n$.

Example 8.4.1 *(Colour dye concentration in ecstasy tablets). A tablet is analyzed ($n = 1$) and the colour dye concentration measured equals 0.155 [%]. Assume distributions of the competing populations C and P are known (Goldmann et al. 2004), that is $X_c \sim N(0.14, 0.01^2)$, and $X_p \sim N(0.3, 0.06^2)$, see Figure 8.4 (a).*

The likelihood ratio for evaluating a linkage to C becomes:

$$V = \frac{f(0.155 \mid 0.14, 0.01^2)}{f(0.155 \mid 0.3, 0.06^2)} = 36.12.$$

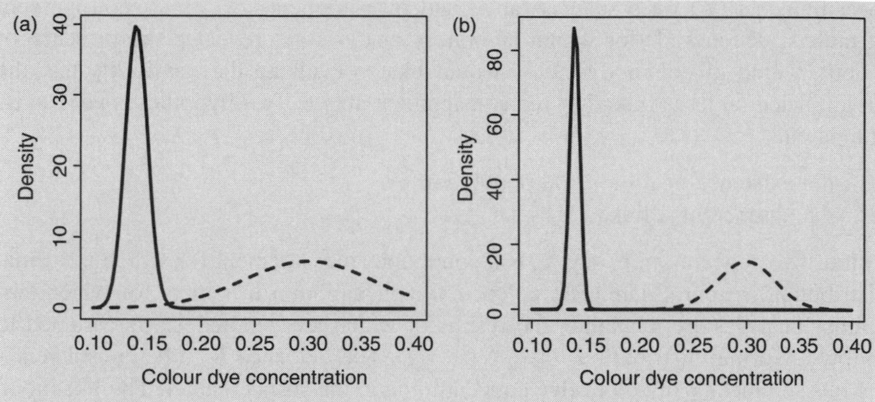

Figure 8.4 (a) (left-hand diagram) Distributions of the competing populations, $X_c \sim N(0.14, 0.01^2)$ (solid curve) and $X_p \sim N(0.3, 0.06^2)$ (dashed curve); (b) (right-hand diagram) Distributions of the sample $y_1, \ldots, y_5$ under the two hypotheses, $X_c \sim N(0.14, 0.01^2/5)$ (solid curve) and $X_p \sim N(0.3, 0.06^2/5)$ (dashed curve). Note the difference in scales on the axes labelled 'Density'.

The observed colour dye concentration in the incriminated tablet may thus be said to be approximately 36 *times more likely if it is linked to C, than if it were linked to a population of unrelated cases.*

Imagine now that a sample of $n = 5$ *incriminated tablets has been observed, and a sample mean* $\bar{y} = 0.155$ *has been obtained. The distributions of the sample mean* $\bar{y}$*, under* H_1 *and* H_2*, are plotted in Figure 8.4 (b). Then,*

$$V = \frac{f(0.155 \mid 0.14, 0.01^2/5)}{f(0.155 \mid 0.3, 0.06^2/5)} = 47451.$$

The uncertainty around the colour concentration measure decreases because five samples have been analyzed. Therefore the value of the evidence is increased by a factor of about 1300*. This is illustrated by the relative heights of the probability density curves under* H_1 *and* H_2 *in Figures 8.4 (a) and 8.4 (b), at the values of 0.155 of colour dye concentration.*

Consider now the case where the means (θ_c, θ_p) of the colour dye concentration distributions are unknown, and prior conjugate distributions are introduced, $\theta_c \sim N(\mu_c, \tau_c^2)$ and $\theta_p \sim N(\mu_p, \tau_p^2)$. The distribution under H_1 is still Normal, but now with $f(y \mid H_1) = N(\mu_c, \sigma_c^2 + \tau_c^2)$, and, equivalently, $f(y \mid H_2) = N(\mu_p, \sigma_p^2 + \tau_p^2)$.

Therefore, the likelihood ratio for evaluating linkages to C becomes:

$$V = \frac{P(E \mid H_1)}{P(E \mid H_2)} = \frac{f(y \mid \mu_c, \sigma_c^2 + \tau_c^2)}{f(y \mid \mu_p, \sigma_p^2 + \tau_p^2)}.$$

Example 8.4.2 *(Colour dye concentration in ecstasy tablets – continued). Assume $\theta_c \sim N(0.14, 0.003^2)$ and $\theta_p \sim N(0.3, 0.016^2)$. Then,*

$$\frac{f(0.155 \mid 0.14, 0.01^2 + 0.003^2)}{f(0.155 \mid 0.3, 0.06^2 + 0.016^2)} = 32.37.$$

Note that the increasing uncertainty due to the prior distributions for θ_c and θ_p reduces the value of the evidence for the concentration of a single tablet from 36.12 to 32.37. For five tablets with a mean concentration of 0.155, the value of the evidence is

$$\frac{f\left(0.155 \mid 0.14, 0.01^2/5 + 0.003^2\right)}{f\left(0.155 \mid 0.3, 0.06^2/5 + 0.016^2\right)} = 5709.$$

8.4.2 A note on the robustness of the likelihood ratio

In the scenario presented in Section 8.4.1, a comparison is wanted between an incriminated tablet and a consignment C of tablets for which laboratory analysis has revealed the presence of a certain kind of colour dye. One would like to evaluate the possibility that the incriminated tablet is linked to C. The two hypotheses of interest – under which the evidence should be compared – are the existence of a link, denoted H_1, and the absence of a link, denoted H_2.

The colour dye concentration, X, is assumed to follow a Normal distribution, $X \sim N(\theta, \sigma^2)$. Using the ideas of Example 8.4.1 with a single measurement now on the incriminated tablet of $y = 0.165$, a likelihood ratio of 3.3 is obtained for evaluating the value of the evidence comparing the presence and absence of a link between the incriminated tablet and C,

$$V = \frac{f(0.165 \mid 0.14, 0.01^2)}{f(0.165 \mid 0.3, 0.06^2)} = 3.3.$$

The scientist may thus report that his evidence slightly supports (see Table 6.2) the prosecution's case by a factor of about 3.3. Such a result may give rise to questions of the following kind (relating to what can be called the *robustness of the likelihood ratio*):

- How often may a forensic scientist obtain such a likelihood ratio for tablets that actually come from consignment C (as assumed by the prosecution's case, H_1)?
- How often may a forensic scientist obtain such a likelihood ratio for tablets unrelated to consignment C that actually come from a population, say P (as assumed by the defence case, H_2)?

Knowledge of the distribution of the likelihood ratio itself is required to answer these questions, irrespective of the observed evidence, but for members of given populations. For the sake of illustration, consider again the colour dye scenario.

A way to approach such questions is to investigate the colour dye concentration on training sets of tablets coming from both populations of interest and to calculate a likelihood ratio for each of these tablets. If this procedure is conducted a certain number of times, a scientist can obtain a likelihood ratio distribution for each of the two studied populations (i.e. consignment C and population P).

Such an investigation can be conducted experimentally but such investigations may be time-consuming and demanding with respect to resources. Another possibility consists in generating data for each of the two target populations, C and P, by simulation. Such an approach may be acceptable, for instance, in settings in which one has well-informed distributions from which appropriate random values can be generated. A simulation of this kind is conducted here for the above-mentioned scenario in which a likelihood ratio of 3.3 was found.

The simulation involves the same distributional assumptions as those mentioned at the beginning of this example, say two Normal distributions. Following the stated assumptions, 10 000 colour dye concentrations are generated for each of the two target populations with a likelihood ratio being calculated for each concentration. Table 8.3 shows – for each of the two target populations – the number of likelihood ratio values obtained for the randomly generated colour dye concentration.

Table 8.3 Counts of values in four apportionnements of 10 000 likelihood ratio values from simulations made in populations C (proposition H_1) and P (proposition H_2), respectively. Repetition of simulations will give small differences in the output.

	< 1	$[1, 10)$	$[10, 100)$	$[100, 1000)$
H_1:	22	161	2099	7718
H_2:	9864	50	59	27

For the simulation conducted here, 161 likelihood ratios with a value between 1 and 10 were obtained for the random colour dye concentration selected from the first population, C. Fifty likelihood ratios with a value between the same range, $[1, 10)$, were found for the random concentrations given from population P.

These figures help answer questions of the kind stated above. The ratio between the two proportions of values in the stated range $[1, 10)$ offers an estimate of the *robustness* of the likelihood ratio. It provides a figure of how many times a likelihood ratio between 1 and 10 will point in the wrong direction.

In the colour dye concentration scenario, a likelihood ratio of 3.3 was calculated; this slightly supports the proposition that the incriminated tablet comes from consignment C rather than coming from the population P. Simulations suggest that about 24% (50/211) of the time that the scientist points (slightly) to the prosecutor's hypothesis, there is a risk of misleading evidence; the decision maker chooses H_1 when H_2 should be chosen.

Note that such a simulation exercise allows a general assessment of false negatives and positives for a given population (proportion of likelihood ratios less than 1 and greater than 1, respectively). Table 8.3 shows that about 0.2% $(22/10000)$ of the simulations offer a likelihood ratio less than 1 when hypothesis H_1 (the incriminated tablet is linked to consignment C) should be supported. In approximately 1.4% $(136/10000)$, a likelihood ratio greater than 1 is (falsely) obtained under hypothesis H_2.

Thus, the scientist can assess the 'discriminating power' of an analytical process (i.e. the colour dye concentration). The flexibility of the likelihood ratio statistic gives it an advantage for discrimination over methods that do not vary with respect to underlying hypotheses. Similar methodology has also been proposed, for example, in DNA evidence evaluation in family searching (Taroni and Hicks 2008) and in gunshot residues evaluation (Biedermann *et al.* 2009).

8.4.3 Normal distribution and questioned documents (case with known variance)

Consider a scenario involving questioned documents which have been printed with a laser-jet printer of unknown type. Imagine, for the sake of simplicity, that the possible printers which might have been used are only two: say printer 1 and printer 2. For forensic purposes, two hypotheses need to be compared:

H_1 : the questioned documents have been printed with printer 1;
H_2 : the questioned documents have been printed with printer 2.

Several variables can be measured from the printed characters (e.g. the area and the diameter, see Mazzella and Marquis 2007). Take for instance the area of a specific printed character; measurements follow a Normal distribution with mean $\theta_{1(2)}$ for printer 1(2). The variance σ^2 is assumed known and equal for both populations of characters (i.e. those documents printed with printer 1 or 2), say

$$X_i \sim N(\theta_i, \sigma^2), \qquad i = 1, 2.$$

Prior conjugate distributions are introduced for the unknown means, $\theta_1 \sim N(\mu_1, \tau_1^2)$ and $\theta_2 \sim N(\mu_2, \tau_2^2)$.

Suppose n characters of the same type are measured from a questioned document, $y_1, \ldots, y_n$, the sample mean $\bar{y}$ is calculated and is the evidence E. The distribution under H_1 is still Normal, that is $f(\bar{y} \mid H_1) = N(\mu_1, \sigma^2/n + \tau_1^2)$, and $f(\bar{y} \mid H_2) = N(\mu_2, \sigma^2/n + \tau_2^2)$. Therefore, the likelihood ratio becomes

$$V = \frac{P(E \mid H_1)}{P(E \mid H_2)} = \frac{f(\bar{y} \mid \mu_1, \sigma^2/n + \tau_1^2)}{f(\bar{y} \mid \mu_2, \sigma^2/n + \tau_2^2)}.$$

Example 8.4.3 *(Questioned document). The area of $n = 10$ characters of type 'a' is measured from a questioned document and a sample mean $\bar{y} = 454888\ \mu m^2$ is obtained. Documents might have been printed with an* Canon model ir 400 *(hypothesis H_1), or with an* HP model 41 *(hypothesis H_2). The variance σ^2 is known and equal to 546187921 (units in μm^2). Available knowledge enables prior distributions for $\theta_1 \sim N(462825, 10000^2)$, and for $\theta_2 \sim N(430350, 10000^2)$ to be specified. The likelihood ratio is*

$$V = \frac{f(454888 \mid 462825, 546187921/10 + 10000^2)}{f(454888 \mid 430350, 546187921/10 + 10000^2)} = 5.71.$$

There is limited evidence in favour of hypothesis H_1.

Since there is no reason to believe one printer to be more likely than the other, it is assumed that $P(H_1) = P(H_2) = 0.5$. Then, 5.71 is also the posterior odds. Given a '0–1' loss function (which seems perfectly reasonable in this scenario since there are no reasons to penalize more or less one type of misclassification error over another), the optimal decision, according to Equation (8.4), will be to classify the observation as originating from printer 1, the Canon model ir400.

Example 8.4.4 *(Questioned document – continued). Consider the same scenario as in Example 8.4.3. As outlined in Section 8.2, the classification procedure can be generalized to several populations. Imagine there is a third printer that might have been used, an* HP model 1300. *Available knowledge on that printer provides a prior distribution for its mean $\theta_3 \sim N(476242, 10000^2)$. Assuming equal prior probabilities $P(p_1) = P(p_2) = P(p_3) = 1/3$, the posterior distribution, Equation (8.6), can be computed for each printer:*

$$P(p_1 \mid 454888) = 0.69$$

$$P(p_2 \mid 454888) = 0.12$$
$$P(p_3 \mid 454888) = 0.19.$$

Given a loss function as in Table 8.2 with $k_{j|i} = 1$ for all i and j, $i \neq j$, the optimal decision is to classify the questioned document as coming from printer 1, since Equation (8.8) holds for $j = 2, 3$.

8.4.4 Normal distribution and sex determination (case with unknown variance)

Imagine a scientist examines skeletal remains with a view to classifying them as remains of females or of males. Analysis and measurements of the sacral base (*basis osseus sacri*) is considered a good determinant of sex (Benazzi *et al.* 2009). The area of the sacral base is assumed to follow a Normal distribution, $X \sim N(\theta, \sigma^2)$, with unknown mean and variance.

Whenever limited information is available prior to data collection (imagine that only a small database is available), a vague prior distribution can be adopted, $\pi(\theta, \sigma) \propto \frac{1}{\sigma}$ (see Section 4.4.2). Suppose that n observations are taken on the recovered skeletal remains: $x = (x_1, \ldots, x_n)$. The predictive distribution of a measurement, Y, is a non-central *Student t* distribution (Bernardo and Smith 2000),

$$(Y \mid x) \sim St\left(n - 1, \bar{x}, \frac{s^2(n+1)}{n}\right), \tag{8.15}$$

with mean $\bar{x} = \frac{1}{n}\sum_{i=1}^{n} x_i$, $(n-1)$ degrees of freedom, and $s^2 = \frac{1}{n-1}\sum_{i=1}^{n}(x_i - \bar{x})^2$. This enables the Bayes factor and the posterior odds to be computed and the standards of classification implemented as discussed in Section 8.2.

Whenever prior information is available, an informative prior distribution can be assumed for the unknown parameters, $\pi(\theta, \sigma^2) = \pi(\theta \mid \sigma^2)\pi(\sigma^2)$, where $(\theta \mid \sigma^2) \sim N(\mu, \sigma^2/n_0)$ and $\sigma^2 \sim IG(\alpha, \beta)$, as in Section 4.4.2. In that case, having observed $x = (x_1, \ldots, x_n)$, the predictive distribution $f(y \mid x)$ is still a non-central *Student t* distribution, with mean as in (4.19) and $2\alpha + n$ degrees of freedom.

Example 8.4.5 *(Sex determination of skeletal remains). Suppose that data concerning the area of the profile of the sacral base are available. In a female population, say F, of $n_F = 38$ individuals the mean of the area is*

$\bar{x}_F = 10.35\,\text{cm}^2$ and the standard deviation is $s_F = 1.42\,\text{cm}^2$. In the alternative population, say M, composed by $n_M = 35$ males, the mean of the area of the sacral base is $\bar{x}_M = 14.09\,\text{cm}^2$ and the standard deviation is $s_M = 1.52\,\text{cm}^2$. A non-informative prior distribution $\pi(\theta, \sigma) \propto \frac{1}{\sigma}$ is adopted. The predictive distributions (8.15) for the two populations are given in Figure 8.5.

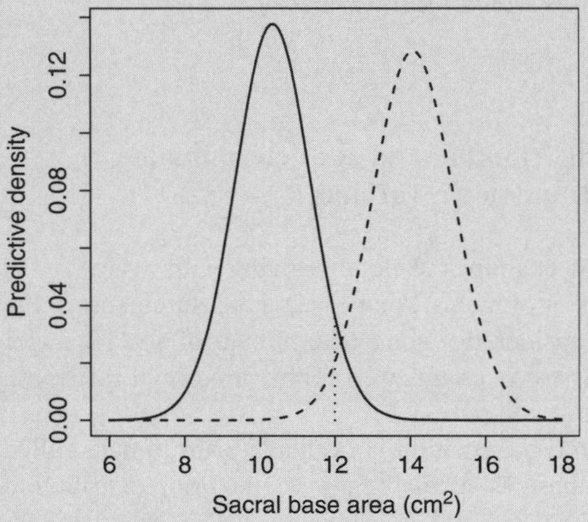

Figure 8.5 Predictive density of the sacral base area for the female (solid line) and the male (dashed line) population. The vertical dotted line indicates the predictive densities for a skeletal remain of area $12\,\text{cm}^2$.

A skeletal remain is recovered and the sacral base area measures $y = 12\,\text{cm}^2$. There are two hypotheses of interest:

H_1 : the skeletal remain belongs to population 1, the female population;
H_2 : the skeletal remain belongs to population 2, the male population.

The value of the evidence is

$$V = \frac{f(y \mid H_1)}{f(y \mid H_2)} = \frac{St\left(y \mid n_F - 1, \bar{x}_F, s_F^2(n_F + 1)/n_F\right)}{St\left(y \mid n_M - 1, \bar{x}_M, s_M^2(n_M + 1)/n_M\right)}$$

$$= \frac{St\,(12 \mid 37, 10.35, 2.07)}{St\,(12 \mid 34, 14.09, 2.38)} = 1.79.$$

As may expected by observing the predictive distributions in Figure 8.5, there is limited evidence in support of hypothesis H_1. The reader is left to develop this scenario, computing the posterior odds and finding the optimal decision according to their state of knowledge (e.g. using conjugate priors).

8.5 NON-NORMAL DISTRIBUTIONS AND COCAINE ON BANK NOTES

Consider the two populations of bank notes described in Section 8.3.1, the first given by bank notes coming from drug trafficking (say population 1), and the second given by bank notes in general circulation (say population 2). It is assumed that the intensities with which a drug is present on the bank notes will be different depending on whether the notes have or have not been involved in drug dealing. Consider a scenario where some bank notes are seized on a suspect: a typical question that may need to be addressed in such a situation is whether or not they have been involved with drug trafficking (see for example Besson 2003, 2004). In Section 8.3.1 a situation was described where the laboratory simply took into account the presence or the absence of traces of drugs on the bank notes. Suppose now the laboratory measures the intensity of drug present on the bank notes, and wishes to determine the probability distribution underlying the generation of the observations. However, not all data have a distribution which is readily modelled by a standard distribution. In particular, not all data are unimodal, symmetric and bell-shaped and able to be modelled by a Normal distribution. The histograms of the measured intensity of contamination with drugs on bank notes of € 200 from population 1, Figure 8.6 (left), and on bank notes of € 200 from population 2, Figure 8.6 (right), illustrate this (Besson 2004). The distribution for bank notes involved in drug trafficking is not unimodal and the distribution for bank notes in general is positively skewed. In such a case the probability distribution may be estimated from data taken from the population. A procedure known as the *kernel density estimation* can be used (see Silverman 1986 for technical details).

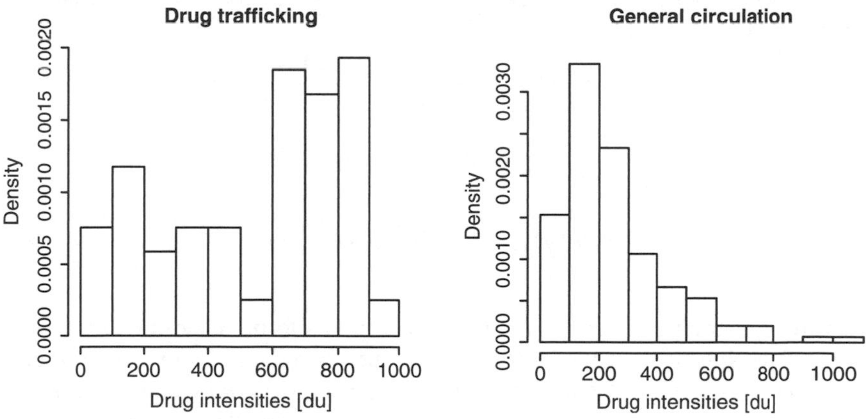

Figure 8.6 Drug intensity measured on bank notes of 200 € in a population of bank notes coming from drug trafficking (left) and in general circulation (right). The unit of measurement (du) is a 'digital unit'.

Consider one of the histograms plotted in Figure 8.6. The procedure is described for a generic set of observations, and is implemented successively for populations 1 and 2 in turn. Consider a histogram as constructed with rectangular blocks, each block corresponding to one observation. The block is positioned according to the interval in which observation lies. The method of kernel density estimation used here replaces the rectangular block by a Normal probability density curve, known in this context as the *kernel function*. The curve is positioned by centring it over the observation to which it relates. The estimate of the probability density curve is obtained by adding the individual curves together over all the observations in the data set and then dividing this sum by the number of observations. Since each component of the sum is a probability density function, each component has area 1. Thus, the sum of the functions divided by the number of observations also has area 1 and is a probability density function.

Mathematically, the kernel density estimate of an underlying probability density function can be constructed as follows (see Aitken and Taroni (2004)). Denote the measurement of the mean concentration of drug in a particular bank note from a generic population by θ. The corresponding probability density function $f(\theta)$ is to be estimated. A dataset $D = \{x_1, x_2, \ldots, x_n\}$ is available to enable this to be done. The variance of the drug concentration from different bank notes is estimated by

$$s^2 = \sum_{i=1}^{n} (x_i - \bar{x})^2 / (n-1),$$

where $\bar{x}$ denote the sample mean. The sample standard deviation s is then multiplied by a parameter, known as the *smoothing parameter*, denoted here by λ, which determines the smoothness of the density estimate. The kernel density function $K(\theta \mid x_i, \lambda)$ for point x_i is then taken to be a Normal distribution with mean x_i and variance $\lambda^2 s^2$,

$$K(\theta \mid x_i, \lambda) = \frac{1}{\lambda s \sqrt{2\pi}} \exp\left\{-\frac{(\theta - x_i)^2}{2\lambda^2 s^2}\right\}.$$

The estimate $\hat{f}(\theta \mid D, \lambda)$ of the probability density function is then given by

$$\hat{f}(\theta \mid D, \lambda) = \frac{1}{n} \sum_{i=1}^{n} K(\theta \mid x_i, \lambda). \tag{8.16}$$

In the construction of a histogram a decision has to be made initially as to the width of the intervals. If the width is wide, the histogram is rather uninformative regarding the underlying distribution. If it is narrow, there may be too much detail with general features of the distribution being lost. Similarly, in kernel density estimation, the spread of the Normal density curves has to be determined. The spread of the curves is represented by the smoothing parameter λ. If the variance is chosen to be large, the resulting estimated curve is very smooth. If the variance is chosen to be small, the resulting curve is very spiky (see Figure 8.7).

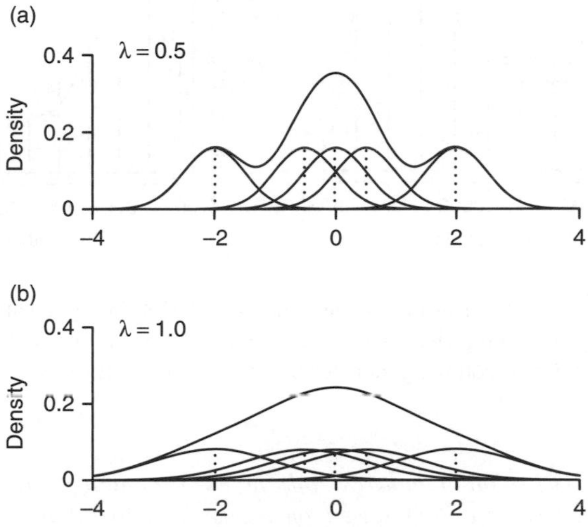

Figure 8.7 Examples of kernel density estimates showing individual kernels. Smoothing parameter values are (a) $\lambda = 0.5$ and (b) $\lambda = 1$. (Figure 10.3 at p. 333 in Aitken CGG and Taroni F 2004 Statistics and the Evaluation of Evidence for Forensic Scientists 2nd edn. John Wiley & Sons, Chichester.)

The smoothing parameter λ has to be chosen. Mathematical procedures exist which enable an automatic choice of λ to be made (see for example Habbema *et al.* (1974) where a so-called *pseudo-maximum likelihood* procedure is proposed). The choice of λ has to be made while bearing in mind that the aim of the analysis is to provide a value V for the evidence in a particular case, as represented by the likelihood ratio. Using the kernel density estimation procedure, an expression for V is derived, see Equation (8.17). An investigation of the variation in V as λ varies is worthwhile (see Aitken and Taroni (2004)). If V does not vary greatly as λ varies then a precise value for λ is not necessary. For example, it is feasible to choose λ subjectively by comparing the density estimate curve $\hat{f}$ obtained for various values of λ with the histogram of the data. The value which provides the best visual fit can then be chosen. A value of λ equal to 0.2 was used to produce the curves in Figure 8.8.

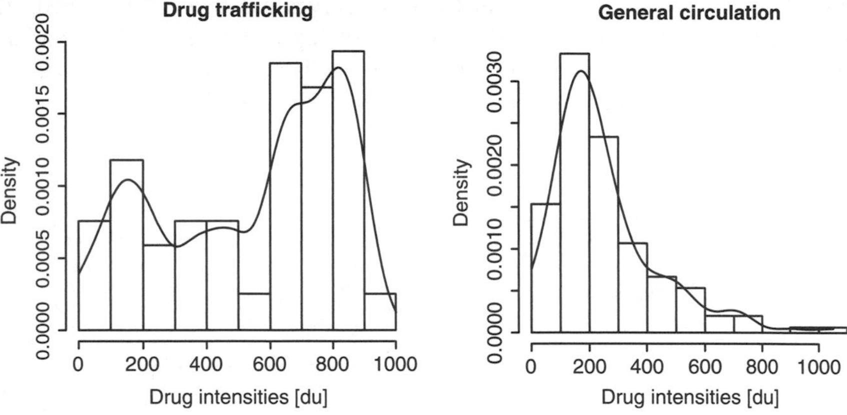

Figure 8.8 Drug intensity measured on bank notes of 200 € in a population of bank notes coming from drug trafficking (left) and in general circulation (right), and associated kernel density estimates with smoothing parameter equal to 0.2 in both cases.

Example 8.5.1 *(Contaminated bank notes - continued from Example 8.3.1). Nine bank notes are seized on a suspect and laboratory measurements show drug intensities [du] equal to* $y = (y_1 = 122, y_2 = 58, y_3 = 184, y_4 = 135, y_5 = 371, y_6 = 835, y_7 = 828, y_8 = 335, y_9 = 625)$*. A dataset containing measurements of concentration of drug in bank notes of 200 € is available for population 1 (drug dealing),* $D_1 = \{x_1^{(1)}, \ldots, x_{n_1}^{(1)}\}$*, and for population 2 (general population),* $D_2 = \{x_1^{(2)}, \ldots, x_{n_2}^{(2)}\}$ *(Besson 2004). The sample mean are* $\bar{x}_1 = 543$ *and* $\bar{x}_2 = 248$ *respectively for populations 1 and 2, while the estimated variances are* $s_1^2 = 78383$ *and* $s_2^2 = 33474$*. The smoothness parameter is taken equal to 0.2 for both populations,* $\lambda_1 = \lambda_2 = 0.2$*. Consider two propositions of interest:*

H_1 *: the bank notes seized on the suspect have been involved in drug dealing;*
H_2 *: the bank notes seized on the suspect are part of general circulation.*

The value of the evidence is given by:

$$V = \frac{\hat{f}(y \mid H_1)}{\hat{f}(y \mid H_2)} = \frac{\prod_{j=1}^{9} \hat{f}(y_j \mid D_1, \lambda_1)}{\prod_{j=1}^{9} \hat{f}(y_j \mid D_2, \lambda_2)}$$

$$= \frac{\prod_{j=1}^{9} \frac{1}{n_1} \sum_{i=1}^{n_1} \frac{1}{\lambda_1 s_1 \sqrt{2\pi}} \exp\left\{ -\frac{\left(y_j - x_i^{(1)}\right)^2}{2\lambda_1^2 s_1^2} \right\}}{\prod_{j=1}^{9} \frac{1}{n_2} \sum_{i=1}^{n_2} \frac{1}{\lambda_2 s_2 \sqrt{2\pi}} \exp\left\{ -\frac{\left(y_j - x_i^{(2)}\right)^2}{2\lambda_2^2 s_2^2} \right\}} = 33.79. \quad (8.17)$$

Table 8.4 Sensitivity of the likelihood ratio in expression (8.17) in Example 8.5.1 given different choices of the smoothness parameter λ.

Smoothness parameter λ	Likelihood ratio V
$\lambda_1 = \lambda_2 = 0.1$	16455
$\lambda_1 = \lambda_2 = 0.2$	33.79
$\lambda_1 = \lambda_2 = 0.3$	5.79
$\lambda_1 = \lambda_2 = 0.4$	2.80
$\lambda_1 = \lambda_2 = 0.5$	1.84

The value 33.79 slightly supports the hypothesis H_1, but it can be observed in Table 8.4 that V is sensitive to the choice of the smoothness parameter λ. One might argue, from a decision-theoretic point of view, that a different choice of the smoothness parameter could lead to a different optimal decision. Consider the '0–k_i' loss function in Table 8.1. The values k_1 and k_2 may be chosen, but the ratio k_1/k_2 is the factor of interest. This ratio is taken to be greater than 1 since the classification of bank notes as coming from drug trafficking, when they do not, is felt a more serious error. Recalling Equation (8.5), one might feel that a value of the likelihood ratio equal to 5.79 ($\lambda_1 = \lambda_2 = 0.3$) is not necessarily followed by decision d_1 (acceptance of H_1) as the optimal decision. In fact, for decision d_1 to be optimal, it is required that

$$5.79 > \frac{k_1}{k_2}\frac{P(p_2)}{P(p_1)}, \tag{8.18}$$

for any $k_1 > 0$, $k_2 > 0$, and any $P(p_1)$, $P(p_2)$. The specific context will suggest values to the decision maker for the losses and for the prior odds. However, a careful reading of Equation (8.18) suggests that to have d_2 as the optimal decision, k_1 should be at least nearly 6 times greater than k_2, if $P(p_1) = P(p_2)$. However the circumstances will often suggest a value of $P(p_1)$ greater than $P(p_2)$, so the ratio between the losses should be even higher. Greater values of λ produce an even smaller likelihood ratio, and more difficult situations to judge, since Equation (8.5) could hold for some values of the losses and of the prior probabilities, and not for some others. However, this is not felt as problematic since it can visually be checked that such smaller values of the likelihood ratio are obtained in association with large choices of λ which in turn produce kernel density estimates that are smooth.

8.6 A NOTE ON MULTIVARIATE CONTINUOUS DATA

Imagine now that multivariate continuous data become available. A piece of evidence, say a handwritten or a printed character in a questioned document, or a fragment of glass recovered at a crime scene or a drug sample, can be described by more than one variable. A statistical model for the evaluation of evidence through the computation of likelihood ratio for multivariate data has been proposed by Aitken and Lucy (2004) in the context of the elemental composition of glass data, and by Bozza *et al.* (2008b) in the context of handwritten questioned documents.

Consider again the populations of printers introduced in Section 8.4.3 and consider the situation where several variables are measured on each printed character (e.g. the area, the box-ratio, the diameter) (Mazzella and Marquis 2007). The background data consist of n_i measurements of p variables for each printer, and are denoted as $\mathbf{x}_{ij} = (x_{ij1}, \ldots, x_{ijp})'$, $i = 1, 2$, $j = 1, \ldots, n_i$.[2] The procedure outlined above for two univariate Normal populations of printers (where only the variable 'area' was considered) will now therefore be extended to the case of multivariate data. Once new documents become available, the problem becomes the classification of them into two multivariate Normal populations, say $N_p(\boldsymbol{\theta}_1, \Sigma)$ and $N_p(\boldsymbol{\theta}_2, \Sigma)$, where $\boldsymbol{\theta}_i = (\theta_{i1}, \ldots, \theta_{ip})'$ is the vector of means of the i-th population, $i = 1, 2$, and Σ is the matrix of variances and covariances of each population.

Denote the recovered measurements to be classified by $\mathbf{y} = (\mathbf{y}_1, \ldots, \mathbf{y}_n)'$, where $\mathbf{y}_j = (y_{j1}, \ldots, y_{jp})'$, $j = 1, \ldots, n$. The probability density of a multivariate Normal variable $N_p(\boldsymbol{\theta}, \Sigma)$ is

$$f(\mathbf{x} \mid \boldsymbol{\theta}, \Sigma) = (2\pi)^{-p/2} \, |\Sigma|^{-1/2} \exp\left[-\frac{1}{2}(\mathbf{x} - \boldsymbol{\theta})' \, \Sigma^{-1} \, (\mathbf{x} - \boldsymbol{\theta})\right],$$

(Appendix B) then, if the population distributions are known exactly (i.e. the mean vectors and the covariance matrix are known), the value V of evidence is given by

$$V = \frac{f(\mathbf{y} \mid \boldsymbol{\theta}_1, \Sigma)}{f(\mathbf{y} \mid \boldsymbol{\theta}_2, \Sigma)} = \frac{\prod_{j=1}^{n} \exp\left[-\frac{1}{2}(\mathbf{y}_j - \boldsymbol{\theta}_1)' \, \Sigma^{-1} \, (\mathbf{y}_j - \boldsymbol{\theta}_1)\right]}{\prod_{j=1}^{n} \exp\left[-\frac{1}{2}(\mathbf{y}_j - \boldsymbol{\theta}_2)' \, \Sigma^{-1} \, (\mathbf{y}_j - \boldsymbol{\theta}_2)\right]} \tag{8.19}$$

$$= \exp\left\{-\frac{1}{2}\left[\sum_{j=1}^{n} (\mathbf{y}_j - \boldsymbol{\theta}_1)' \, \Sigma^{-1} \, (\mathbf{y}_j - \boldsymbol{\theta}_1) - \sum_{j=1}^{n} (\mathbf{y}_j - \boldsymbol{\theta}_2)' \, \Sigma^{-1} \, (\mathbf{y}_j - \boldsymbol{\theta}_2)\right]\right\}.$$

[2]The superscript $'$ at the end of a vector denotes the transpose, that is $\mathbf{x}_{ij}$ is written as

$$\mathbf{x}_{ij} = \begin{pmatrix} x_{ij1} \\ x_{ij2} \\ \vdots \\ x_{ijp} \end{pmatrix}.$$

Generally, population distributions will not be completely known. A simple criterion of classification consists in estimating $\boldsymbol{\theta}_1$, $\boldsymbol{\theta}_2$ and Σ from the background data and substituting them into (8.19). The mean vector $\boldsymbol{\theta}_i$ can be estimated by

$$\bar{\mathbf{x}}_i = \frac{1}{n_i}\sum_{j=1}^{n_i} \mathbf{x}_{ij}, \qquad i = 1, 2.$$

The covariance matrix Σ can be estimated by

$$S = \frac{1}{n_1 + n_2 - 2}\left[\sum_{j=1}^{n_1}\left(\mathbf{x}_{1j} - \bar{\mathbf{x}}_1\right)\left(\mathbf{x}_{1j} - \bar{\mathbf{x}}_1\right)' + \sum_{j=1}^{n_2}\left(\mathbf{x}_{2j} - \bar{\mathbf{x}}_2\right)\left(\mathbf{x}_{2j} - \bar{\mathbf{x}}_2\right)'\right]. \tag{8.20}$$

Example 8.6.1 *(Questioned document – continued). The area and the box-ratio of $n = 2$ characters of type 'a' are measured from a questioned document, $y_1 = (461044.2, 1.27)$, $y_2 = (469538.4, 1.2)$. The document might have been printed by a* Canon model ir*400 (hypothesis H_1), or by an* HP model 41 *(hypothesis H_2). The mean vectors $\bar{\mathbf{x}}_1$ and $\bar{\mathbf{x}}_2$, which are used as estimates of $\boldsymbol{\theta}_1$ and $\boldsymbol{\theta}_2$, are $\bar{\mathbf{x}}_1 = (462550.6, 1.243889)'$ and $\bar{\mathbf{x}}_2 = (430350.164, 1.3205)'$. The covariance matrix Σ is estimated by S, from (8.20), and is*

$$S = \begin{pmatrix} 237525100 & -211.4275 \\ -211.4275 & 0.0009978549 \end{pmatrix}.$$

Substituting $\bar{\mathbf{x}}_i$, $i = 1, 2$, and S in (8.19) gives

$$V = \frac{f(\mathbf{y} \mid \bar{\mathbf{x}}_1, S)}{f(\mathbf{y} \mid \bar{\mathbf{x}}_2, S)} = 6074.$$

A Bayesian criterion of classification would require the introduction of a prior distribution for $\boldsymbol{\theta}_1$ and $\boldsymbol{\theta}_2$, and it is briefly sketched below. Consider the simplest case where the prior distribution of the $\boldsymbol{\theta}_i$ is taken to be Normal, say $\boldsymbol{\theta}_i \sim N(\boldsymbol{\mu}_i, C)$, $i = 1, 2$.

The marginal distribution under H_1 (the questioned document has been printed with printer 1), $f(\mathbf{y} \mid \boldsymbol{\mu}_1, C, \Sigma, H_1)$, is given by

$$\int_{\boldsymbol{\theta}_1} f(\mathbf{y} \mid \boldsymbol{\theta}_1, \Sigma) f(\boldsymbol{\theta}_1 \mid \boldsymbol{\mu}, C) d\boldsymbol{\theta}_1 = \int_{\boldsymbol{\theta}_1} \prod_{j=1}^{n} |2\pi|^{-p/2} |\Sigma|^{-1/2} \times$$
$$\times \exp\left\{-\frac{1}{2}\left(\mathbf{y}_j - \boldsymbol{\theta}_1\right)' \Sigma^{-1} \left(\mathbf{y}_j - \boldsymbol{\theta}_1\right)\right\} |2\pi|^{-p/2} |C|^{-1/2}$$
$$\times \exp\left\{-\frac{1}{2}\left(\boldsymbol{\theta}_1 - \boldsymbol{\mu}_1\right)' C^{-1} \left(\boldsymbol{\theta}_1 - \boldsymbol{\mu}_1\right)\right\} d\boldsymbol{\theta}_1$$

and can be shown to be equal to[3]

$$f(\mathbf{y} \mid \boldsymbol{\mu}_1, \Sigma, C, H_1) = |2\pi\,\Sigma|^{-n/2} |2\pi C|^{-1/2} \left|2\pi\left(n\Sigma^{-1} + C^{-1}\right)^{-1}\right|^{1/2}$$

$$\exp\left\{-\frac{1}{2}\left[tr\left(S\,\Sigma\right)^{-1}\right) + (\bar{\mathbf{y}} - \boldsymbol{\mu}_1)'\left(\frac{1}{n}\Sigma + C\right)^{-1}(\bar{\mathbf{y}} - \boldsymbol{\mu}_1)\right]\right\},$$

where $\bar{\mathbf{y}} = \frac{1}{n}\sum_{j=1}^{n}\mathbf{y}_j$, and $S = \sum_{j=1}^{n}\left(\mathbf{y}_j - \bar{\mathbf{y}}\right)\left(\mathbf{y}_j - \bar{\mathbf{y}}\right)'$. In the same way the marginal distribution under H_2 (the questioned document has been printed with printer 2), is given by $f\,(\mathbf{y} \mid \boldsymbol{\mu}_2, \Sigma, C, H_2)$. The value of the evidence is given by

$$V = \frac{f\,(\mathbf{y} \mid \boldsymbol{\mu}_1, \Sigma, C, H_1)}{f\,(\mathbf{y} \mid \boldsymbol{\mu}_2, \Sigma, C, H_2)}. \tag{8.21}$$

However, the prior elicitation of a multivariate Normal distribution, in particular of a covariance matrix, can be a difficult problem and it will not be pursued anymore. Some suggestions are given in O'Hagan *et al.* (2006). Moreover, a multivariate Normal distribution for $\boldsymbol{\theta}$ may not always necessarily be a reasonable assumption. The assumption of normality may be removed by considering a kernel-density estimation procedure (see, for example, Aitken and Lucy 2004). Consideration of a prior distribution for Σ adds another level of complexity and is beyond the scope of this book.

8.7 R CODE

A symbol '∗', '+', ',' and so on at the end of a line indicates that the command continuous to the following line. The absence of such a symbol indicates the end of a command.

Example 8.3.1

Data and prior parameters

```
n=100
xtilde=76
n1=462
n2=992
```

[3]The symbol *tr* denotes the *trace of a matrix*; given an $(m \times m)$ square matrix $A = [a_{ij}]$, the trace is defined as the sum of the diagonal elements:

$$tr(A) = \sum_{i=1}^{m} a_{ii}\,.$$

```
x1=382
x2=562
p1=x1/n1
p2=x2/n2
alpha1=round(p1*(n1-1))
alpha2=round(p2*(n2-1))
beta1=round((1-p1)*(n1-1))
beta2=round((1-p2)*(n2-1))
```

Bayes factor

```
library(VGAM)
lp1=dbetabin.ab(xtilde,n,alpha1,beta1,log=TRUE)
lp2=dbetabin.ab(xtilde,n,alpha2,beta2,log=TRUE)
B=exp(lp1-lp2)
print(paste('Bayes factor =',round(B)))
```

Impact of different prior probabilities

```
pp1=seq(0.001,0.9,0.001)
pp2=1-pp1
pp1post=log(pp1)+dbetabin.ab(xtilde,n,alpha1,beta1,log=TRUE)
pp2post=log(pp2)+dbetabin.ab(xtilde,n,alpha2,beta2,log=TRUE)
post_odds=exp(pp1post-pp2post)
pp1[which(post_odds>1)][1]
```

Example 8.3.2

Observation to be classified

```
y=5
```

Populations' parameters known

```
lambda=c(3.91,1.32)
f=dpois(y,lambda)
print(paste('Value of evidence:',round(f[1]/f[2],1)))
```

Populations' parameters unknown

```
alpha=c(125,7)
beta=c(32,5)
pred=(1/factorial(y))*beta^alpha/gamma(alpha)*
gamma(alpha+y)/(beta+1)^(alpha+y)
print(paste('Value of evidence:',round(pred[1]/pred[2],2)))
```

Influence of populations' prior probabilities

```
p1=seq(.01,.9,.01)
p2=1-p1
post_odds=(pred[1]/pred[2])*(p1/p2)
p1[which(post_odds>=1)][1]
```

Examples 8.4.1 and 8.4.2

Observations to be classified

```
y=0.155
n=5
ybar=0.155
```

Populations' parameters known

```
thetac=0.14
sigmac=0.01
thetap=0.3
sigmap=0.06
v=dnorm(y,thetac,sigmac)/dnorm(y,thetap,sigmap)
print(paste('Value of evidence:',round(v,2)))
vn=dnorm(ybar,thetac,sigmac/sqrt(n))/dnorm(ybar,thetap,
sigmap/sqrt(n))
print(paste('Value of evidence ( n =',n,'):',round(vn)))
```

Populations' parameters unknown one sample

```
sigmac2=sigmac^2
sigmap2=sigmap^2
muc=0.14
tauc2=0.003^2
mup=0.3
taup2=0.016^2
vp=dnorm(y,muc,sqrt(sigmac2+tauc2))/
dnorm(y,mup,sqrt(sigmap2+taup2))
print(paste('Value of evidence:',round(vp,2)))
```

more samples

```
n=5
vpn=dnorm(ybar,muc,sqrt(sigmac2/n+tauc2))/
dnorm(y,mup,sqrt(sigmap2/n+taup2))
print(paste('Value of evidence ( n =',n,'):',round(vpn)))
```

8.4.2 A note on the robustness of the likelihood ratio

Prior parameters

```
thetac=0.14
sigmac=0.01
thetap=0.3
sigmap=0.06
```

Simulations

```
n=10000
valc=rnorm(n,thetac,sigmac)
valp=rnorm(n,thetap,sigmap)
```

```
lrt=dnorm(valc,thetac,sigmac)/dnorm(valc,thetap,sigmap)
lrp=dnorm(valp,thetac,sigmac)/dnorm(valp,thetap,sigmap)

length(which(lrt<1))
length(which(lrt>=1 & lrt<10))
length(which(lrt>=10 & lrt<100))
length(which(lrt>=100 & lrt<1000))

length(which(lrp<1))
length(which(lrp>=1 & lrp<10))
length(which(lrp>=10 & lrp<100))
length(which(lrp>=100 & lrp<1000))
```

Examples 8.4.3 and 8.4.4

Observations to be classified

```
n=10
bary=454888
```

Two populations

```
sigma2=546187921
mu1=462825
mu2=430350
tau2=10000^2
v=dnorm(bary,mu1,sqrt(sigma2/n+tau2))/
dnorm(bary,mu2,sqrt(sigma2/n+tau2))
```

Several populations

```
mu3=476242
p1=1/3
p2=1/3
p3=1/3
p=c(p1,p2,p3)
post=matrix(0,nrow=1,ncol=length(p))
mu=c(mu1,mu2,mu3)
tau=c(10000^2,10000^2,10000^2)
for (i in 1:length(p)){
post[i]=p[i]*dnorm(bary,mu[i],sqrt(sigma2/n+tau[i]))/
(p1*dnorm(bary,mu1,sqrt(sigma2/n+tau[1]))+
p2*dnorm(bary,mu2,sqrt(sigma2/n+tau[2]))+
p3*dnorm(bary,mu3,sqrt(sigma2/n+tau[3])))
}
print(post)
```

Example 8.4.5

Observation to be classified

```
y=12
```

Classification into female or male population

```
n=c(38,35)
xbar=c(10.35,14.09)
s=c(1.42,1.52)

lambda=s^2*(n+1)/n
alpha=n-1
c=(gamma((alpha+1)/2)/gamma(alpha/2))/
((lambda*alpha)^(1/2)*gamma(.5))
v=c*(1+((y-xbar)^2)/(lambda*alpha))^-(alpha+1)/2
print(paste('Value of the evidence =',round(v[1]/v[2],2)))
```

Example 8.5.1

Note. This routine needs an available database to be implemented.

Observations to be classified

```
banknotes=c(122,58,184,135,371,835,828,335,625)
```

Loading data (populations 1 and 2)

```
population=data.frame(read.csv('namedatabase.csv',header=TRUE))
attach(population)
```

The name of the database containing the population data should be inserted in place of

```
namedatabase.csv
```

Note that the variables are comma separated (*csv*). Rows correspond to individual bank notes, columns to separate variables. In this example columns 4 and 5 contain the drug intensities (*du*) corresponding to bank notes of 200 € from populations 1 and 2, respectively.

```
d1=population[,4]
d2=population[,5]
```

Estimate of the probability density function and plots.
Population 1

```
n1=length(d1)
m1=mean(d1)
s1=sum((d1-m1)^2)/(n1-1)
lambda1=0.3
sk1=lambda1*sqrt(s1)
x=seq(0,1000,1)
fhat1=0
for (i in 1:n1){
zi=d1[i]
k=dnorm(x,zi,sk1)
fhat1=fhat1+k
```

```
}
dfhat1=fhat1/n1
hist(b1,freq=FALSE,xlab='Drug intensities [du]',
main='Drug trafficking')
lines(dfhat1,type='l')
```

Population 2

```
n2=length(d2)
m2=mean(d2)
s2=sum((d2-m2)^2)/(n2-1)
lambda2=lambda1
sk2=lambda2*sqrt(s1)
x=seq(0,1050,1)
fhat2=0
for (i in 1:n2){
zi=d2[i]
k=dnorm(x,zi,sk2)
fhat2=fhat2+k
}
dfhat2=fhat2/n2
```

Value of the evidence

```
prod(dfhat1[banknotes])/prod(dfhat2[banknotes])
```

Example 8.6.1

Note. This routine needs an available database to be implemented.

Observations to be classified

```
recovered=c(...,...)
```

Loading data (example)

```
name.data=data.frame(read.csv('namedatabase.csv',header=TRUE))
attach(name.data)
```

The name of the database containing the population data should be inserted in place of

```
namedatabase.csv
```

Note that the variables are comma separated (*csv*). Rows correspond to individual characters, columns to separate variables. In this example columns 2, 6 and 7 contains the measurements of the variables *Area* and *Box-ratio*, and the type of printer (*Printer* 1, 2).

```
variables=c(2,6,7)
population=as.matrix(name.data[,variables])
```

Estimation of the mean vectors and the covariance matrix

```
x1bar=apply(population[which(population[,3]==1),],2,mean)[1:2]
x2bar=apply(population[which(population[,3]==2),],2,mean)[1:2]
n1=length(which(population[,3]==1))
n2=length(which(population[,3]==2))

p=population[which(population[,3]==1),-3]
pbar1=cbind(p[,1]-x1bar[1],p[,2]-x1bar[2])
p=population[which(population[,3]==2),-3]
pbar2=cbind(p[,1]-x2bar[1],p[,2]-x2bar[2])
S=(t(pbar1)%*%pbar1+t(pbar2)%*%pbar2)/(n1+n2-2)
```

Value of the evidence

```
library(mvtnorm)
V=exp(sum(log(dmvnorm(recovered,mean=x1bar,sigma=S)))-
sum(log(dmvnorm(recovered,mean=x2bar,sigma=S))))
```

In the specific example

```
recovered=matrix(c(461044.2,469538.4,1.27,1.2),nrow=2)
x1bar=c(462550.6,1.243889)
x2bar=c(430350.164,1.3205)
S=matrix(c(237525100,-211.4275,-211.4275,0.0009978549),nrow=2)
V=exp(sum(log(dmvnorm(recovered,mean=x1bar,sigma=S)))-
sum(log(dmvnorm(recovered,mean=x2bar,sigma=S))))
```

9

Bayesian Forensic Data Analysis: Conclusions and Implications

9.1 INTRODUCTION

There is a growing community in forensic science that has spent many years applying Bayesian ideas to inference from evidence applied to legal contexts. A main reason for this is, probably, that Bayes' theorem is helpful in explaining the concept of evidence and proof in science and that formal procedures are easier to explain and justify than informal ones (Edwards 1988). But yet, if one looks at science more generally, the interest in Bayesian methods to solve problems relating to data analysis appears to be relatively under-utilized (D'Agostini 2004b), with forensic science being no exception.

With regard to this, a principal motivation for writing this book was to bridge the gap between research and operational data-analytic problems in forensic science and the current methodological backup to approach such issues. Part I of this book thus focused on presenting the Bayesian methodology for – summarized in simple terms here – the consistent choice among available options (which may be hypotheses according to the context) when the available information (i.e. evidence) is incomplete. The second part of this book, in particular Chapters 4 to 8, then focused on the study of that methodology for forensic problems in data analysis, gravitating, in one respect or another, around the question of how a scientist may arrive at sensible conclusions. Let us notice again at this point that the term

Data Analysis in Forensic Science: A Bayesian Decision Perspective Franco Taroni, Silvia Bozza, Alex Biedermann, Paolo Garbolino and Colin Aitken

'conclusion' was not interpreted in the sense of a decision made by a judge or court (although the application of the methodology to such issues is perfectly feasible). Rather, attention was concentrated on the idea of 'assessments' or 'judgements' that the scientist may face during the analytical process (e.g. estimating a proportion or choosing a sample size).

At this juncture, the sceptical reader may ask 'Why should I consider the formalisms advocated in this book?'. We feel that this is a critical question because experience shows that many scientists, when confronted with the discipline of statistics, tend to adopt and practise an approach much like the following of a recipe, however uncomfortable or not that may be. This is precisely the attitude that we wish that the reader would *not* take with respect to the concepts presented in this book. Ideally, the aim is that the reader has an understanding of the rationale behind the proposed methods and that the reader will use them because they are *convinced* that the methods are sound and address the practical needs appropriately. It thus seems useful to conclude this book with a review of the principal topics and issues, offered in a 'question and answer' style. The reason for this choice is that, even though this mode of presentation differs somewhat from the rest of this book, it is one that forensic scientists may find more natural, principally because their profession involves much activity where the interplay between questioning and answering holds an essential role. For example, scientists may need to explain and justify a particular method for data analysis, how they arrived at some specified conclusion and why, on the whole, some chosen methodology is one that can be trusted. Notice further that all of this also stands in close relation to the precepts of evidential assessment set forth in Chaper 1.

There is much foundational and philosophical literature available on a broad range of aspects of the Bayesian decision-theoretic approach. In what follows hereafter, there is no claim of an exhaustive or entirely representative account. Some references will be given to where the interested reader may find discussion that goes into further details than those that can be given here.

9.2 WHAT IS THE PAST AND CURRENT POSITION OF STATISTICS IN FORENSIC SCIENCE?

Since the early 1960s, forensic science openly faced the problem of interpretation and evaluation of data, as is critically noted by a famous quote of the pioneer forensic scientists Kirk and Kingston:

> When we claim that criminalistics is a science, we must be embarrassed, for no science is without some mathematical background, however meagre. This lack must be a matter of primary concern to the educator [. . .]. Most, if not all, of the amateurish efforts of all of us to justify our own evidence interpretations have been deficient in mathematical exactness and philosophical understanding. (Kirk and Kingston 1964, pp. 435–436)

However, interpretation and data evaluation are still among the most neglected areas in the entire field of physical evidence (National Research Council 2009). In fact, many evaluations are simply subjective. Note that this usage of the term 'subjective', contrary to the remainder of this book, should be understood as 'arbitrary'. As mentioned by Kirk and Kingston (1964), it was indeed rare (and often, currently, still is rare) that an opinion is based on a statistical or probabilistic study. This is so even though the presentation of a general model which remedies these deficiencies was put forward by Darboux *et al.* (1908) and rediscovered by Kingston (1965a,b) through the application of the Bayesian model to forensic science.

Given that forensic science as well as science, in general, should abandon the idea of certainty, as was argued in Chapter 1, it becomes a logical necessity to determine the degree of belief that may be assigned to a particular event or proposition[1]. In this context, statistics can offer the most valuable approach there is to science. When uncertainty does exist, and a statistical approach is possible, then this approach is the best one available since it offers to assess and measure the uncertainty based upon a precise and logical line of reasoning.

9.3 WHY SHOULD FORENSIC SCIENTISTS CONFORM TO A BAYESIAN FRAMEWORK FOR INFERENCE AND DECISION MAKING?

Above all, the fundamental problem of scientific progress, and also a fundamental problem of everyday life, is that of learning from experience. In part, knowledge obtained in this way is merely a description of what has already been observed. Another part, however, consists of making inferences from past experience to predict future experience. This is the part that may be referred to and has been presented here as induction (see Chapter 2). Stated otherwise, on the basis of what one sees, one seeks to evaluate the uncertainty associated to an event of interest, the kind of reasoning used to learn new things. Forensic scientists need a defensible approach to this challenging issue.

Bayesian inference uses the likelihood function to convert a prior probability distribution that characterizes an observer's beliefs about a continuous population parameter θ, for instance, into a posterior distribution that takes account of information (data) (see, for instance, Sections 2.3 and 3.3). The principal relevance of this resides in the fact that classical (or frequentist) inference restricts itself to sample data and does not lead to direct probability statements for the possible values of a parameter. In contrast to Bayesian analysis, where unobserved

[1]It should be emphasised here again that 'event' and 'proposition' are taken as equivalent terms with only a different emphasis on the fact and on the formulation expressing it (de Finetti 1968).

parameters are treated as random and observed data as fixed, classical analysis does just the opposite: it treats the data as random even after observation and considers the parameter as a fixed unknown constant not having a probability distribution.

The Bayesian method has a clearly stated objective, that of using data to revise the state of knowledge about a hypothesis (e.g. a proposition of the kind 'the suspect is the source of the recovered bloodstain') or, more generally, a parameter of interest.

As such, Bayesian analysis can be regarded as a framework to capture how one ought to make up one's mind. This also stems from the fact that it pertains to an interpretation of probability as a particular measure of the opinions of (ideally) consistent people, wherein probability is a coherent opinion and inference from data is nothing other than the revision of such opinion in the light of relevant new information (Edwards *et al.* 1963; Schlaifer 1959).

A further argument in favour of Bayesian inference is its connection to decision theory, which includes an additional feature by incorporating into the analysis the consequence of actions or decisions. Bayes' theorem relates naturally to a rational decision calculus. The posterior probabilities are exactly what is needed to find optimal decisions, based on elicited utilities (or losses). This relationship is a major strength of the Bayesian method (Kadane 1995). As observed by de Finetti (1970, p. 139)

> All information leads to inference on a new distribution, and this gives an adequate basis for decision. Though logically independent, inference and decision are connected because the results of the former are the point of departure of the latter.

Finally, in that it provides a standard, the Bayesian argument is normative. The argument prescribes how an individual who is faced with a decision problem involving uncertainty should proceed in the choice of a course of action that is consistent with their personal basic judgements and preferences. In the sense explained in Section 2.1, it is mandatory to check consciously the consistency of one's personal inputs and to calculate their implications for action. The point is well made, even with regards to concessions in practice, by de Finetti (1972, p. 150), according to which

> this is the behaviour that 'ought to be' followed by a reasonable person, if he is to avoid incoherence, and it may further be said to tend to agree with the actual behaviour of people, though distortions and errors, especially in intricate questions, must be expected and do occur. But how frequent and serious these anomalies are is entirely irrelevant to the theory of probability.

9.4 WHY REGARD PROBABILITY AS A PERSONAL DEGREE OF BELIEF?

Part of an argument in favour of viewing probability as a degree of belief stems from the fact that other interpretations of probability encounter applicability problems. As an illustration of this point, observe again the meaning of probability concerning the occurrence of events or phenomena. The so-called frequentist definition of probability postulates a long sequence of repetitions of a given situation assuming identical conditions. For the purpose of illustration, consider a sequence of n repetitions in which an event E occurs X times. X may take some value greater than or equal to 0 and less than or equal to n. In such a setting, X/n is the relative frequency. It could vary in different sequences of n repetitions. It is, however, supposed that, in a sequence where the number n of repetitions grows indefinitely under identical conditions, the relative frequency tends to a definite limiting value. In a frequentist framework the probability of event E is defined to be the limiting value of the relative frequency.

Consider the foregoing now from a practical point of view. In reality, it is difficult, if not impossible, to maintain identical conditions between trials. Consequently, in anything other than idealized situations, such a definition of probability proves unworkable. As an example, consider the prediction of the rate of unsolved crimes for the following year. It is simply inappropriate to use the frequentist definition to determine the probability that the rate will lie for example between 20% and 25% of the total number of crimes investigated, essentially because it is not possible to consider crime as a sequence of repetitions under identical conditions. Unsolved crime rate in the following year is a unique, one-time event (Berger 1988).

Frequentist probabilities are sometimes referred to as 'objective' probabilities. They are said to be objective in the sense that there is a well-defined set of circumstances for the long-run repetition of the trials, such that the corresponding probabilities are well defined. One's personal or subjective views will not alter the value of the probabilities. Each person considering these circumstances will provide the same values for the probabilities. The frequentist model strictly refers to a relative frequency obtained in a long sequence of trials that are assumed to be performed in an identical manner, physically independent of each other. Such a circumstance encounters certain difficulties. It follows from this point of view, for instance, that no statement of probability is allowed for any situation that does not happen to be embedded, at least conceptually, in a long sequence of events giving equally likely outcomes (Gelman *et al.* 1997; Press and Tanur 2001).

In many situations, law being a prime example, one cannot assume equally likely outcomes any more than one can count past occurrences of events to determine relative frequencies.

> What is the chance that the defendant is guilty? Are we to imagine a sequence of trials in which the judgements, 'guilty' or 'not guilty', are made and the frequency of the former found? It will not work because it confuses the judgement of guilt, but, more importantly, because it is impossible to conceive of a suitable sequence. Do we repeat the same trial with a different jury; or with the same jury but different lawyers; or do we take all Scottish trials; or only Scottish trials for the same offence? The whole idea of chance is preposterous in this context. (Lindley 1991, p. 48)

Contrast what has been noted so far in this section with the idea of subjective probability which regards the probability of an event to happen as a measure of *personal* belief in the occurrence of that event. For example, a person may have a personal feeling that the unsolved crime rate will be between 20% and 25%, even though no frequency probability can be assigned to the event. This should not appear to be surprising. It is actually very common to think in terms of personal probabilities all the time, such as when betting on the outcome of a football game or when stating the probability of rain tomorrow.

More specifically, a subjective probability amounts to a personal degree of belief, as actually held by someone based on his whole knowledge, experience and information, with respect to the truth of a given statement or event, the truth or falsity of which is, for whatever reason, unknown to that person. It follows from this that such an assessment:

- depends on information;
- may change as the information changes;
- may vary amongst individuals because different individuals may have different information or assessment criteria.

It may readily be seen that this view of probability closely relates to the kind of events and parameters encountered in such diverse fields such as history, economy, law, forensic science and many other contexts. Most importantly, in these contexts events and parameters are usually not the result of repetitive or replicable processes. On the contrary, they are singular and unique. It is not conceivable to play the world over and over again to tabulate the number of occasions on which some past event actually occurred. The way ahead for the determination of probabilities for such situations is through subjective probabilities. Interestingly, the previously mentioned obstacles that hinder the applicability of the frequency concept of probability are just the very field of the *mise en œuvre* of the personalist interpretation of probability.

Personal probabilities are sometimes viewed cautiously, however, as a concept that may appear abstract and nontrivial to capture. In the same context, people may also have preconceptions about expressing probabilities numerically, suggesting that this interpretation of probability is arbitrary and, from a practical point of view, an inaccessible concept. Such perceptions are unnecessarily restrictive because they disregard the fact that personal degrees of belief can actually be elicited and investigated empirically. One possibility of effecting this is in terms of bets that an individual is willing to accept (see Chapter 2). For instance, the probability maintained by an individual in the truth or otherwise of a proposition can be compared with the probability of drawing a black ball from an urn, a setting that can be represented in terms of two gambles involving the same prize (with no stakes). An idealized method for measuring the probability the individual entertains in the truth of a proposition would be to propose the choice between a gamble which offers a certain prize if the proposition is true and a gamble which offers the same prize if a black ball is drawn from an urn of known composition: if the individual chooses the first game, this means that, for him, the probability of the proposition at issue is greater than the proportion p of black balls in the urn; if the individual chooses the second game, this means that, for him, the probability of the proposition is smaller than the proportion of black balls. Ideally, one can vary the proportion p up to the point when the individual declares himself to be indifferent between the two gambles: that value p will be his probability for the proposition.

An illustration of this idea can also be found, for example, in Lindley (1985, pp. 25–26):

> A chemical engineer realized that there was a chance of the process for which he was responsible failing but was reluctant to assess it numerically. However he knew the monetary consequences of failure and so I asked him: suppose I was able to offer you a device which would make the process certain, how much would you pay me for it, a thousand dollars, ten thousand? He laughed at the latter figure as being ridiculously high but contemplated the former more seriously. After some bargaining we settled for 750, a figure which can be converted into a probability.

One might expect that numbers should be familiar to scientists because, as noted by Lindley (1985, p. 13),

> our aim is to describe the concept of uncertainty numerically: for number is the essence of the scientific method and it is by measuring things that we know them.

However, scientists may have preconceptions about particular numerical values and argue, as a consequence, that the concept as a whole is to be rejected. With regards to this, it is worth recognizing that a rough qualitative appreciation of probabilities is often sufficient for practical purposes (de Finetti 1974). Actually, one could even take a step further and argue that probability is not really about numbers but about the structure of reasoning. Implementation of the personalist interpretation of probability requires an individual to assign numbers to the likelihood of events. What really matters is the fact that numbers allow one to use powerful rules of reasoning which can be set into effect practically, such as by computer programs. The very important issue is not whether the specified numbers are 'precise', whatever the meaning of 'precision' may be in reference to subjective degrees of belief based upon personal knowledge. What is really crucial is that one is enabled to use sound rules of reasoning that allow one to check the logical consequences of propositions and that one is enabled to answer questions like: what are the consequences with respect to the degree of belief in A of assuming that the degree of belief in B is high? And how does the degree of belief in A change, if we lower the degree of belief in B? (Taroni *et al.* 2006a).

More generally, it should also be reminded that science is a human activity and, as with every human activity, it is a product of thought. Probability, as such, is an aid in dealing with uncertainty which is pervasive in everyday life. On this point, de Finetti, (1989, p. 170) wrote:

> [...] no science will permit us to say: this fact will come about, it will be thus and so because it follows from a certain law, and that law is an absolute truth. Still less will it lead us to conclude negatively: the absolute truth does not exist, and so this fact might or might not come about, it may go like this or in a totally different way, I know nothing about it. What we can say is this: *I foresee* that such a fact will come about, and that it will happen in such and such a way, because past experience and its scientific elaboration by human thought make this forecast seem reasonable to me.

Probability is precisely what makes it possible to attach uncertainty to a forecast. Since a forecast is always made in reference to a subject, shaped by its experience and convictions, the logical instrument that is needed is the subjective theory of probability. It is in this context that Bruno de Finetti provocatively labelled his probabilism as *subjectivist* and expressed this position with the memorable phrase for which he is now well known: 'Probability does not exist'[2]. Savage wrote in the same line of thought, but more diplomatically, 'probabilities are states of mind and not states of nature' (Savage 1981).

With regard to judicial applications of both probability theory and its subjective interpretation, scholars in probability have often found examples involving uncertain reasoning in legal contexts as particularly convincing. Consider, for example, de Finetti (1993a, p. 293).

> The forecast and assumptions we continuously make, constitute the usual object of our thinking in all the practical circumstances of life, more than the much rarer judgements which are logically certain. Depending on situations, we feel likely to rely more or less on the reliability of such forecasts and assumptions. In combining these judgements on the reliability degree of our different forecasts and assumptions, we in fact reason in accordance with probability theory, although without awareness and in a rather approximate way. One of the most effective examples is that one of the judicial or police investigations, where procedure is always on the basis of clues and suppositions, where the work is never based on certainty, but always and only on the probable.

As in the example of the legal investigation, it is not sensible to consider probabilities in a frequentist format or to consider that the circumstances of the

[2]For a discussion on de Finetti's subjectivism and the suspicious attitude embraced by some scientists regarding the use of subjective probabilities, see Dawid and Galavotti (2009).

investigation are such that the frequentist probabilities would be able to answer the questions of interest. What is at issue is always a specific set of circumstances, relating to the situation of a given individual. An individual can consider the set most close to their actual needs by adopting a personalist view of probability, rather than by a frequentist perspective which requires assumptions to be made that – as has been seen – are known not to apply in real life situations.

A further recurrent issue in discussions on subjective probability is encountered in relation with the assessment of prior probabilities. Within a personalist regime, prior probabilities are assigned according to the actual beliefs held by the subject of interest. The formation of a prior probability thus amounts to a choice which is the result of a largely context-dependent procedure, a point concisely made by Cornfield (1967, p. 47):

> The inability to assign in a unique way prior probabilities either from experience or from principles like that of ignorance or invariance has led to a favourable re-examination by some of the nineteenth-century doctrine as expressed by De Morgan (1847) that probability is a degree of belief. De Morgan held that a probability is not an objective characteristic of the external world, but a subjective attitude towards it, which can and does vary from individual to individual.
>
> [. . .]
>
> Furthermore if one accepts such a view of probability one must reject the idea that an outcome must lead to, in Fisher's words, 'a rigorous and unequivocal' (i.e. a unique) conclusion. I must confess that although I once entertained obiections somewhat like this I now regard the subjective view as inescapable.

The shift from prior to posterior probabilities thus means passing from one subjective assessment of probability to another. What is involved here, in essence, is a change of opinion in response to new information.

It seems that concerns about this view are the result of a confusion between *subjectivity* (in the sense of arbitrary) and the rather difficult technical question of how probabilities ought to be assigned. A popular argument states that if a probability represents a degree of belief, then it must be subjective in the sense of arbitrary, because the belief of one individual could be different from that of another. Following Sivia (1996), the Bayesian view is that a probability does indeed represent how much one believes that something is true, but that this belief should be based on all the relevant information available. That latter requirement is important because the information at one individual's disposal may not be the same as that accessible to another. This is not the same as arbitrary. It solely means that probabilities are always conditional, and this conditioning must be stated explicitly (Berry 1997).

The fact that Bayesian analysis requires an explicit prior distribution offers an excellent opportunity for an actor to state publicly beliefs about a phenomenon at hand – with the sole obvious restriction that, in the case of a scientist, the phenomenon of interest is one that differs from an issue that is in the exclusive province of the court, such as a proposition of the kind 'the suspect is the source of the crime stain'. More generally, however, as noted by Kadane (1995), the exhibition of prior probabilities is a highly desirable property because it can be taken as a 'step toward honesty in analysis'. It is also one of the desiderata of evidential assessment – transparency – outlined in Chapter 1.

Given the preceding, one can see that it is of little help to talk about prior distributions without placing them into the framework to which they belong. An awareness is required of their role in Bayes' theorem, as well of the role of the theorem itself, in order to develop a constructive relationship with prior probabilities. If this can be assured, then one may choose the priors most suitable for a specific problem or ignore them if they are irrelevant. Alternatively, it may also be that one decides that only the likelihood ratio can be provided. On yet other occasions, one may even skip Bayes' theorem altogether, or use it in a reverse mode to discover which kind of priors might give rise to the final beliefs that one unconsciously has (D'Agostini 1999).

9.5 WHY SHOULD SCIENTISTS BE AWARE OF DECISION ANALYSIS?

Substantial theory and practice demonstrates that there is a lot of scope for the use of statistical ideas in law. In part this stems from the fact that the criminal justice system works towards reducing uncertainty, as in a court of law, by the provision of scientific evidence, notably using the analysis of forensic data. This may be readily illustrated by recalling, for instance, the use of measurements made on a sample of a suspect's blood in order to estimate (blood) alcohol concentration.

Law distinguishes two main separate functions: inference and decision making (Lindley 2006). Forensic scientists have, however, not always recognized that they can make a contribution also to the latter, that is explaining how data may be used to assist a judiciary action, rather than just presenting data informatively, such as through a likelihood ratio.

This is pointed out, for instance, through the topic discussed in Chapter 4, point estimation, where the process of inference is represented by the assessment of the posterior probability distribution on the parameter of interest, θ. The decision process consists of the choice of a point estimate that best respects the scientist's personal preferences (as represented by the utility or loss function). Stated otherwise, the inference process in which the scientist engages will help him to answer a question of the kind 'What should I believe?', whereas the decision process would answer a query of the kind 'What should I do?' (Royall 1997). The two questions are closely related. In fact, as presented in Chapter 3 and Chapter 4, inference and decision are closely linked. 'All information leads to inference on a new distribution, and this gives an adequate basis for decision' (de Finetti 1970).

In Bayesian statistics, the measurement of both uncertainty and value play an essential role. Judicial literature is well aware of this and has proposed formal (Bayesian) decision models to solve, for instance, litigation cases (see Section 2.4.3).

On their side, the forensic sciences did not take account of advantages of such models, probably because of a reluctance to approach the concept of value and lacking procedures for coping with it. Notwithstanding, decision – like uncertainty – is an omnipresent issue. Forensic scientists, as much as scientists in general, are systematically faced with problems involving decision making. They may need to inquire on how they are to judge an estimate, how to decide among competing hypotheses or how to judge the relative appropriateness of two (or more) competing models.

According to the view advocated here, the optimal choice is the option for which the expected value of the utility function is largest (or the expected value of the loss function is the smallest). But since utility (loss) also relates to numbers, which may be viewed cautiously, it is useful to remember – as noted by Lindley (1992, p. 7) – that '[a]ll action within the Bayesian view is based on maximization of expected utility (or minimization of expected loss) where the maximum (or the minimum) value is irrelevant, all that matters is that it is the largest (or the smallest) in comparison with everything else. We do something not because it is 'good' but because it is 'better' than anything else we can think of.' The perception is that of an optimal strategy not in the sense of a universally 'best' choice, but that of a strategy which is best for a situation at hand (Raiffa 1968).

Decision analysis represents a relevant topic of interest for forensic scientists because it allows the quality of one's decision to be transparent. More formally, decision analysis is unavoidable since otherwise there remains the question of how the quality of a decision ought to be measured in the absence of decision theory. Decision theory provides the framework for exploring actions. Its value resides in the fact that it forces the decision maker (who may be a scientist) to recognize that they may err either by taking an unnecessary action or by failing to take a necessary action. It helps to formalize and categorize thinking to make sure that all relevant possibilities have been considered.

A concept only briefly mentioned in Chapter 4 (Section 4.2.2), but not treated in further detail, is that of *decision hierarchy* through a temporal coherency. As an example to illustrate this idea, consider once again the problem of estimating blood alcohol concentration through the analysis of breath. One can see that, temporally speaking, the first decision focuses on the estimate of alcohol concentration. The scientist can use a model (as illustrated in Chapter 4) which puts forth the scientist's own worth. For sake of illustration, the scientist may prefer to penalize the underestimation of the true alcohol concentration more than the overestimation. This may be so because falsely concluding a low alcohol concentration in an individual with increased blood alcohol is regarded as a more serious error. Based on such a loss function and taking advantage of the posterior distribution of the alcohol concentration given the data from the laboratory (i.e. measurements or data), the scientist decides on a point estimate. Later in time, the case may come to court where a judge decides on a guilty or not guilty verdict. Here, the judge should specify their costs for a false acquittal and for a false conviction. In other words, decisions should be (and generally are) sequential. Decision theory places the works of decision makers in a coherent and transparent framework.

A summary definition of decision analysis is due to Lindley (2000a, p. 75):

> Life is a balance between reason and emotion. Maximisation of expected utility reflects this in the combination of probability as the reasoning tool, and utility as the numerical expression of emotion.

A question may however arise from this statement. Technical concepts may be involved in the choice of utility or loss functions, associating many different features, some tangible, such as money, but also others, like pleasure, intangible. It may be objected that it is not sensible to reduce such a collection of disparate aspects to a single number in the form of utility or loss. In reply to this, one can consider that in any situation of decision analysis, it is necessary to deal with several consequences that have to be contrasted and combined. Representation of these consequences by numbers enables the consequences to be contrasted and combined more easily than other concepts related to decisions, and in accordance with strict rules which thus ensures coherence of the results. There is no claim that utility amounts to a complete description of a consequence. It is only a summary that is adequate for the stated purpose, that is the choice of action in a particular context. Ultimately, it overcomes the particular difficulty, not only of combining beliefs, or of contrasting preferences in the form of consequences, but of combining beliefs with preferences.

9.6 HOW TO IMPLEMENT BAYESIAN INFERENCE AND DECISION ANALYSIS?

Throughout this book, the proposed inference and decision analyses have been illustrated by examples accompanied by code for R, a widely used, highly flexible statistical software environment for quantitative analyses, statistics and graphics. For MCMC methods the BUGS project is also useful, see Section 4.2.4.

As the discussion here is a more general one, the intention is to draw the reader's attention to methods for the implementation of different aspects of Bayesian analyses for inference and decision, and, where possible, in a combined fashion. One such method is that offered by Bayesian networks. Their use has already been illustrated for some examples in earlier chapters of this book (see Sections 2.2.4, 3.2.2, 4.3.3 and 7.3). The flexibility of Bayesian networks and their relevance for forensic inference and decision problems is pointed out by two additional examples below. The first example, Example 9.6.1, illustrates the use of continuous random variables for inference about a Normally distributed parameter (with variance assumed known). This example contrasts examples of Bayesian networks presented earlier in the book where continuous variables were approximated through discretization. The next example, Example 9.6.2, emphasizes that Bayesian networks can be extended to Bayesian decision networks (also called influence diagrams, see Section 2.2.4 and Example 3.2.4) in order to co-ordinate inference and decision analysis within a single model. The proposed examples focus on models that involve a limited number of nodes, sufficient to illustrate the principle, which also applies to more complex models. Actually, the examples represent local network fragments which can be viewed as components within more complex graphical models, as described for forensic applications in Taroni *et al.* (2006a), for example.

Example 9.6.1 *(Alcohol concentration in blood – continued). Consider again the problem of inference on a Normally distributed variable with unknown mean θ and known variance σ^2, presented earlier in Section 4.4.1. Assume that experimental data are available as outlined in Example 4.4.1, where a blood sample from an individual is analyzed for alcohol concentration in a laboratory using two independent methods, denoted HS and ID. Two measurements are performed with each of these procedures. Denote by X_1 and X_2 the two readings obtained with the HS method and by X_3 and X_4 the two readings obtained with the ID method. Let θ denote the true, but unknown, level of alcohol. The distributional assumptions are as follows:*

$$\theta \sim N(1, \tau^2 = 0.3^2), \quad X_i \sim N(\theta, \sigma^2) \quad (\text{for } i = 1, \ldots, 4)$$

As in Example 4.4.1, the variance σ^2 for the HS method is set to 0.0229^2 whereas that of the ID method is set to 0.0463^2.

The dependency of the X_i on θ is translated into a Bayesian network as shown in Figure 9.1. The double borders of the nodes indicate the continuous natures of the nodes, as opposed to the discrete nodes with single borders used in earlier sections of this book. Detailed accounts on the theory of continuous Bayesian networks are available, for instance, in Cowell et al. (1999), and Kjærulff and Madsen (2008).

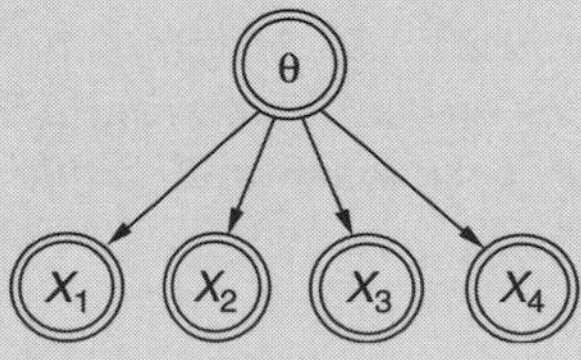

Figure 9.1 Bayesian network with five continuous nodes. The node θ acts as a parental variable for the nodes X_i.

In the model shown in Figure 9.1, the absence of directed edges between the X_i is an expression of the assumption that their distributions are independent of knowledge of θ. In particular, the conditional distributions of the X_i are $N(\theta, \sigma^2)$. The unconditional distribution of the X_i is $N(\theta, \sigma^2 + \tau^2)$.

This result can be traced by inspecting the compiled Bayesian network in its initial state (without any evidence entered), shown in Figure 9.2(i). The mean in the case here is 1 *and the variance equals* $0.0229^2 + 0.3^2 = 0.09524$. *For economy of space, only the results for the HS method are displayed. The dotted edge indicates that the results for the ID method may be obtained analogously.*

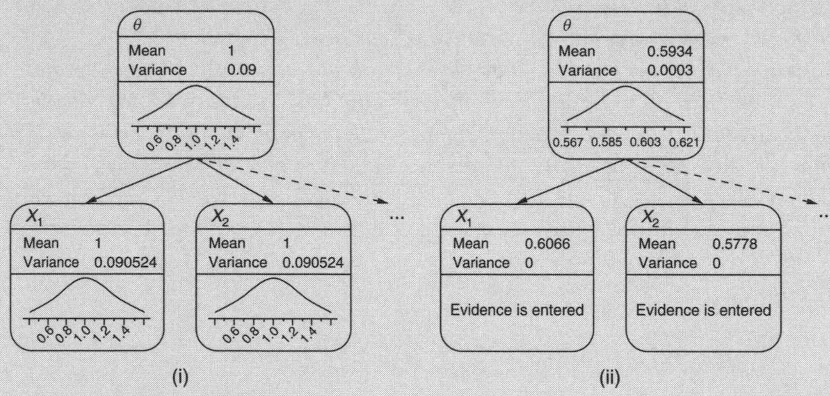

Figure 9.2 Expanded representation of the Bayesian network for inference on a Normally distributed parameter θ. The figure on the left shows the Bayesian network in its initial state whereas the figure on the right displays the posterior distribution for θ given data entered for the observed variables X_1 and X_2. For economy of space, only the results for the HS method are displayed. The dotted edge indicates that the results for the ID method may be obtained analogously.

Next, recall from Example 4.4.1 the two measurements obtained with the HS method: 0.6066 *and* 0.5778. *These data can be entered at the nodes X_1 and X_2, as shown in Figure 9.2(ii). In this figure, the node θ now displays the posterior distribution for the unknown parameter. The result can be found to agree with what has earlier been found in Example 4.4.1.*

***Example 9.6.2** (Deciding among discrete propositions). It has been mentioned at several instances in this and earlier chapters that inference and* *decision are closely related, notably, that the former represents the starting point of the latter. In particular, it is part of the principal argument put forward in this book that the decision to be made, such as about a discrete proposition, is a function of the probabilities of the various propositions as well as the losses (or utilities) that measure the consequences given by the combination of choices and individual propositions (that is, actual states of nature).*

This 'philosophy of statistics', as it is called by Lindley (2000b), can be translated in terms of a Bayesian decision network. For the ease of argument, consider two discrete, mutually exclusive and exhaustive hypotheses, denoted H_0 and H_1, but the argument extends by analogy to multiple propositions. Hypotheses H_0 and H_1 could represent the hypotheses of the defence and the prosecution, respectively. In turn, the available decisions may thus be denoted by D_0 and D_1, respectively. In a Bayesian decision network for this setting (Figure 9.3), the set of hypotheses is represented by a discrete chance node H, in the form of a circle, whereas the set of available decisions is represented by a node D, in the form of a square. The two nodes H and D share a common, diamond shaped child node L. This node contains the loss function. Assume that an incorrect decision for a defendant (i.e. a false conviction) is ten times as serious as an incorrect decision for a prosecutor (i.e. a false acquittal). Assume further that correct decisions, that is choosing D_0 when H_0 holds and D_1 when H_1 holds, incur a zero loss. This loss function, with k representing the loss associated with a false acquittal (i.e. choosing D_0 when H_1 is true), is specified in Table 9.1. It emphasizes the importance of node L. In combination, the three nodes D, H and L can be regarded as the decision-theoretic core of the network.

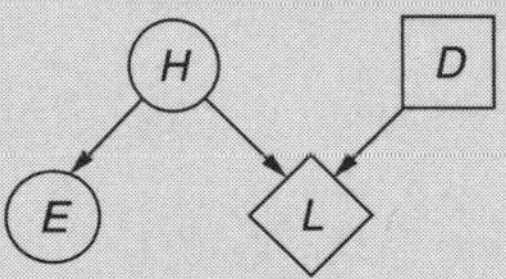

Figure 9.3 Bayesian decision network for analysing the problem of deciding between competing discrete propositions (node H), given evidence E. The node D represents the available decisions (the number of states equals the number of competing propositions assumed by the node H) whereas the node L accounts for the losses associated for each decision D given each state of the node H.

Figure 9.3 also contains a node E, a discrete chance node which represents evidence that a decision maker judges relevant for judging the truthstate

Table 9.1 Loss function for the Bayesian decision network shown in Figure 9.3.

H:	H_0		H_1	
D:	D_0	D_1	D_0	D_1
L:	0	$10k$	k	0

of H. This part, $H \rightarrow E$, of the proposed graphical model accounts for the Bayesian inference. Based on evidence on the node E, a posterior distribution for H is obtained, given by Bayes' theorem:

$$P(H \mid E) = [P(E \mid H)P(H)]/P(E)$$

The fact that the node H is part of the network that accounts for the Bayesian inference (in terms of the fragment $H \rightarrow E$) as well as part of the network that covers the decision-theoretic core (i.e. $H \rightarrow L \leftarrow D$), makes explicit that the topics of inference and decision are related.

This connection is also apparent in the expected decision losses. In particular, the values stored for the node H enter calculations of the expected losses of the decisions D_0 and D_1 which, in general, are given by:

$$L(D_0, P(H \mid E)) = L(D_0, H_0)P(H_0 \mid E) + L(D_0, H_1)P(H_1 \mid E)$$

and

$$L(D_1, P(H \mid E)) = L(D_1, H_0)P(H_0 \mid E) + L(D_1, H_1)P(H_1 \mid E)$$

Thus, for a loss function as defined by Table 9.1 one has:

$$L(D_0, P(H \mid E)) = L(D_0, H_1)P(H_1 \mid E) = kP(H_1 \mid E) \quad (9.1)$$

and

$$L(D_1, P(H \mid E)) = L(D_1, H_0)P(H_0 \mid E) = 10kP(H_0 \mid E). \quad (9.2)$$

Figure 9.4 shows the influence diagram (Figure 9.3) in a partially expanded form, with a value of $k = 1$ and an illustrative current belief

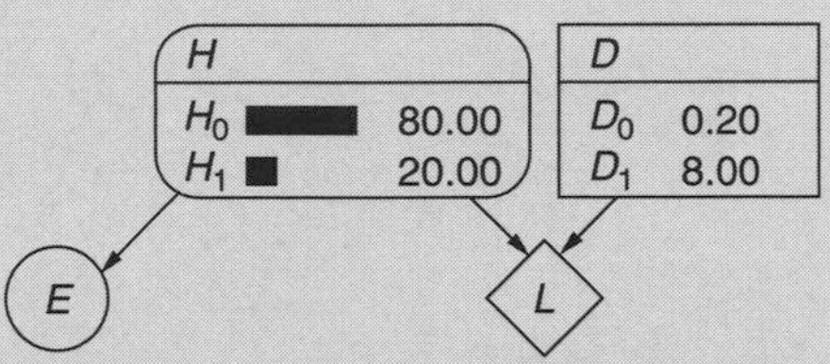

Figure 9.4 Partially expanded Bayesian decision network for analyzing the problem of deciding between competing discrete propositions (node H), given evidence E. The node D displays the expected decision losses for the available options D_0 and D_1, respectively. The node L contains the loss function defined in Table 9.1, assuming the value of $k = 1$.

state for the node H of $\{0.8, 0.2\}$. *The node D displays the expected losses associated with the decisions* D_0 *and* D_1*, which agree with the values obtained by Equations (9.1) and (9.2).*

Notice, as was done for the previous example, that the proposed network is chosen primarily for its simplicity, in order to illustrate the underlying principle and assumptions. Detailed accounts on the computational background of Bayesian decision networks are available, for instance, in Cowell et al. (1999), Jensen (2001) and, more recently, in Kjærulff and Madsen (2008).

The influence diagram shown in Figure 9.3 is a generic one, capable of representing a wide variety of situations involving a choice among discrete propositions (many classification problems, as discussed in Chapter 8, conform to this set-up as well as the problem of forensic identification presented in Sections 3.2.1 and 3.2.2). In particular, the nodes E and H serve merely as a placeholders for evidence and target propositions in general. Actually, E and H can be replaced by a set of nodes forming a sub-network that organizes a particular inference task. For the purpose of illustration, consider the Bayesian decision network shown in Figure 9.5, where the target propositions focus on whether a given individual (the putative father), who is not available for DNA testing (but a brother and a sister are), is or is not the father of a given child c. In such complex scenarios of disputed paternity, an influence diagrammatic analysis offers valuable assistance since decision- and inference-aspects of the problem can be handled in a modular, transparent and coherently connected way. The part of the network covering discrete chance nodes is described, for instance, in Dawid et al. (2002) and Taroni et al. (2006a).

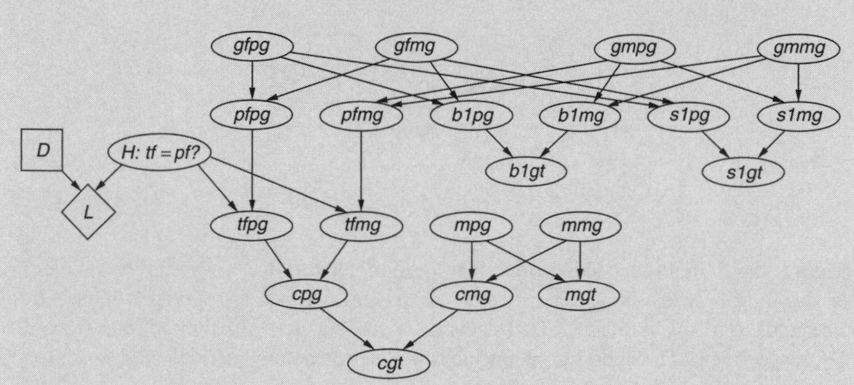

Figure 9.5 A Bayesian decision network for assessing a case of disputed paternity where the putative father is not available, but a brother and a sister of the putative father are available. The network describes the way in which evidence for a single genetic marker may be used to infer something about the major proposition (disputed paternity). Nodes *gfpg*, *gfmg*, *gmpg*, *gmmg* denote the grandfather (*gf*) and grandmother (*gm*) paternal *p* and maternal *m* genes. Nodes *pfpg*, *pfmg*, *b1pg*, *b1mg*, *s1pg* and *s1mg* denote the putative father, *pf*, the brother *b1*, and sister *s1* paternal *p* and maternal *m* genes. Nodes *tfpg*, *tfmg*, *mpg*, *mmg*, *cpg* and *cmg* denote the true father *tf*, mother *m* (in first place) and child *c* paternal *p* and maternal *m* (in second place) genes. Nodes *b1gt*, *s1gt*, *mgt* and *cgt* denote the brother *b1*, sister *s1*, mother *m* and child *c* genotypes. Node $H: tf = pf?$ takes two values: $H_0: tf \neq pf$, the true father is not the putative father, and $H_1: tf = pf$, the true father is the putative father. The node L accounts for the decision loss whereas the node D covers the available conclusions, that is D_0, the putative father is not the true father, and D_1, the putative father is the true father.

Appendix A

Discrete Distributions

Bernoulli, $Br(\theta)$

$f(x \mid \theta) = \theta^x(1-\theta)^{1-x}$; $x = 0, 1$; $0 \leq \theta \leq 1$
$E(x) = \theta$
$Var(x) = \theta(1-\theta)$

Bayesian learning process:

$x = (x_1, \ldots, x_n)$; $y = \sum_{i=1}^n x_i$
$\pi(\theta) = Be(\alpha, \beta)$
$\pi(\theta \mid x) = Be(\alpha + y, \beta + n - y)$
$f(x_{n+1} \mid x) = Bb(x_{n+1} \mid \alpha + y, \beta + n - y, 1)$

Binomial, $Bin(n, \theta)$

$f(x \mid n, \theta) = \binom{n}{x}\theta^x(1-\theta)^{n-x}$; $x = 0, 1, \ldots, n$; $0 \leq \theta \leq 1$

$E(x) = n\theta$

$Var(x) = n\theta(1-\theta)$

Negative-binomial, $Nb(r, \theta)$

$f(x \mid r, \theta) = \binom{r+x-1}{x}\theta^r(1-\theta)^x$; $x = 0, 1, \ldots$; $0 \leq \theta \leq 1$
$E(x) = \frac{r(1-\theta)}{\theta}$
$Var(x) = \frac{r(1-\theta)}{\theta^2}$

Beta-binomial, $Bb(\alpha, \beta, n)$

$f(x \mid \alpha, \beta, n) = \binom{n}{x}\frac{\Gamma(\alpha+\beta)}{\Gamma(\alpha)\Gamma(\beta)}\frac{\Gamma(\alpha+x),\ \Gamma(\beta+n-x)}{\Gamma(\alpha+\beta+n)}$, $x = 0, 1, \ldots, n$; $\alpha, \beta > 0$

Data Analysis in Forensic Science: A Bayesian Decision Perspective Franco Taroni, Silvia Bozza, Alex Biedermann, Paolo Garbolino and Colin Aitken

$E(x) = n\frac{\alpha}{\alpha+\beta}$
$Var(x) = \frac{n\alpha\beta}{(\alpha+\beta)^2}\frac{(\alpha+\beta+n)}{(\alpha+\beta+1)}$

Poisson, $Pn(\lambda)$

$f(x \mid \lambda) = \frac{\lambda^x e^{-\lambda}}{x!}\ x = 0, 1, 2, \ldots ; \lambda > 0$
$E(x) = \lambda$
$Var(x) = \lambda$

Bayesian learning process:

$x = (x_1, \ldots, x_n),\ y = \sum_{i=1}^n x_i$
$\pi(\lambda) = Ga(\alpha, \beta)$
$\pi(\lambda \mid x) = Ga(\alpha + y, \beta + n)$
$f(x_{n+1} \mid x) = Pg(x_{n+1} \mid \alpha + y, \beta + n, 1)$

Poisson-Gamma, $Pg(\alpha, \beta, n)$

$f(x \mid n, \alpha, \beta) = \frac{\beta^\alpha}{\Gamma(\alpha)}\frac{\Gamma(\alpha+x)}{x!}\frac{n^x}{(\beta+n)^{\alpha+x}}\ x = 0, 1, 2, \ldots ; \alpha, \beta > 0$
$E(x) = n\frac{\alpha}{\beta}$
$Var(x) = \frac{n\alpha}{\beta}(1 + \frac{n}{\beta})$

Hypergeometric, $Hy(N, M, n)$

$f(x \mid N, M, n) = \frac{\binom{M}{x}\binom{N-M}{n-x}}{\binom{N}{n}}; N, M, n \geq 0 ; x = 0, 1, \ldots, n$
$E(x) = \frac{nM}{N}$
$Var(x) = \frac{nM}{N}\frac{(N-M)(N-n)}{N(N-1)}$

Multinomial, $M_k(n, \theta_1, \ldots, \theta_k)$

$f(x_1, \ldots, x_k \mid \theta_1, \ldots, \theta_k) = \binom{n}{x_1 \cdots x_k}\theta_1^{x_1} \ldots \theta_k^{x_k},\ x_i = 0, 1, \ldots,\ \sum_{i=1}^k x_i \leq n;$
$0 \leq \theta_i \leq 1,$
$\sum_{i=1}^k \theta_i = 1$
$E(x_i) = n\theta_i$
$Var(x_i) = n\theta_i(1 - \theta_i)$

Bayesian learning process:

$x = (x_1, \ldots, x_k)$
$\pi(\theta_1, \ldots, \theta_k) = D_k(\alpha_1, \ldots, \alpha_k)$
$\pi(\theta_1, \ldots, \theta_k \mid x) = D_k(\alpha_1 + x_1, \ldots, \alpha_k + x_k)$

Appendix B

Continuous Distributions

Uniform, U(a,b)

$f(x \mid a, b) = \frac{1}{b-a}, \; b > a, \; a \leq x \leq b$

$E(x) = \frac{a+b}{2}$

$Var(x) = \frac{(b-a)^2}{12}$

Beta, Be(α, β)

$f(x \mid \alpha, \beta) = \frac{1}{B(\alpha,\beta)} x^{\alpha-1}(1-x)^{\beta-1}, \; 0 \leq x \leq 1, \; \alpha, \beta > 0$

$E(x) = \frac{\alpha}{\alpha+\beta}$

$Var(x) = \frac{\alpha\beta}{(\alpha+\beta)^2(\alpha+\beta+1)}$

The constant in the beta probability density function can be defined in terms of gamma functions, $B(\alpha, \beta) = \frac{\Gamma(\alpha)\Gamma(\beta)}{\Gamma(\alpha+\beta)}$.

Exponential, Ex(θ)

$f(x \mid \theta) = \theta e^{-\theta x} x > 0, \; \theta > 0$

$E(x) = \frac{1}{\theta}$

$Var(x) = \frac{1}{\theta^2}$

Bayesian learning process

$x = (x_1, \ldots, x_n); \; y = \sum_{i=1}^{n} x_i$

Data Analysis in Forensic Science: A Bayesian Decision Perspective Franco Taroni, Silvia Bozza, Alex Biedermann, Paolo Garbolino and Colin Aitken

$\pi(\theta) = Ga(\alpha, \beta)$

$\pi(\theta \mid x) = Ga(\alpha + n, \beta + y)$

Gamma, $Ga(\alpha, \beta)$

$f(x \mid \alpha, \beta) = \frac{\beta^\alpha}{\Gamma(\alpha)} x^{\alpha-1} e^{-\beta x}$, $x > 0$, $\alpha, \beta > 0$

$E(x) = \frac{\alpha}{\beta}$

$Var(x) = \frac{\alpha}{\beta^2}$

The gamma function satisfies the following relationships:

$$\Gamma(x+1) = \Gamma(x),$$

and, for any integer $x > 0$,

$$\Gamma(x) = (x-1)!$$

Also $\Gamma(1/2) = \pi^{\frac{1}{2}}$

Inverse gamma, $IG(\alpha, \beta)$

$f(x \mid \alpha, \beta) = \frac{\beta^\alpha}{\Gamma(\alpha)} x^{-(\alpha+1)} e^{-\beta/x}$, $x > 0$, $\alpha, \beta > 0$

$E(x) = \frac{\beta}{\alpha-1}$ $(\alpha > 1)$

$Var(x) = \frac{\beta^2}{(\alpha-1)^2(\alpha-2)}$ $(\alpha > 2)$

This distribution is the distribution of X^{-1} when $X \sim Ga(\alpha, \beta)$.

Normal distribution, $N(\theta, \sigma^2)$

$f(x \mid \theta, \sigma^2) = \frac{1}{\sqrt{2\pi\sigma^2}} \exp\left(-\frac{1}{2\sigma^2}(x-\theta)^2\right)$, $-\infty < x < \infty$, $-\infty < \theta < \infty$, $\sigma > 0$

$E(x) = \theta$

$Var(x) = \sigma^2$

Bayesian learning process (known variance):

$x = (x_1, \ldots, x_n)$, $\bar{x} = \frac{1}{n}\sum_{i=1}^n x_i$

$\pi(\theta) = N(\mu, \tau^2)$

$\pi(\theta \mid x) = N(\mu(x), \tau^2(x))$, $\mu(x) = \frac{\frac{\sigma^2}{n}}{\frac{\sigma^2}{n}+\tau^2}\mu + \frac{\tau^2}{\frac{\sigma^2}{n}+\tau^2}\bar{x}$, $\tau^2(x) = \frac{\frac{\sigma^2}{n}\tau^2}{\frac{\sigma^2}{n}+\tau^2}$

$f(x_{n+1} \mid x) = N(x \mid \mu(x), \sigma^2 + \tau^2(x))$

Bayesian learning process (unknown variance):

$x = (x_1, ..., x_n),\ \bar{x} = \frac{1}{n}\sum_{i=1}^{n} x_i,\ s^2 = \frac{1}{n-1}\sum_{i=1}^{n}(x_i - \bar{x})^2$

$\pi(\theta \mid \sigma) = 1, \pi(\sigma) = \sigma^{-1}$

$\pi(\theta \mid \sigma, \bar{x}, s^2) = N(\bar{x}, \sigma^2/n)$

$\pi(\sigma^2 \mid \bar{x}, s^2) = IG\left(\frac{n-1}{2}, \frac{n-1}{2}s^2\right)$

$\pi(\theta \mid \bar{x}, s^2) = St\left(n-1, \bar{x}, s^2/n\right)$

$f(x_{n+1} \mid x) = St\left(n-1, \bar{x}, \frac{s^2(n+1)}{n}\right)$

$\pi(\theta, \sigma) = N\left(\mu, \sigma^2/n_0\right) IG(\alpha, \beta)$

$\pi(\theta \mid \sigma^2, \bar{x}, s^2) = N\left(\mu_n, \tau_n^2\right),\ \mu_n = (n+n_0)^{-1}(n\bar{x} + n_0\mu),\ \tau_n^2 = \sigma^2(n+n_0)^{-1}$

$\pi(\sigma^2 \mid \bar{x}, s^2) = IG\,(\alpha_n, \beta_n),$
$\quad \alpha_n = \alpha + n/2,\ \beta_n = \beta + \frac{1}{2}\left[(n-1)s^2 + (n_0+n)^{-1}n_0 n(\bar{x} - \mu)^2\right]$

$\pi(\theta \mid x) = St\left(2\alpha + n, \mu_n, (n+n_0)^{-1}(\alpha + n/2)^{-1}\beta_n\right)$

$f(x_{n+1} \mid x) = St\left(2\alpha_n + n, \mu_n, (n+n_0)^{-1}(n+n_0+1)(\alpha + n/2)^{-1}\beta_n\right)$

Student t distribution, $St(\alpha, \mu, \lambda)$

$f(x \mid \alpha, \mu, \lambda) = \frac{\Gamma((\alpha+1)/2)/\Gamma(\alpha/2)}{\lambda^{1/2}(\alpha\pi)^{1/2}}\left[1 + \frac{(x-\mu)^2}{\lambda\alpha}\right]^{-(\alpha+1)/2},\ -\infty < 0 < \infty,\ \alpha > 0,\ \lambda > 0$

$E(x) = \mu$

$Var(x) = \lambda\frac{\alpha}{\alpha-2}$

The distribution is symmetrical about $x = \mu$. If $Y = (X - \mu)\lambda^{-1/2}$, where $X \sim St$ (α, μ, λ), then Y has a standard Student t distribution $St(\alpha, 0, 1)$.

Chi squared, $\chi^2(\alpha)$

$f(x \mid \alpha) = \frac{2^{-\alpha/2}}{\Gamma(\alpha/2)}x^{\alpha/2-1}e^{-x/2},\ 0 < x < \infty$

$E(x) = \alpha$

$Var(x) = 2\alpha$

Snedecor's F, $F(\alpha_1, \alpha_2)$

$f(x \mid \alpha_1, \alpha_2) = \frac{\Gamma\left(\frac{\alpha_1+\alpha_2}{2}\right)}{\Gamma\left(\frac{\alpha_1}{2}\right)\Gamma\left(\frac{\alpha_2}{2}\right)}\alpha_1^{\alpha_1/2}\alpha_2^{\alpha_2/2}\frac{x^{(\alpha_1-2)/2}}{(\alpha_2+\alpha_1 x)^{(\alpha_1+\alpha_2)/2}},\ x > 0,\ \alpha_1, \alpha_2 > 0$

$E(x) = \frac{\alpha_2}{\alpha_2 - 2}, \alpha_2 > 2$

$Var(x) = 2\left(\frac{\alpha_2}{\alpha_2-2}\right)^2 \frac{\alpha_1+\alpha_2-2}{\alpha_1(\alpha_2-4)}, \quad \alpha_2 > 4$

X has an $F(\alpha_1, \alpha_2)$ distribution if X has the same distribution as

$$\frac{W_1/\alpha_1}{W_2/\alpha_2},$$

where W_1 and W_2 are independent and $W_1 \sim \chi^2(\alpha_1)$, $W_2 \sim \chi^2(\alpha_2)$.

Fisher's z distribution, $z(\alpha_1, \alpha_2)$

X has a z distribution on α_1 and α_2 degrees of freedom, denoted $X \sim z(\alpha_1, \alpha_2)$, if $Y = \exp(2X) \sim F(\alpha_1, \alpha_2)$.

Unless α_1 and α_2 are very small, the distribution of z is approximately Normal.

Dirichlet distribution, $D_k(\alpha_1, \ldots, \alpha_k)$

$f(x_1, \ldots, x_k \mid \alpha_1, \ldots, \alpha_k) = \frac{\Gamma(\alpha_1 + \cdots + \alpha_k)}{\Gamma(\alpha_1)\ldots\Gamma(\alpha_k)} \prod_{j=1}^k x_j^{\alpha_j - 1}, \ \alpha_1, \ldots, \alpha_k > 0, \ \alpha_0 = \sum_{j=1}^k \alpha_j$

$E(x_i) = \frac{\alpha_i}{\alpha_0}$

$Var(x_i) = \frac{\alpha_i(\alpha_0 - \alpha_i)}{\alpha_0^2(\alpha_0+1)}, \ Cov(x_i, x_j) = -\frac{\alpha_i \alpha_j}{\alpha_0^2(\alpha_0+1)}$

Multivariate Normal, $N_k(\boldsymbol{\theta}, \boldsymbol{\Sigma})$

$x = (x_1, \ldots, x_k), \ x \in R^k$

$f(\mathbf{x}) = (2\pi)^{-k/2} |\Sigma|^{-1/2} \exp\left[-\frac{1}{2}(\mathbf{x} - \boldsymbol{\theta})' \Sigma^{-1} (\mathbf{x} - \boldsymbol{\theta})\right]$

$E(\mathbf{x}) = \boldsymbol{\theta}$

$Var(\mathbf{x}) = \Sigma$

Bibliography

Aitken CGG 1995 *Statistics and the Evaluation of Evidence for Forensic Scientists*. John Wiley & Sons, Ltd, Chichester.

Aitken CGG 1999 Sampling – how big a sample?. *Journal of Forensic Sciences* **44**, 750–760.

Aitken CGG and Lucy D 2002 Estimation of the quantity of a drug in a consignment from measurements on a sample. *Journal of the Forensic Science Society* **47**, 968–975.

Aitken CGG and Lucy D 2004 Evaluation of trace evidence in the form of multivariate data. *Applied Statistics* **53**, 109–122.

Aitken CGG and Stoney DA 1991 *The Use of Statistics in Forensic Science*. Ellis Horwood, Chichester.

Aitken CGG and Taroni F 2004 *Statistics and the Evaluation of Evidence for Forensic Scientists*, 2nd edn. John Wiley & Sons, Ltd, Chichester.

Aitken CGG, Bring J, Leonard T and Papasouliotis O 1997 Estimation of quantities of drugs handled and the burden of proof. *Journal of the Royal Statistical Society A* **160**, 333–350.

Aitken CGG, Taroni F and Garbolino P 2003 A graphical model for the evaluation of cross-transfer evidence in DNA profiles. *Theoretical Population Biology* **63**, 179–190.

Albert J 1992 Teaching statistical inference using Bayes. Technical report, Department of Mathematics and Statistics, Bowling Green State University.

Albert J 2007 *Bayesian Computation with R*. Springer, New York.

Allais M 1953 Le comportement de l'homme rationnel devant le risque: critique des postulats et axiomes de l'école Américaine. *Econometrica* **21**, 503–546.

Allais M and Hagen O 1979 *Expected Utility and the Allais Paradox*. Reidel, Dordrecht.

Altham P 1969 Exact Bayesian analysis of a 2×2 contingency table and Fisher's "exact" significance test. *Journal of the Royal Statistical Society, Series B* **31**, 261–269.

Anderson T 2003 *An Introduction to Multivariate Statistical Analysis*, 3rd edn. John Wiley & Sons, Ltd, Hoboken.

Antelman GR 1997 *Elementary Bayesian Statistics*. Edward Elgar, Cheltenham.

Antelman GR 1972 Interrelated Bernoulli processes. *Journal of the American Statistical Association* **67**, 831–841.

Armendt B 1980 Is there a Dutch Book argument for probability kinematics?. *Philosophy of Science* **47**, 583–588.

Balding DJ 2005 *Weight-of-Evidence for Forensic DNA Profiles*. John Wiley & Sons, Ltd, Chichester.

Data Analysis in Forensic Science: A Bayesian Decision Perspective Franco Taroni, Silvia Bozza, Alex Biedermann, Paolo Garbolino and Colin Aitken

Balding DJ and Nichols RA 1994 DNA profile match probability calculation: how to allow for population stratification, relatedness, database selection and single bands. *Forensic Science International* **64**, 125–140.

Baron J 2000 *Thinking and Deciding*, 3rd edn. Cambridge University Press, Cambridge.

Bayes T 1763 An essay toward solving a problem in the doctrine of chances In *Studies in the History of Statistics and Probability. Vol.1* (ed. Pearson ES and Kendall MG), Griffin, London, pp. 134–153.

Benazzi S, Maestri C, Parisini S, Vecchi F and Gruppioni G 2009 Sex assessment from the sacral base by means of image processing. *Journal of Forensic Sciences* **54**, 249–254.

Berger JO 1988 *Statistical Decision Theory and Bayesian Analysis*, 2nd edn. Springer-Verlag, New York.

Berger JO and Berry DA 1988 Statistical analysis and the illusion of objectivity. *American Scientist* **76**, 159–165.

Berger JO and Delampady M 1987 Testing precise hypotheses. *Statistical Science* **2**, 317–352.

Bernardo JM 1997 Statistical inference as a decision problem: the choice of sample size. *The Statistician* **46**, 151–153.

Bernardo JM and Smith AFM 2000 *Bayesian Theory*, 2nd edn. John Wiley & Sons, Ltd, Chichester.

Berry DA 1991 Experimental design for drug development: a Bayesian approach. *Journal of Biopharmaceutical Statistics* **1**, 81–101.

Berry DA 1997 Teaching elementary Bayesian statistics with real applications in science. *The American Statistician* **51**, 241–246.

Berry DA and Stangl DK 1996 Bayesian methods in health-related research. In *Bayesian Biostatistics* (ed. Berry DA and Stangl DK), Marcel Dekker, Inc., New York, pp. 3–66.

Besson L 2003 Contamination en stupéfiants des billets de banque en Euro. Technical report, Institut de police scientifique, Université de Lausanne.

Besson L 2004 Détection des stupéfiants par IMS. Technical report, Institut de police scientifique, Université de Lausanne.

Biedermann A, Bozza S and Taroni F 2008 Decision theoretic properties of forensic identification: underlying logic and argumentative implications. *Forensic Science International* **177**, 120–132.

Biedermann A, Bozza S and Taroni F 2010 Probabilistic evidential assessment of gunshot residue particle evidence (part II): Bayesian parameter estimation for experimental count data. *Forensic Science International*. In press.

Biedermann A, Taroni F and Garbolino P 2007a Equal prior probabilities: can one do any better? *Forensic Science International* **172**, 85–93.

Biedermann A, Taroni F, Bozza S and Aitken CGG 2007b Analysis of sampling issues using Bayesian networks. *Law, Probability & Risk* **7**, 35–60.

Birnbaum A 1962 On the foundations of statistical inference. *Journal of the American Statistical Association* **57**, 269–306.

Bolstad WM 2004 *Introduction to Bayesian Statistics*. John Wiley & Sons, Ltd, Hoboken.

Bovens L and Hartmann S 2003 *Bayesian Epistemology*. Clarendon Press, Oxford.

Box GEP and Tiao GC 1973 *Bayesian Inference in Statistical Analysis*. John Wiley & Sons, Ltd, New York.

Bozza S and Taroni F 2009 Making decisions in forensic science, European Meeting of Statisticians (EMS 2009). Toulose (France), July 20–24 2009.

Bozza S, Taroni F, Biedermann A and Aitken CGG 2008a A decision-theoretic approach to the choice of sample size in forensic science applications. Technical report, Institut de police scientifique, Université de Lausanne.

Bozza S, Taroni F, Marquis R and Schmittbuhl M 2008b Probabilistic evaluation of handwriting evidence: likelihood ratio for authorship. *Applied Statistics* **57**, 329–341.

Buckleton JS, Triggs CM and Walsh SJ 2004 *Forensic DNA Evidence Interpretation*. CRC Press, Boca Raton.

Bunch S 2000 Consecutive matching striations criteria: a general critique. *Journal of Forensic Sciences* **45**, 955–962.

Carlin BP and Louis TA 1998 *Bayes and Empirical Bayes Methods for Data Analysis*. Chapman & Hall/CRC, Boca Raton.

Casella G and Berger RL 2002 *Statistical Inference*, 2nd edn. Duxbury Press, Pacific Grove.

Casella G, Hwang JTG and Robert CP 1993a Loss function for set estimation. In *Statistical Decision Theory and Related Topics V* (ed. Berger JO and Gupta SS), Springer-Verlag, New York, pp. 237–252.

Casella G, Hwang JTG and Robert CP 1993b A paradox in decision-theoretic interval estimation. *Statistica Sinica* **3**, 141–155.

Chib S and Greenberg E 1995 Understanding the Metropolis-Hastings algorithm. *The American Statistician* **49**, 327–335.

Cohen J 1977 *The Probable and the Provable*. Claredon Press, Oxford.

Congdon P 2001 *Bayesian Statistical Modelling*. John Wiley & Sons, Ltd, Chichester.

Cook R, Evett IW, Jackson G, Jones PJ and Lambert JA 1998 A hierarchy of propositions: deciding which level to address in casework. *Science & Justice* **38**, 231–239.

Cornfield J 1967 Bayes theorem. *Review of the International Statistical Institute* **35**, 34–49.

Cournot AA 1843 *Exposition de la théorie des chances et des probabilités*. Librairie de L. Hachette, Paris.

Cowell RG, Dawid AP, Lauritzen SL and Spiegelhalter DJ 1999 *Probabilistic Networks and Expert Systems*. Springer, New York.

Cox DR 1958 Some problems connected with statistical inference. *The Annals of Mathematical Statistics* **29**, 357–372.

Curran JM, Hicks TN and Buckleton JS 2000 *Forensic Interpretation of Glass Evidence*. CRC Press, Boca Raton.

D'Agostini G 1999 Overcoming priors anxiety In *Bayesian Methods in the Sciences* (ed. Bernardo JM), Revista de la Real Academia de Ciencias, Madrid.

D'Agostini G 2000 Teaching Bayesian statistics in the scientific curricula. *The ISBA Newletter* **7**, 18.

D'Agostini G 2004a *Bayesian Reasoning in Data Analysis*. World Scientific Publishing Co., Singapore.

D'Agostini G 2004b From observations to hypotheses. Probabilistic reasoning versus falsificationism and its statistical variations *Vulcano Workshop on Frontier Objects in Astrophysics and Particle Physics*, Vulcano, Italy.

Darboux J, Appell P and Poincaré J 1908 Examen critique des divers systèms ou études graphologiques auxquels a donné lieu le bourdereau. *L'affaire Dreyfus – La révision du procès de Rennes – enquête de la chambre criminelle de la Cour de Cassation*. Ligue française des droits de l'homme et du citoyen Paris pp. 499–600.

Dawid AP and Galavotti MC 2009 De Finetti's subjectivism, objective probability, and the empirical validation of probability assessment In *Bruno de Finetti – Radical probabilist* (ed. Galavotti MC), College Publications, London, pp. 97–114.

Dawid AP, Mortera J, Pascali VL and van Boxel D 2002 Probabilistic expert systems for forensic inference from genetic markers. *Scandinavian Journal of Statistics* **29**, 577–595.

de Finetti B 1930 Funzione caratteristica di un fenomeno aleatorio. *Memorie Accademia Nazionale dei Lincei* **4**, 86–133.

de Finetti B 1937 La prévision: ses lois logiques, ses sources subjectives. Annales de l'Institut Henri Poincaré 7 1-68 (English translation) In *Studies in Subjective Probability (1980)* (ed. Kyburg HE and Smokler HE), 2nd edn, Dover Publications, Inc., New York, pp. 93–158.

de Finetti B 1968 Probability: the subjectivistic approach In *La philosophie contemporaine* (ed. Klibansky R), La Nuova Italia, Firenze, pp. 45–53.

de Finetti B 1970 Logical foundations and measurement of subjective probability. *Acta Psychologica* **34**, 129–145.

de Finetti B 1972 *Probability, Induction and Statistics*. John Wiley & Sons, Ltd, London.

de Finetti B 1974 Bayesianism: its unifying role for both the foundations and applications of statistics. *International Statistical Review* **42**, 117–130.

de Finetti B 1989 Probabilism. *Erkenntnis* **31**, 169–223.

de Finetti B 1993a On the subjective meaning of probability (Paper originally published in the 'Fundamenta mathematicae', 17, 1931, pp. 298–329) In *Probabilità e induzione* (ed. Monari P and Cocchi D), CLUEB, Bologna, pp. 291–321.

de Finetti B 1993b Recent suggestions for the reconciliation of theories of probability (Paper originally published in the "Proceedings of the Second Berkely Symposium on Mathematical Statistics and Probability", held from July 31 to August 12, 1950, University of California Press, 1951, pp. 217–225) In *Probabilità e induzione* (ed. Monari P and Cocchi D) CLUEB Bologna pp. 375–387.

de Groot MH 1970 *Optimal Statistical Decisions*, 2nd edn. McGraw-Hill, New York.

de Morgan A 1847 *Formal Logic: the Calculus of Inference Necessary and Probable*, London.

Diaconis P and Zabell S 1982 Updating subjective probability. *Journal of the American Statistical Association* **77**, 822–830.

Dickey J 1973 Scientific reporting and personal probabilities: Student's hypothesis. *Journal of the Royal Statistical Society B* **35**, 285–305.

Domotor Z, Zanotti M and Graves H 1980 Probability kinematics. *Synthese* **44**, 421–442.

Duda R, Hart P and Stork D 2001 *Pattern Classification*, 2nd edn. John Wiley & Sons, Ltd, New York.

Earman J 1992 *Bayes or bust?*. MIT Press, Cambridge (Mass.).

Edwards W 1988 Summing up: the Society of Bayesian Trial Lawyers In *Probability and Inference in the Law of Evidence, The Uses and Limits of Bayesianism (Boston Studies in the Philosophy of Science)* (ed. Tillers P and Green ED), Springer, Dordrecht, pp. 337–342.

Edwards W 1991 Influence diagrams, Bayesian imperialism, and the Collins case: an appeal to reason. *Cardozo Law Review* **13**, 1025–1073.

Edwards W, Lindman H and Savage LJ 1963 Bayesian statistical inference for psychological research. *Psychological Review* **70**, 193–242.

Egglestone R 1983 *Evidence, proof and probability*, 2nd edn. Weidenfeld & Nicolson, London.

ENFSI 2004 Guidelines on representative drug sampling. Technical report, European Network of Forensic Science Institutes, Drug Working Group, Den Haag, The Netherlands.

Evett IW 1987 Bayesian inference and forensic science: problems and perspectives. *The Statistician* **36**, 99–105.

Evett IW 1990 The theory of interpreting scientific transfer evidence. *Forensic Science Progress*, vol. 4, Springer-Verlag, Berlin, pp. 141–179.

Evett IW 1991 Interpretation: a personal odissey In *The Use of Statistics in Forensic Science* (ed. Aitken CGG and Stoney DA), Ellis Horwood, New York, pp. 9–22.

Evett IW 1993 Establishing the evidential value of a small quantity of material found at the crime scene. *Journal of the Forensic Science Society* **33**, 83–86.

Evett IW 1996 Expert evidence and forensic misconceptions of the nature of exact science. *Science & Justice* **36**, 118–122.

Evett IW and Weir BS 1998 *Interpreting DNA Evidence*. Sinauer Associates Inc., Sunderland.

Evett IW, Jackson G, Lambert JA and McCrossan S 2000 The impact of the principles of evidence interpretation and the structure and content of statements. *Science & Justice* **40**, 233–239.

Fienberg SE 2006 Does it make sense to be an 'Objective Bayesian'? (comment on articles by Berger and by Goldstein). *Bayesian Analysis* **1**, 429–432.

Fienberg SE and Finkelstein MO 1996 Bayesian statistics and the law In *Bayesian Statistics 5* (ed. Bernardo JM, Berger J, Dawid AP and Smith AFM), Oxford University Press, Oxford, pp. 129–146.

Fienberg SE and Kadane JB 1983 The presentation of Bayesian statistical analyses in legal proceedings. *The Statistician* **32**, 88–98.

Fienberg SE and Schervish MJ 1986 The relevance of Bayesian inference for the presentation of statistical evidence and for legal decision making. *Boston University Law Review* **66**, 771–798.

Finkelstein MO and Fairley WB 1970 A Bayesian approach to identification evidence. *Harvard Law Review* **83**, 489–517.

Foreman LA, Smith AFM and Evett IW 1997 Bayesian analysis of DNA profiling data in forensic identification applications. *Journal of the Royal Statistical Society A* **160**, 429–469.

Frank RS, Hinkley SW and Hoffman CG 1991 Representative sampling of drug seizures in multiple containers. *Journal of Forensic Sciences* **36**, 350–357.

Gamerman D 2004 *Markov Chain Monte Carlo: Stochastic Simulation for Bayesian Inference*. Chapman & Hall, London.

Gamerman D and Lopes HF 2006 *Markov Chain Monte Carlo: Stochastic Simulation for Bayesian Inference*. Chapman & Hall, London.

Gelman A, Carlin JB, Stern HS and Rubin DB 1997 *Bayesian Data Analysis*. Chapman & Hall, London.

Gelman A, Carlin JB, Stern HS and Rubin DB 2004 *Bayesian Data Analysis*, 2nd edn. Chapman & Hall, London.

Gigerenzer G and Hoffrage U 1995 How to improve Bayesian reasoning without instruction: frequency formats. *Psychological Review* **102**, 684–704.

Gigerenzer G, Todd PM and The ABC Research Group 1999 *Simple Heuristics that Make Us Smart*. Oxford University Press, New York.

Gilks WR, Richardson S and Spiegelhalter DJ 1996 *Markov Chain Monte Carlo in practice*. Chapman & Hall, London.

Girotto V and Gonzalez M 2001 Solving probabilistic and statistical problems: a matter of question form and information structure. *Cognition* **78**, 247–276.

Girotto V and Gonzalez M 2002 Chances and frequencies in probabilistic reasoning. *Cognition* **84**, 353–359.

Glymour C 2001 *The Mind's Arrows*. MIT Press, Cambridge (Mass.).

Goldmann T, Taroni F and Margot P 2004 Analysis of dyes in illicit pills (amphetamine and derivates). *Journal of Forensic Sciences* **49**, 716–722.

Goldstein WM and Hogarth RM 1997 *Judgment and Decision Making. Currents, Connections, and Controversies*. Cambridge University Press, Cambridge.

Good IJ 1950 *Probability and the weighting of the evidence*. Charles Griffin, London.

Good IJ 1962 How rational should a manager be?. *Management Science* **8**, 383–393.

Good IJ 1979 A. M. Turing's statistical work in World War 2. *Biometrika* **66**, 393–396.

Good IJ 1985 Weight of evidence: a brief survey (with discussion) In *Bayesian Statistics 2* (ed. Bernardo JM, De Groot MH, Lindley DV and Smith AFM), North-Holland and Valencia University Press, Amsterdam, pp. 249–269.

Good IJ 1988 The interface between statistics and philosophy of science. *Statistical Science* **4**, 386–397.

Goodman SN 1999 Toward evidence-based medical statistics. 1: The p-value fallacy. *Annals of Internal Medicine* **130**, 995–1004.

Goodman SN 2005 Introduction to Bayesian methods 1: measuring the strength of evidence. *Clinical Trials* **2**, 282–290.

Goodman SN and Royall R 1988 Evidence and scientific research. *American Journal of Public Health* **78**, 1568–1574.

Habbema J, Hermans J and van den Broek K 1974 A stepwise discrimination program using density estimation In *Compstat 1974* (ed. Bruckman G), pp. 100–110. Physica Verlag, Vienna.

Halmos PR and Savage LJ 1949 Applications of the Radon-Nikodym theorem to the theory of sufficient statistics. *The Annals of Mathematical Statistics* **57**, 225–241.

Hoffrage U, Gigerenzer G, Krauss S and Martignon L 2002 Representation facilitates reasoning; what natural frequencies are and what they are not. *Cognition* **84**, 343–352.

Horwich P 1982 *Probability and evidence*. Cambridge University Press, New York.

Howard JV 1998 The 2×2 table: a discussion from a Bayesian viewpoint. *Statistical Science* **13**, 351–367.

Howard R and Matheson J 1984 Influence diagrams In *Readings on the Principles and Applications of Decision Analysis. Vol. 2* (ed. Howard RA and Matheson JE), Strategic Decisions Group, Menlo Park, pp. 719–762.

Howson C 2002 Bayesianism in statistics In *Bayes's Theorem* (ed. Swinburne R), pp. 39–69, Proceedings of the British Academy. Oxford University Press, Oxford.

Howson C and Urbach P 1996 *Scientific Reasoning: The Bayesian Approach*, 2nd edn. Open Court Publishing Company, Chicago.

Hwang JTG, Casella G, Robert CP, Wells MT and Farrel R 1992 Estimation of accuracy in testing. *Annals of Statistics* **20**, 490–509.

Hyndman RJ 1996 Computing and graphing highest density regions. *The American Statistician* **50**, 120–126.

Izenman AJ 2001 Statistical and legal aspects of the forensic study of illicit drugs. *Statistical Science* **16**, 35–57.

Jackson G 2000 The scientist and the scales of justice. *Science & Justice* **40**, 81–85.

Jackson G, Jones S, Booth G, Champod C and Evett IW 2006 The nature of forensic science opinion – a possible framework to guide thinking and practice in investigations and in court proceedings. *Science & Justice* **46**, 33–44.

Jaynes ET 2003 *Probability Theory: the Logic of Science*. Cambridge University Press, Cambridge.

Jeffrey RC 1975 Probability and falsification: critique of the Popper program. *Synthese* **30**, 95–117.

Jeffrey RC 1983 *The Logic of Decision*, 2nd edn. University of Chicago Press, New Chicago.

Jeffrey RC 1988 Conditioning, kinematics and exchangeability In *Causation, Chance and Credence. Vol. 1* (ed. Skyrms B and Harper WL), Kluwer Academic Publishers, Amsterdam, pp. 221–255.

Jeffrey RC 2004 *Subjective Probability: The Real Thing*. Cambridge University Press, Cambridge.

Jeffreys H 1961 *Theory of Probability*, 3rd edn. Clarendon Press, Oxford.

Jensen FV 2001 *Bayesian Networks and Decision Graphs*. Springer, New York.

Jensen FV and Nielsen TD 2007 *Bayesian Networks and Decision Graphs*, 2nd edn. Springer, New York.

Johnson NL, Kotz S and Balakrishnan N 1995 *Continuous Univariate Distributions*, vol. 2, 2nd edn. John Wiley & Sons, Ltd, New York.

Joyce H 2005 Career story: Consultant forensic statistician. Communication with Ian Evett. *Significance* **2**, 34–37.

Kadane JB 1995 Prime time for Bayes. *Controlled Clinical Trials* **16**, 313–318.

Kahneman D and Tversky A 2000 *Choices, Values, and Frames*. Cambridge University Press, New York.

Kahneman D, Slovic P and Tversky A 1982 *Judgment under Uncertainty: Heuristics and Biases*. Cambridge University Press, Cambridge.

Kaplan J 1968 Decision theory and the factfinding process. *Stanford Law Review* **20**, 1065–1092.

Kass RE and Raftery AE 1995 Bayes Factors. *Journal of the American Statistical Association* **90**, 773–795.

Kaye DH 1986a Commentary on 'Proficiency of professional document examiners in writer identification'. *Journal of Forensic Sciences* **39**, 13–44.

Kaye DH 1986b Quantifying probative value. *Boston University Law Review* **66**, 761–766.

Kaye DH 1987a Apples and oranges: confidence coefficients and the burden of persuasion. *Cornell Law Review* **73**, 54–77.

Kaye DH 1987b The validity of tests: Caveant omnes. *Jurimetrics Journal* pp. 349–361.

Kaye DH 1988 What is Bayesianism? In *Probability and Inference in the Law of Evidence, The Uses and Limits of Bayesianism (Boston Studies in the Philosophy of Science)* (ed. Tillers P and Green ED), Springer, Dordrecht, pp. 1–19.

Kaye DH 1999 Clarifying the burden of persuasion: what Bayesian decision rules do and do not do. *The International Journal of Evidence and Proof* **3**, 1–29.

Keeney R and Raiffa H 1976 *Decisions with Multiple Objectives: Preferences and Value Trade-offs*. John Wiley & Sons, Ltd, New York.

Kingston CR 1965a Application of probability theory in criminalistics. *Journal of the American Statistical Association* **60**, 70–80.

Kingston CR 1965b Application of probability theory in criminalistics – 2. *Journal of the American Statistical Association* **60**, 1028–1034.

Kirk PL and Kingston CR 1964 Evidence evaluation and problems in general criminalistics. *Journal of Forensic Sciences* **9**, 434–444.

Kjærulff U and Madsen A 2008 *Bayesian Networks and Influence Diagrams: A Guide to Construction and Analysis*. Springer, New York.

Koehler JJ 1992 Probabilities in the courtroom: an evaluation of the objection and policies In *Handbook of Psychology and Law* (ed. Kagehiro DK and Laufer WS), Springer, New York, pp. 167–184.

Kolmogorov AN 1933 *Grundbegriffe der Wahrscheinlichkeitsrechnung. English translation (1950): Foundations of the theory of probability. Chelsea, New York*. Springer, Berlin.

Krantz D, Luce R, Suppes P and Tversky A 1971 *Foundations of Measurement. Vol.1*. Academic Press, New York.

Krawczak M 2001 Forensic evaluation of Y-STR haplotype matches: a comment. *Forensic Science International* **118**, 114–115.

Kuhn TS 1970 *The Structure of Scientific Revolutions*, 2nd edn. University of Chicago Press, Chicago.

Laudan L 2006 *Truth, Error, and Criminal Law. An Essay in Legal Epistemology*. Cambridge University Press, Cambridge.

Lavine M and Schervish MJ 1999 Bayes factors: what they are and what they are not. *The American Statistician* **53**, 119–122.

Lee PM 2004 *Bayesian Statistics*, 3rd edn. Hodder Arnold, London.

Lempert R 1988 The New Evidence Scholarship In *Probability and Inference in the Law of Evidence, The Uses and Limits of Bayesianism (Boston Studies in the Philosophy of Science)* (ed. Tillers P and Green ED), Springer, Dordrecht, pp. 61–102.

Lempert R 1977 Modeling relevance. *Michigan Law Review* **75**, 1021–1057.

Leonard T and Hsu JSJ 1999 *Bayesian methods*. Cambridge University Press, Cambridge.

Lindley DV 1961 The use of prior probability distributions in statistical inferences and decisions *Proceedings of the Fourth Berkeley Symposium on Mathematics and Probability. Vol. 1*, University of California Press, Berkeley, pp. 453–468.

Lindley DV 1965 *Introduction to Probability and Statistics. Part 1 – Probability, and Part 2 – Inference*. Cambridge University Press, Cambridge.

Lindley DV 1977a Probability and the law. *The Statistician* **26**, 203–220.

Lindley DV 1977b A problem in forensic science. *Biometrika* **64**, 207–213.

Lindley DV 1982 Scoring rules and the inevitability of probability. *International Statistical Review* **50**, 11–26.

Lindley DV 1985 *Making Decisions*, 2nd edn. John Wiley & Sons, Ltd, Chichester.

Lindley DV 1990 The 1988 Wald Memorial Lectures: The present position in Bayesian statistics. *Statistical Science* **5**, 44–89.

Lindley DV 1991 Probability In *The use of statistics in forensic science* (ed. Aitken CGG and Stoney DA), Ellis Horwood, New York, pp. 27–50.

Lindley DV 1992 Is our view of Bayesian statistics too narrow? In *Bayesian statistics 4* (ed. Bernardo JM, Berger JO, Dawid AP and Smith AFM), Oxford University Press, Oxford, pp. 1–15.

Lindley DV 1997 The choice of sample size. *The Statistician* **46**, 129–138.

Lindley DV 2000a Bayesian thoughts. *Significance* **1**, 73–75.

Lindley DV 2000b The philosophy of statistics. *The Statistician* **49**, 293–337.

Lindley DV 2006 *Understanding Uncertainty*. John Wiley & Sons, Ltd, Hoboken, New Jersey.

Locard E 1920 *L'enquête Criminelle et les méthodes scientifiques*. Flammarion, Paris.

Lütkepohl H 1996 *Handbook of Matrices*. John Wiley & Sons, Ltd, Chichester.

Maher P 1993 *Betting on Theories*. Cambridge University Press, Cambridge.

Marden JI 2000 Hypothesis testing: from p-values to Bayes factor. *Journal of the Amercian Statistical Association* **95**, 1316–1320.

Marschak J 1950 Rational behavior, uncertain prospects, and measurable utility. *Econometrica* **18**, 111–141.

Mavridis D and Aitken CGG 2009 Sample size determination for categorical responses. *Journal of Forensic Sciences* **54**, 135–151.

Mazzella WD and Marquis R 2007 Forensic image analysis of laser-printed documents. *Journal of the Amercian Society of Questioned Documents* **10**, 19–24.

National Research Council 2009 *Strengthening Forensic Science in the United States: a Path Forward*. The National Academies Press, Washington D.C.

Neal R 1996 *Bayesian learning for neural networks*. Springer, New York.

Neyman J and Pearson ES 1928a On the use and the interpretation of certain test criteria for purposes of statistical inference, part 1. *Biometrika* **20A**, 175–240.

Neyman J and Pearson ES 1928b On the use and the interpretation of certain test criteria for purposes of statistical inference, part 2. *Biometrika* **20A**, 263–294.

O'Hagan A 1994 *Kendall's Advanced Theory of Stastistics, Vol. 2B, Bayesian inference*. Edward Arnold, London.

O'Hagan A 2004 Bayesian statistics: principles and benefits In *Bayesian statistics and quality modelling in agro-food production chain (Wageningen UR Frontis Series)* (ed. van Boekel MAJS, Stein A and van Bruggen A H C), Kluwer Academic Publishers, Dordrecht, pp. 31–45.

O'Hagan A 2006 Science, subjectivity and software. *Bayesian Analysis* **1**, 445–450.

O'Hagan A, Buck CE, Daneshkhah A, Eiser JR, Garthwaite PH, Jenkinson DJ, Oakley JE and Rakow T 2006 *Uncertain Judgements: Eliciting Experts' Probabilities*. John Wiley & Sons, Ltd, Hoboken, NY.

Parmigiani G 2002 *Modeling in Medical Decision Making*. John Wiley & Sons, Ltd, Chichester.

Parmigiani G and Inoue L 2009 *Decision theory*. John Wiley & Sons, Ltd, Chichester.

Pearl J 2000 *Causality: Models, Reasoning, and Inference*. Cambridge University Press, Cambridge.

Pham-Gia T and Turkkan N 1993 Bayesian analysis of the difference of two proportions. *Communications in Statistics – Theory and Methods* **22**, 1755–1771.

Pham-Gia T and Turkkan N 2003 Determination of exact sample sizes in the Bayesian estimation of the difference of two proportions. *The Statistician* **52**, 131–150.

Popper KR 1959 *The Logic of Scientific Discovery*. Hutchinson, London.

Press SJ 1989 *Bayesian Stastistics: Principles, Models, and Applications*. John Wiley & Sons, Ltd, New York.

Press SJ 2003 *Subjective and Objective Bayesian Stastistics*. John Wiley & Sons, Ltd, Hoboken.

Press SJ and Tanur JM 2001 *The Subjectivity of Scientists and the Bayesian Approach*. John Wiley & Sons, Ltd, New York.

Putnam H 1975 Probability and confirmation In *Mathematics, Matter and Method* (ed. Putnam H), Cambridge University Press Cambridge (Mass.), pp. 293–304.

R 2003 *R: A language and environment for statistical computing* R Foundation for Statistical Computing Vienna, Austria. ISBN 3-900051-00-3.

Raiffa H 1968 *Decision Analysis – Introductory Lectures on Choices Under Uncertainty*. Addison-Wesley, Reading.

Raiffa H and Schlaifer R 1961 *Applied Statistical Decision Theory*. MIT Press, Cambridge (Mass.).

Ramsey F 1931 Truth and probability In *The Foundations of Mathematics and Other Logical Essays* (ed. Braithwaite R), Routledge & Kegan Paul Ltd., London, pp. 156–198.

Redmayne M 2001 *Expert Evidence and Criminal Justice*. Oxford University Press, Oxford.

Robert CP 1994 *The Bayesian Choice. A Decision-Theoretic Motivation*. Springer-Verlag, New York.

Robert CP 2001 *The Bayesian Choice*, 2nd edn. Springer, New York.

Robertson B and Vignaux GA 1993 Probability – the logic of the law. *Oxford Journal of Legal Studies* **13**, 457–478.

Robertson B and Vignaux GA 1995 *Interpreting Evidence. Evaluating Forensic Science in the Courtroom*. John Wiley & Sons, Ltd, Chichester.

Robertson B and Vignaux GA 1998 Explaining evidence logically. *New Law Journal, Expert Witness Supplement* **148**, 159–162.

Robinson N, Taroni F, Saugy M, Ayotte C, Mangin P and Dvorak J 2001 Detection of nandrolone metabolites in urine after a football game in professional and amateur players: a Bayesian comparison. *Forensic Science International* **122**, 130–135.

Roseveare N 1982 *Mercury's Perihelion from Leverrer to Einstein*. Oxford University Press, Oxford.

Royall R 1997 *Statistical Evidence: A Likelihood Paradigm*. Chapman & Hall, London.

Saks MJ and Koehler JJ 2005 The coming paradigm shift in forensic identification science. *Science* **309**, 892–895.

Salmon WC 1990 Rationality and objectivity in science or Tom Kuhn meets Tom Bayes In *Scientific Theories, Minnesota Studies in the Philosophy of Science. Vol.14* (ed. Savage C), University of Minnesota Press, Minneapolis, pp. 175–204.

Salmon WC 1967 *The Foundations of Scientific Inference*. University of Pittsburgh Press, Pittsburgh.

Savage LJ 1972 *The Foundations of Statistics*. Dover Publications, Inc., New York.

Savage LJ 1981 *The Writings of Leonard Jimmie Savage – A Memorial Selection*. American Statistical Association and The Institute of Mathematical Statistics, Washington, D.C.

Schervish MJ 1989 A general method for comparing probability assessors. *Annals of Statistics* **17**, 1856–1879.

Schlaifer R 1959 *Probability and Statistics For Business Decisions – An Introduction to Managerial Economics Under Uncertainty*. McGraw-Hill Book Company, New York.

Schmittbuhl M, Le Minor JM, Schaaf A and Mangin P 2002 The human mandible in lateral view: elliptical Fourier descriptors of the outline and their morphological analysis. *Annals of Anatomy* **184**, 199–207.

Schmittbuhl M, Le Minor JM, Taroni F and Mangin P 2001 Sexual dimorphism of the human mandible: demonstration by elliptical Fourier analysis. *International Journal of Legal Medicine* **115**, 100–101.

Schmittbuhl M, Rieger J, Le Minor J, Schaaf A and Guy F 2007 Variations of the mandibular shape in extant hominoids: generic, specific, and subspecific quantification using elliptical Fourier analysis in lateral view. *American Journal of Physical Anthropology* **132**, 119–131.

Schum DA 1994 *Evidential Foundations of Probabilistic Reasoning*. John Wiley & Sons, Ltd, New York.

Shachter RD 1986 Evaluating influence diagrams. *Operations Research* **34**, 871–882.

Shakespeare W 1994 Julius Caesar *Complete Works of William Shakespeare* Harper Collins Publishers Glasgow pp. 1019–1048.

Silverman BW 1986 *Density Estimation*. Chaptman & Hall, London.

Simonoff JF 1996 *Smoothing Methods in Statistics*. Springer, New York.

Singpurwalla ND 2006 *Reliability and Risk. A Bayesian Perspective*. John Wiley & Sons, Ltd, Chichester.

Sivia DS 1996 *Data Analysis – a Bayesian Tutorial*. Clarendon Press, Oxford.

Skyrms B 1987 Dynamic coherence and probability kinematics. *Philosophy of Science* **54**, 1–20.

Sloman S 2005 *Causal Models. How People Think about the World and Its Alternatives*. Oxford University Press, New York.

Smith J 1988 *Decision Analysis. A Bayesian Approach*. Chapman & Hall, London.

Sokal A and Bricmont J 1998 *Fashionable Nonsense. Postmodern Intellectuals' Abuse of Science*. Picador, New York.

Stoney DA 1991 Transfer evidence In *The Use of Statistics in Forensic Science* (ed. Aitken CGG and Stoney DA) Ellis Horwood New York pp. 107–138.

Stoney DA 1992 Reporting of highly individual genetic typing results: a practical approach. *Journal of the Forensic Science Society* **37**, 373–386.

Stoney DA 1994 Relaxation of the assumption of relevance and an application to one-trace and two-trace problems. *Journal of the Forensic Science Society* **34**, 17–21.

Swinburne R 2003 *Bayes' Theorem. Proceedings of the British Academy*. Oxford University Press, Oxford.

Taroni F, and Aitken CGG 1998 Probabilistic reasoning in the law, part 1: assessment of probabilities and explanation of the value of DNA evidence. *Science & Justice* **38**, 165–177.

Taroni F, Aitken CGG, Garbolino P and Biedermann A 2006a *Bayesian networks and probabilistic inference in forensic science*. John Wiley & Sons, Ltd, Chichester.

Taroni F and Aitken CGG 1999 The likelihood approach to compare populations: a study on DNA evidence and pitfalls of intuitions. *Science & Justice* **39**, 213–222.

Taroni F and Hicks TN 2008 Exploitation des profils adn partiels et les recherches parmi les familiers. Journée Romande de Médecine Légale, Genéve.

Taroni F, Bozza S and Biedermann A 2006b Two items of evidence, no putative source: an inference problem in forensic intelligence. *Journal of the Forensic Sciences* **51**, 1350–1361.

Taroni F, Champod C and Margot P 1998 Forerunners of Bayesianism in early forensic science. *Jurimetrics Journal* **38**, 183–200.

Teller P 1973 Conditionalization and observation. *Synthese* **26**, 218–258.

Thompson WC and Schumann EL 1987 Interpretation of statistical evidence in criminal trials: The prosecutor's fallacy and the defense attorney's fallacy. *Law and Human Behaviour* **11**, 167–187.

Thompson WC, Taroni F and Aitken CGG 2003 How the probability of a false positive affects the value of DNA evidence. *Journal of Forensic Sciences* **48**, 47–54.

Tribe LH 1971 Trial by mathematics: precision and ritual in the legal process. *Harvard Law Review* **84**, 1329–1393.

Tversky A and Kahneman D 1986 Rational choice and the framing of decisions. *Journal of Business* **59**, 251–278.

Tzidony D and Ravreboy M 1992 A statistical approach to drug sampling: a case study. *Journal of the Forensic Sciences* **37**, 1541–1549.

van Fraassen B 1989 *Laws and Symmetry*. Clarendon Press, Oxford.

von Neumann J and Morgenstern O 1953 *Theory of Games and Economic Behavior*, 3rd edn. Princeton University Press, Princeton.

Wald A 1947 *Sequential Analysis*. John Wiley & Sons, Inc., New York.

Wand MP and Jones MC 1995 *Kernel Smoothing*. Chapman & Hall, London.

Weisberg HI 1972 Bayesian comparison of two ordered multinomial populations. *Biometrics* **28**, 859–867.

Wetherill GB and Glazebrook KD 1986 *Sequential Methods in Statistics*. Chapman & Hall, London.

Williams PM 1980 Bayesian conditionalisation and the principle of minimum information. *British Journal for the Philosophy of Science* **31**, 131–144.

Williamson J 2004 *Bayesian Nets and Causality: Philosophical and Computational Foundations*. Clarendon Press, Oxford.

Winkler RL 2001 Why Bayesian analysis hasn't caught on in healthcare decision making. *International Journal of Technology Assessment in Health Care* **17**, 56–66.

Author Index

Data Analysis in Forensic Science: A Bayesian Decision Perspective Franco Taroni, Silvia Bozza, Alex Biedermann, Paolo Garbolino and Colin Aitken

Subject Index

Data Analysis in Forensic Science: A Bayesian Decision Perspective Franco Taroni, Silvia Bozza, Alex Biedermann, Paolo Garbolino and Colin Aitken

STATISTICS IN PRACTICE

Human and Biological Sciences

Berger – Selection Bias and Covariate Imbalances in Randomized Clinical Trials
Berger and Wong – An Introduction to Optimal Designs for Social and Biomedical Research
Brown and Prescott – Applied Mixed Models in Medicine, Second Edition
Chevret (Ed) – Statistical Methods for Dose-Finding Experiments
Ellenberg, Fleming and DeMets – Data Monitoring Committees in Clinical Trials: A Practical Perspective
Hauschke, Steinijans & Pigeot – Bioequivalence Studies in Drug Development: Methods and Applications
Lawson, Browne and Vidal Rodeiro – Disease Mapping with WinBUGS and MLwiN
Lesaffre, Feine, Leroux & Declerck – Statistical and Methodological Aspects of Oral Health Research
Lui – Statistical Estimation of Epidemiological Risk
Marubini and Valsecchi – Analysing Survival Data from Clinical Trials and Observation Studies
Molenberghs and Kenward – Missing Data in Clinical Studies
O'Hagan, Buck, Daneshkhah, Eiser, Garthwaite, Jenkinson, Oakley & Rakow – Uncertain Judgements: Eliciting Expert's Probabilities
Parmigiani – Modeling in Medical Decision Making: A Bayesian Approach
Pintilie – Competing Risks: A Practical Perspective
Senn – Cross-over Trials in Clinical Research, Second Edition
Senn – Statistical Issues in Drug Development, Second Edition
Spiegelhalter, Abrams and Myles – Bayesian Approaches to Clinical Trials and Health-Care Evaluation
Walters – Quality of Life Outcomes in Clinical Trials and Health-Care Evaluation
Whitehead – Design and Analysis of Sequential Clinical Trials, Revised Second Edition
Whitehead – Meta-Analysis of Controlled Clinical Trials
Willan and Briggs – Statistical Analysis of Cost Effectiveness Data
Winkel and Zhang – Statistical Development of Quality in Medicine

Earth and Environmental Sciences

Buck, Cavanagh and Litton – Bayesian Approach to Interpreting Archaeological Data
Glasbey and Horgan – Image Analysis in the Biological Sciences
Helsel – Nondetects and Data Analysis: Statistics for Censored Environmental Data
Illian, Penttinen, Stoyan, H and Stoyan D–Statistical Analysis and Modelling of Spatial Point Patterns
McBride – Using Statistical Methods for Water Quality Management
Webster and Oliver – Geostatistics for Environmental Scientists, Second Edition
Wymer (Ed) – Statistical Framework for Recreational Water Quality Criteria and Monitoring

Industry, Commerce and Finance

Aitken and Taroni – Statistics and the Evaluation of Evidence for Forensic Scientists, Second Edition
Balding – Weight-of-evidence for Forensic DNA Profiles
Brandimarte – Numerical Methods in Finance and Economics: A MATLAB-Based Introduction, Second Edition
Brandimarte and Zotteri – Introduction to Distribution Logistics
Chan – Simulation Techniques in Financial Risk Management
Coleman, Greenfield, Stewardson and Montgomery (Eds) – Statistical Practice in Business and Industry
Frisen (Ed) – Financial Surveillance
Fung and Hu – Statistical DNA Forensics
Gusti Ngurah Agung – Time Series Data Analysis Using EViews
Jank and Shmueli (Ed.) – Statistical Methods in e-Commerce Research
Lehtonen and Pahkinen – Practical Methods for Design and Analysis of Complex Surveys, Second Edition
Ohser and Mücklich – Statistical Analysis of Microstructures in Materials Science
Pourret, Naim & Marcot (Eds) – Bayesian Networks: A Practical Guide to Applications
Taroni, Aitken, Garbolino and Biedermann – Bayesian Networks and Probabilistic Inference in Forensic Science